Pathology & Parasitology

for Veterinary Technicians

Second Edition

Pathology & Parasitology

for Veterinary Technicians

Second Edition

Dr. Leland S. Shapiro
Contributing Author: Patricia Mandel

DELMAR
CENGAGE Learning™

**Pathology and Parasitology for
Veterinary Technicians, Second Edition**
Dr. Leland S. Shapiro

Vice President, Career and Professional
    Editorial: Dave Garza
Director of Learning Solutions:
    Matthew Kane
Acquisitions Editor: Benjamin Penner
Managing Editor: Marah Bellegarde
Senior Product Manager: Darcy M. Scelsi
Editorial Assistant: Scott Royael
Vice President, Career and Professional
    Marketing: Jennifer McAvey
Marketing Director: Debbie Yarnell
Marketing Manager: Erin Brennan
Marketing Coordinator: Jonathan Shee-
    han
Production Director: Carolyn Miller
Production Manager: Andrew Crouth
Content Project Manager: Katie Wachtl
Senior Art Director: Dave Arsenault

For product information and technology assistance, contact us at
Cengage Learning Customer & Sales Support, 1-800-354-9706

For permission to use material from this text or product, submit
all requests online at **www.cengage.com/permissions.**
Further permissions questions can be e-mailed to
permissionrequest@cengage.com

Library of Congress Control Number: 2009925002

ISBN-13: 978-1-4354-3855-2
ISBN-10: 1-4354-3855-8

Delmar
5 Maxwell Drive
Clifton Park, NY 12065-2919
USA

Cengage Learning is a leading provider of customized learning solutions
with office locations around the globe, including Singapore, the United Kingdom,
Australia, Mexico, Brazil, and Japan. Locate your local office at:
international.cengage.com/region

Cengage Learning products are represented in Canada by Nelson Education, Ltd.

To learn more about Delmar, visit **www.cengage.com/delmar**
Purchase any of our products at your local college store or at our preferred online
store **www.ichapters.com**

NOTICE TO THE READER
Publisher does not warrant or guarantee any of the products described herein or
perform any independent analysis in connection with any of the product information contained herein. Publisher does not assume, and expressly disclaims, any
obligation to obtain and include information other than that provided to it by the
manufacturer. The reader is expressly warned to consider and adopt all safety
precautions that might be indicated by the activities described herein and to avoid
all potential hazards. By following the instructions contained herein, the reader will-
ingly assumes all risks in connection with such instructions. The publisher makes no
representations or warranties of any kind, including but not limited to, the warranties of fitness for particular purpose or merchantability, nor are any such representations implied with respect to the material set forth herein, and the publisher takes
no responsibility with respect to such material. The publisher shall not be liable for
any special, consequential, or exemplary damages resulting, in whole or part, from
the readers' use of, or reliance upon, this material.

Printed in the United States of America
5 6 7 8 9 10 11   18 17 16 15 14

Dedication

To my parents (Murray and Shirley), who gave me life; to my children (Ilana and Aaron), who showed me life and taught me how to live; and to Lori, who taught me how to feel alive.

Table of Contents

Part One: Pathology

Chapter 1: Introduction to Pathology 2

What Is Pathology? 3

History and Development 5

Review 7

Chapter 2: Cell Injury and Cell Death 8

Disease in Relation to Cell Injury 9

Necrosis and Patterns of Necrosis 10

Adaptations of Cells in Response to Injury 12

Pathological Mineralization and Pigmentation 14

Review 20

Chapter 3: Inflammation and Healing 22

Inflammation 23

The Acute Inflammatory System 25

Chronic and Granulomatous Inflammation 28

Healing of Injured Tissue 30

Review 35

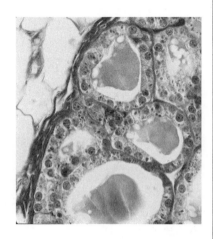

Chapter 4: Fluid and Circulatory Disturbances 37

Review of the Circulatory System   38

Blood   40

Complete Blood Count and Differential WBC Count   41

Chemistry Profile   42

Fluid Homeostasis   42

Hemodynamic Disturbances   43

Fluid Imbalances   48

Review   52

Chapter 5: Neoplasia: An Introduction to Tumors 55

Classification of Tumors   56

Behavior of Tumors   56

Chemical Carcinogenesis   58

Malignant versus Benign   60

The Morphology of Tumors   61

Cancer Research   63

Review   65

Chapter 6: Immunological Diseases 67

Immunity   68

Autoimmunity   68

Mechanisms of Tissue Damage in Autoimmune Disease   69

Disturbances in Immune Function   71

Neoplasms of Lymphoid Cells   72

Review   76

**Chapter 7: Mechanisms of Infections
and Causes of Infectious Disease** 78

Infection   79

The Fate of the Infecting Organism   79

Sources of Infection   79

Infection and Contagion   81

Properties of Pathogenic Organisms   82

*Mechanisms of Disease Production by Pathogenic
    Organisms   82*

Pathogenesis of Parasitic Infections   84

Tolerance in the Host-Parasite Relationship   86

Review   88

Chapter 8: Genetic Disorders 90

    Genetic Technology 91

    Categories of Genetic Disorders 92

    Molecular Diagnosis 94

    Review 98

Chapter 9: Environmental and Nutritional Diseases 100

    Environmental Impact on Health 101

    Air Pollution 102

    Chemical and Drug Injury 102

    Physical Injury 103

    Nutritional Disease 104

    Nutritional Excesses and Imbalances 106

    Review 107

Chapter 10: Zoonoses and Safety on the Job 109

    Zoonoses 110

    Responsibility of the Veterinary Staff and Public Health 112

    Parasitic Diseases 113

    Bacterial Diseases 113

    Viral Diseases 117

    Protozoal Diseases 118

    Tickborne Diseases 118

    Review 121

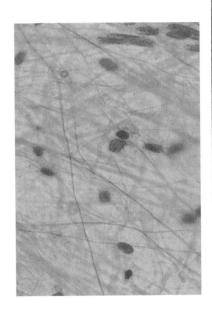

Part Two: Clinical Parasitology

Chapter 11: The Variety of Parasites 124

    Infestation versus Infection 125

    Types of Parasites 125

    Review 128

Chapter 12: Ectoparasites of Large Animals 129

    Common Ectoparasites 130

    Ectoparasites of Cattle 134

    Ectoparasites of Sheep 142

    Ectoparasites of Swine 145

    Ectoparasites of Horses 146

    Review 150

Chapter 13: Ectoparasites of Small Animals 151

Ectoparasites in Companion Animals   152

Ectoparasites of Dogs and Cats   152

Ectoparasites of Rabbits and Rodents   160

Ectoparasites of Ferrets, Foxes, and Mink   166

Ectoparasites of Reptiles   167

Ectoparasites of Chickens, Turkeys, and Other Birds   168

Review   172

Chapter 14: Endoparasites of Large Animals 174

Endoparasites of Laboratory and Farm Animals   175

Endoparasites of Ruminants (Cattle, Sheep, and Goats)   177

Endoparasites of Swine and Pet Pigs   185

Endoparasites of Horses   190

Review   196

Chapter 15: Endoparasites of Small Animals   198

Endoparasites of Dogs and Cats   199

Endoparasites of Ferrets   212

Endoparasites of Rodents and Rabbits   212

Endoparasites of Reptiles   213

Endoparasites of Poultry and Other Birds   215

Review   221

**Appendix I: Important Techniques
for Veterinary Technicians** 223

*How to Do a Staining for Cytologic Preparations
and Bacteria Examination   224*

*How to Prepare an Ear Cytology
and Microbiological Examination   225*

How to Do a Fine-Needle Aspiration   226

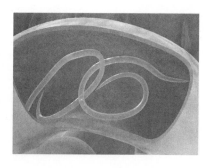

How to Do a Cellophane Tape Preparation for Lice or Mites 226

How to Do a Skin Scraping 227

How to Do a Direct Smear for a Fecal Exam 227

How to Do a Fecal Flotation 228

How to Use the Modified Knott's Technique for Heartworm (Dirofilaria immitis *and* Dipetalonema reconditum) 229

How to Use the Filter Technique for Heartworm (Dirofilaria immitis *and* Dipetalonema reconditum) 229

How to Use the Buffy Coat Method for Microfilariae 232

How to Do a Fungal Culture Using Dermatophyte Testing Media 232

How to Do a Blood Smear 233

How to Use a Hemacytometer for a White Blood Cell Count 234

How to Do a Urinalysis 234

How to Do a Quantitative Fecal Egg Count 236

Appendix II: Common Diseases 242

Mice and Rats 242

Guinea Pigs 244

Rabbits 246

Cats 248

Dogs 250

Primates 253

Cattle, Sheep, Goats, and Horses 255

Appendix III: Narrative Guide for the Image Library 260

Internal Parasites of Domestic Animals 261

Identification of External Parasites 268

Types of Animal Diseases 276

Glossary 285

Index 295

Preface

This book is intended for students studying veterinary technology, animal production, and pre-veterinary sciences. A basic understanding of animal anatomy and physiology will assist the student in identifying sites and mechanisms of injurious agents and predicting the effects of the injury.

Veterinarians attempt to prevent disease or make diagnoses of disease every day of their professional lives. A disease occurs because the normal functioning of a cell or organ is affected by some injury. In other words, all diseases are due to cellular injury. The study of cell injury and its result is the science of pathology. You must become familiar with pathology in order to appreciate the effects different diseases have on animals. The information you'll learn pertaining to how cells respond to injury can help you make predictions about how the affected animals will act. It's this understanding of injury and response that will aid you while working as a veterinary technician.

The veterinary technician should be able to:

- Explain what happens to cells and tissues when the cells die, and how injured tissues heal.

- Describe how tissues protect themselves using the inflammatory response.

- State the effects on organs when their blood supply increases or decreases.

- Explain how a single cell can become "immortal" and cause cancer.

- Describe the mechanisms of infection and the causes of infectious diseases.

The purpose of this text is not to make its reader a diagnostician of animal disease. That is the job of the veterinarian. However, it is teamwork, in and out of the veterinary setting, that will help promote overall quality animal care and health. The basic understanding of veterinary pathology and parasitology will enable this teamwork to be accomplished and will increase the competency of the veterinary technician in working directly with the veterinarian. With this second edition we incorporated several of the recommendations of previous users of our text, both faculty and students, to make it more "user friendly" and provide additional hands-on techniques practiced in veterinary clinics.

L.S. Shapiro

2009

Acknowledgments

The writing of this text could not have been undertaken without the assistance of many who shared in their knowledge, experience, and critique. I want to especially thank Patty Mandel, a former student and graduate of our registered veterinary technician program at Pierce College. Patty's assistance in writing the case studies and in editing my first edition of this text was instrumental in creating a very hands-on oriented guide to the veterinary technician. Dr. Fret Lucero, also a former student, lent his expertise in veterinary pathology and current procedures performed in the veterinary clinic and hospital. Dr. Frank Lavac, Wilshire Animal Hospital; Dr. Jim Peddie, professor emeritus of Moorpark College Exotic Animal Training and Management Program; Dr. Monica Silvers, a Pierce alumnus and currently a practitioner in Austin, Texas; and Drs. Kalie Pasek, Rebecca Yates, and Jana Smith of Los Angeles Pierce College's RVT Program, and Elizabeth White, director of Los Angeles Pierce College's RVT Program, all played a part in providing review, critique, and valuable information for this text.

In addition, I would like to thank the following individuals at Delmar Cengage Learning who helped in the development of this manuscript: Darcy Scelsi, Senior Product Manager; David Arsenault, Art Director; Katie Wachtl, Content Project Manager; Andrew Crouth, Production Manager; and Scott Royael, Editorial Assistant. The author and Delmar would like to thank those individuals who reviewed the manuscript and offered suggestions, feedback, and assistance. Their work is greatly appreciated: Freya Burnett, MS, Wilson College, Chambersburg, Pennsylvania; Tanya LeRoith, DVM, PhD, Virginia Tech, Blacksburg; Mary O'Horo Loomis, DVM, State University of New York, Canton; Frances Turner, RVT, McLennan Community College, Waco, Texas; Brenda R. Woodard, DVM, Director of Veterinary Technology, Northwestern State University of Louisiana, Natchitoches; Darwin R. Yoder, MS, DVM, Director of Veterinary Technology, Sul Ross State University, Alpine, Texas; Betsy A. Krieger, DVM, Front Range Community College, Fort Collins, Colorado; John L. Robertson, DVM, PhD, VA-MD Regional College of Veterinary Medicine, Blacksburg, Virginia; and Kay M. Bradley, CVT, Madison Area Technical College, Madison, Wisconsin.

Finally, I want to thank my children (Ilana and Aaron) and wife, Lori, who sacrificed countless hours away from their father and husband, while I researched, wrote, and edited this text.

About the Author

Dr. Leland S. Shapiro is the director of the pre-veterinary science program at Los Angeles Pierce College. He has been a professor of animal sciences for 33 years and is a member of the American Dairy Science Association, Dairy Shrine Club, Gamma Sigma Delta Honor Society of Agriculture, and Association of Veterinary Technician Educators, Inc. Dr. Shapiro was a dairy farmer for almost two decades, holds a California State pasteurizer's license, and for 14 years was a certification instructor for artificial insemination. Professor Shapiro has actively served on the college's ethics and work environment committees and has completed two post-doctoral studies in Bio-Ethics. Dr. Shapiro is a University of California Davis "Mentor of Veterinary Medicine" and the recipient of several local college teaching awards as well as the prestigious National Institute for Staff and Organizational Development Community College Leadership Program (NISOD) Excellence in Teaching Award, in Austin, Texas. Dr. Shapiro has authored 16 books in the veterinary and animal sciences including *Applied Animal Ethics* 2nd edition.

About the Contributing Author

Patricia Ann Mandel is a registered veterinary technician and licensed veterinary technician. In addition, she is a member of the Much Love animal rescue organization and a member of the California Veterinary Medical Association. She is the author and editor of veterinary books and educational manuals.

PART ONE

Pathology

CHAPTER 1

Introduction to Pathology

OBJECTIVES

Upon completion of this chapter, the reader should be able to:

- Explain the basic areas of pathology.

- Describe the main characteristics of an injurious agent allowing the prediction of the effects of the injury.

- Identify common causes and classes of cell and tissue lesion.

KEY TERMS

biopsy

diagnosis

differential diagnoses

etiology

humors

lesions

necropsy

pathology

pathogenesis

virology

What Is Pathology?

Pathology is the scientific study of disease. The word *pathology* is derived from two root words: *pathos*, which means "suffering" or "disease," and *logos*, which means "the study of." This study of disease may be divided into four main parts:

1. The study of the cause, or **etiology** of disease, which may be genetic or acquired. Genetic etiologies arise from changes within genes, whereas acquired etiologies refer to disease caused by outside sources, such as bacterial or viral infections or metabolic or nutritional disorders.

2. The study of mechanisms in the development of disease, or **pathogenesis**. This describes the sequence of events—the progression of changes from their inception—at the cell and tissue level as a disease expresses itself. Pathogenesis includes factors that influence the development of a disease, such as the immune status of an animal.

3. The study of **lesions**, or morphological alterations in tissues that occur with disease. Lesions give rise to functional disturbances. These disturbances serve to distinguish one disease from another and occasionally may be diagnostic of an etiological agent. The alteration of tissue is studied by gross examination and by microscopic examination (Figure 1-1). To appreciate changes in tissues or organs that are a result of disease, you must know the normal condition of those tissues or organs. Changes are described according to size, color, shape, consistency, and weight (if possible). The surgical removal of lesions within organs and tissues, primarily for the purpose of examination, is known as **biopsy**.

4. The study of the functional consequences of these lesions, which give rise to physical signs or symptoms of disease. A **diagnosis** is achieved when the precise nature of the lesions causing the symptoms is recognized. A list of observations based on a **necropsy**, which is the examination of an animal's body after death, can help lead to diagnosis.

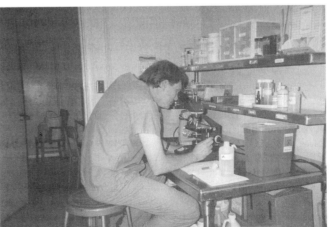

Delmar/Cengage Learning

FIGURE 1-1

Lesions are often studied microscopically to aid the diagnosis of disease.

Our understanding of normal tissues and organs can help in the determination of disease. Although a great number of things can cause cell injury, cells have a limited number of ways they can respond to injury, regardless of its cause. The proportions of cells and cell materials determine the makeup of tissue.

Different types of tissues will adapt to injury as a result of their compositions. Any agent that acts upon tissue and exceeds this ability to adapt causes disease. If you know the mechanism and site of something that injures tissue, you can predict the effects of the injury. For example, the chemical carbon tetrachloride is known to damage the DNA of liver cells.[1] The damage to DNA is the mechanism, and because it's targeting liver cells, the liver is the site. Genetic damage of liver cells suggests that the effect of an injury from carbon tetrachloride will be the development of liver cancer. Knowing the structural and functional changes of the tissue allows the veterinarian to predict the clinical manifestations, or the signs, that an animal will exhibit. Animals can breathe in carbon tetrachloride at or near manufacturing plants or waste sites, drink contaminated water near the same, or eat contaminated soil at waste sites.

Table 1-1 lists some of the most common causes and classes of cell and tissue lesions. Becoming acquainted with these categories will be crucial to your understanding of animal pathology in general.

In our example of carbon tetrachloride, an animal will exhibit signs of liver disease because of the damage the cancer causes in the liver—specifically, the structural and functional changes the liver undergoes. It's not always possible to make a specific diagnosis for disease, as many agents can cause similar signs. Liver cancer can

TABLE 1–1   Cell and Tissue Lesions at a Glance

Causes of Cell or Tissue Lesions	Classes of Cell or Tissue Lesions
1. Genetic	1. Disruptive
2. Congenital	2. Degenerative
3. Toxic	3. Blood vessel/circulatory defects
4. Infection	4. Inflammatory
5. Trauma	5. Growth and differentiation defects
6. Degenerative	6. Developmental
7. Immunologic	
8. Neoplastic	
9. Nutritional	
10. Metabolic	
11. Iatrogenic	
12. Hypoxia (lack of oxygen) Anoxia (absence of oxygen in the blood or tissues) Ischemia (deficiency of blood in an area due to obstruction or constriction of a blood vessel)	

[1]Iwai, S., et al. (2002, May 8). *Cancer Letters, 179*(1), 15–24.

occur from things besides carbon tetrachloride, and yet the signs of liver cancer are almost always the same. Clinical manifestations and the complaints that are presented will allow one to make **differential diagnoses**, or a list of possible diseases possessing the same or similar symptoms.

History and Development

From the fossil record, we've learned that even prehistoric creatures fell prey to disease (Figure 1-2). For instance, arthritis was common in cave bears and dinosaurs. Veterinary practice was known to occur in ancient Mesopotamia. It may have been handled by lower-class barbers or the upper-class asu, or priests, whose title is derived from the biblical king, Asa, whose name meant "healer of God." The oldest medical text is a cuneiform tablet from Mesopotamia. Yet Egyptian medical writings from about 3,000 years ago refer to far older texts.

Special schools for training physicians were commonly attached to temples. By the time of Hippocrates (ca. 460–370 B.C.E.), the Greeks had developed a system of illness based on the four cardinal **humors**—blood, phlegm, yellow bile (choler), and black bile (melancholy). Any imbalance of these resulted in disease. A healthy person, ideally, had a balance of these four humors. Hippocrates wrote many medical books, covering such subjects as anatomy, physiology, surgery, therapy, and general pathology, where disease was determined to be the result of either

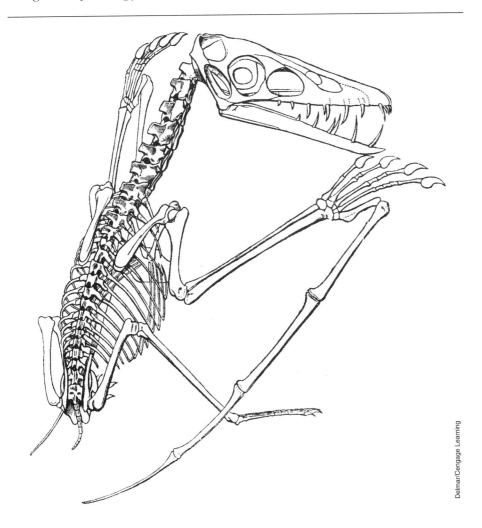

FIGURE 1-2

Prehistoric remains have aided scientists in establishing the types of disease that animals suffered from in ancient times.

Delmar/Cengage Learning

internal difficulties or outside influences that upset the humors. Hippocrates is particularly remembered for the Hippocratic oath and the Hippocratic method. The Hippocratic oath contains affirmations, or pledges to the gods, teachers, and future students, and prohibitions, or rules, such as the following:

- Do not harm the patient.

- Do not administer deadly drugs.

- Keep confidences concerning the patient.

In a similar way, the Hippocratic method outlines a philosophy that has become the basis of modern medicine: Observe all, study the patient (not the disease), evaluate honestly, and assist nature.

Although physicians throughout history have relied on autopsies for information, the procedure didn't become a major focus of medicine until the 19th century. The development of treatment based on pathological changes in tissues occurred at this time, and the idea of "humors" was dismissed. In addition, the 19th century brought about the discovery of bacteria as the cause of disease. Louis Pasteur (1822–1895) demolished the theory of spontaneous generation and established the germ theory. Robert Koch (1843–1910) further revolutionized the science of bacteriology.

In the 20th century, techniques and concepts of autopsy pathology became clinically practical. Tissue samples could be removed from patients and interpreted to determine a diagnosis, a course of therapy, and a prognosis. Pathologists also began to study the chemicals of the body, which led to the development of clinical and experimental pathology.

It wasn't until the 19th and 20th centuries that veterinarians became recognized, certified professionals. Many advancements in veterinary medicine have had a direct impact on human medicine. Among these was the establishment of the field of **virology** in 1898, when Friedrich Loeffler and Paul Frosch demonstrated that foot-and-mouth disease in cattle was caused by a filterable virus. With the development of modern technologies such as electron microscopy, the science of pathology has continued to further our understanding of the disease process in both humans and animals.

Summary

Pathology, the scientific study of disease, includes the study of the cause, the development, the morphological changes, and the functional consequences of disease. A basic understanding of anatomy, physiology, chemistry, and microbiology is essential to the understanding of pathology. Veterinary pathology is used to understand human disease, and the applications in the study and treatment of human disease have allowed us to make considerable advancements in veterinary medicine. Veterinary pathologists specialize in recognizing the cause of disease not only by examining the animal but also by studying the organs, tissues, cells, chemicals, molecules, and genes. Laboratory findings, surgical biopsies, and necropsies help the pathologist determine what the disease is and how it is transmitted and prevented. Thus, it is essential for the veterinary technician to have a basic understanding of pathology to assist the veterinarian in the diagnosis, prevention, and treatment of animal disease.

REVIEW

FILL IN THE BLANKS

1. The prefix *pathos* is Greek for _____ .

2. The suffix *logos* is Greek for _____ .

3. The field of _____ was established in 1898 from research conducted in cattle.

4. The surgical removal of lesions within organs and tissues is known as _____ .

5. _____ is the study of the cause of disease.

6. The study of the mechanisms in the development of disease is known as _____ .

7. _____ is the examination of the organs, tissues, and cells of a dead animal to determine the cause of death or to study the pathological changes.

DISCUSSION QUESTIONS

1. What are the four areas of pathology?

   _____

   _____

   _____

2. What two characteristics of an injurious agent allow you to predict the effects of the injury?

   _____

   _____

   _____

3. List three prohibitions contained in the Hippocratic oath.

   _____

   _____

   _____

4. Research and discuss with your teacher examples of causes of cell or tissue lesions listed in Table 1–1.

   _____

   _____

   _____

CHAPTER 2

Cell Injury and Cell Death

OBJECTIVES

Upon completion of this chapter, the reader should be able to:

- Describe the differences between repair and regeneration of tissue/cells.

- State the main components of the nucleus and the cytoplasm.

- List the main causes of cell injury.

- Define and differentiate between hypertrophy and hyperplasia and give examples of each.

- Define and differentiate between metaplasia and dysplasia and give examples of each.

- Describe how cells respond to injury.

KEY TERMS

agenesis

aplasia

apoptosis

atrophy

calcification

cell degeneration

cell membrane

cellular morphology

dysplasia

dystrophic calcification

endoplasmic reticulum

hypercalcemia

hyperplasia

hypertrophy

hypoplasia

ions

ischemia

labile cells

lethal

lysosomes

metaplasia

metastatic calcification

mitochondria

necrosis

neoplasia

nucleus

pathological mineralization

permanent cells

proliferation

proliferative responses

sodium pump

stable cells

sublethal

Disease in Relation to Cell Injury

The underlying principle of pathology is that all disease processes have a cellular basis, or, stated more simply, all disease is due to cell injury. Changes in the arrangement, structure, or products of cells are of fundamental importance in diagnostic pathology.

Cellular injury can be **lethal** (fatal), causing **necrosis**, or cell death. It can also be **sublethal**, meaning the cell or tissue function is altered but the cell isn't destroyed. Deterioration of cell function can occur following nonlethal cell injury. This alteration is referred to as **cell degeneration**.

Cellular degenerations can be reversible, in which case the cells either return to normal or, if the injury is progressive, the cells degenerate until the injury becomes lethal. Cells that undergo nonlethal or degenerative injury experience proliferative changes. In nonlethal injury there is a balance between degenerative changes and proliferative changes or repair. **Proliferation** can refer to an increase in cell size or number.

As you learned earlier, there are many causes of cellular injury. Although the specific mechanisms are numerous, most injuries are the result of one or more of the following occurrences:

- Interference with membrane function

- Interference with energy production

- Interference with protein synthesis

- Lysosomal changes

- Interference with operation of the genome

Charged particles, or **ions**, exist in different concentrations and proportions inside and outside a cell. **Cell membranes** keep the environment inside and outside a cell balanced and separate by means of a **sodium pump**. This mechanism forces sodium ions out of a cell to maintain water balance. The pump can fail if the cell isn't getting enough oxygen (anoxia), or energy (adenosine triphosphate (ATP)), in which case water rushes in and the cell swells up. Although such swelling may be reversible it can also prove to be fatal and result in necrosis. In addition, if the cell membrane is damaged from free radical attack or from exposure to a toxin or heavy metal (lead poisoning), the water that leaks in can also cause other structures within the cell to swell. These structures include the **nucleus**, where deoxyribonucleic acid (DNA) is stored and converted to ribonucleic acid (RNA); the **mitochondria**, which produce ATP for energy; and the **endoplasmic reticulum**, which makes proteins (Figure 2-1).

Lysosomal change can also result in cellular damage. **Lysosomes** contain chemicals that are able to dissolve food particles, foreign invaders, and even the cells themselves. If lysosomes break open because of damage done to (1) the membrane, (2) the energy production mechanism, or (3) the protein synthesis mechanism, the cell will self-destruct.

FIGURE 2–1

Among the structures commonly affected by damage to the cellular membrane are the nucleus, the mitochondria, and the endoplasmic reticulum.

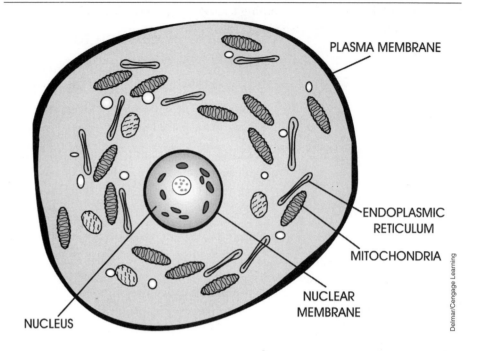

Reactions to cell injury tend to follow a limited set of patterns that can be observed microscopically. Since these are degenerations, the injuries are reversible if the cell is able to heal itself. These degenerations include:

- *Cloudy swelling*—the result of a defective sodium pump

- *Hydropic degeneration (cell swelling)*—a more severe form of swelling, due to the cells absorbing excess fluid in the cell cytoplasm

- *Fatty degeneration or fatty change (steatosis)*—the accumulation of fat in cells

- *Hyaline degeneration (hyalinosis)*—referring to processes that affect tissues and various cells, resulting in droplets that are acidophilic, homogeneous, and translucent

- *Mucoid degeneration*—a conversion of any of the connective tissues into a gelatinous or mucoid substance (such as a mucopolysaccharide)

- *Amyloid degeneration or amyloidosis*—an accumulation of amyloid between cells and fibers of tissues and organs

A cell that's swollen is in danger of having additional internal machinery malfunction or even bursting. Remember, any foreign material that accumulates inside or outside a cell can physically interfere with the normal function of that cell or tissue.

Necrosis and Patterns of Necrosis

Necrosis, as mentioned previously, is the final stage of an irreversible degenerative process. The term is usually applied to one or more cells, a portion of a tissue, or an organ, resulting in pathological death. Necrosis should be differentiated from **apoptosis**, which is endogenous programmed cell death that occurs normally in both adult and developing tissues. You should understand that studying a dying cell is like examining a single snapshot, when in fact you could learn a lot more from a video or DVD recording.

There are many things that will cause necrosis:

- **Ischemia** (involving a deficiency of blood to an area and therefore a lack of oxygen)

- Toxicity (a result of poison)

- Physical damage (a result of heat, cold, radiation, ultraviolet light, trauma)

- Infection and infectious agents

- Genetic abnormalities

- Altered metabolism

Damaged cells, which undergo necrosis, exhibit changes in their appearance, also called their **cellular morphology**. Among these changes are three types of nuclear changes that are easy to detect (Figure 2-2):

- *Pyknosis.* The nucleus shrinks and becomes very dense.

- *Karyolysis.* The nucleus dissolves with a loss of its affinity for basic stains. This occurs more commonly in necrosis.

- *Karyorrhexis.* The nucleus breaks into small fragments and the chromatin breaks up into unstructured granules.

Cytoplasmic changes are more difficult to detect. Here, ribosomes detach from the endoplasmic reticulum, and the staining properties of the cell change.

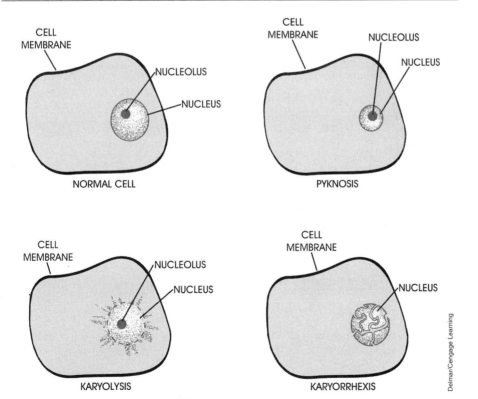

FIGURE 2–2

Types of nuclear changes within a cell.

Delmar/Cengage Learning

Cells are known to undergo several types of necrosis:

- *Coagulation necrosis.* In this type of necrosis, the cytoplasm of the cell co-agulates (thickens or gels), and details within the cell are lost. Affected tissues feel firm and dry. The cell membranes remain intact and, micro-scopically, you can see "ghost outlines of cells."

- *Liquefaction necrosis.* This type is more commonly known as pus. However, there are two types of liquefaction necrosis: (1) Regular pus consists of the dead and dying tissue cells that are partially liquefied and the white blood cells (primarily neutrophils) the body makes to "clean up the mess." Pus is usually associated with bacteria. An abscess occurs when the pus develops in pockets and puts pressure on the surrounding tissue. The tissue responds by walling off the pus with fibrous connective tissue. (2) Liquefaction necrosis of the central nervous system is more watery because of the higher water and fat content of nerve cells.

- *Caseous necrosis.* Due to the slow, progressive destruction of cells during this type of necrosis, a high response of white blood cells never occurs. Caseous necrosis results from the effects of hydrolytic enzymes and other substances released from the macrophages and other inflammatory cells in response to the persistence of microorganisms. This results in an outpour-ing of proteinaceous fluid without large numbers of white blood cells. All of the structural details of the tissue are lost with the production of a dry, white, cheesy, granular material. Caseous necrosis is common in rabbits, birds, reptiles, and animals with tuberculosis.

- *Fat necrosis.* This type usually occurs as a result of injury to fat as in pan-creatitis. Cellular injury to the pancreatic acini leads to the release of en-zymes that damage fat. Fatty acids are converted into soaps by reactions with calcium, magnesium, and sodium. The digested fat attracts a large amount of white blood cells, resulting in pain and discomfort.

- *Gangrenous necrosis.* This type is a combination of ischemia and a super-imposed bacterial infection. There are two types of gangrenous necrosis: (1) Dry gangrene consists of the mummified skin from coagulation necrosis. Tissues are usually dark because the dead blood cells release pigments which become iron sulfides, or rust. (2) Wet gangrene consists of liquefaction necrosis and bacteria that invade to feed on the dead tissue. The bacteria produce an abundance of gases and poisons that can kill an affected animal. Severe wet gangrene is also called *gas gangrene* because of the large amount of gas bubbles that are produced.

Adaptations of Cells in Response to Injury

The factors that control the growth and differentiation of cells are both environ-mentally and genetically controlled. Abnormalities of these two functions are frequent, forming an important part of many disease processes. Cellular injury and tissue adaptation often result in hyperplasia or growth abnormalities. We call these abnormalities **proliferative responses**.

Disorders of growth are based on the growth potential of cells. With regard to growth potential, cells are divided into three types—labile cells, stable cells, and permanent cells. **Labile cells** divide frequently to replace lost cells. These cells are found in the bone marrow, skin, and gut. **Stable cells** are fully differentiated cells that don't divide often (unless they are induced to do so). These cells are found in liver and kidney tissue. **Permanent cells** lose the ability to divide after birth. Muscles and neurons are examples of permanent cells.

Abnormalities of cell and tissue growth and differentiation are common responses to cellular injury. This facet of pathology also introduces you to a unique vocabulary with which you must become familiar. (Remember, growth can mean a decrease or diminution as well as an excess.)

Hypertrophy is an increase in cell size without an increase in cell numbers. Hypertrophic cells show an increase in size of the cytoplasm and in the numbers of cellular organelles and occasionally an increase in the size of the nucleus. An example is what happens to skeletal muscle when you exercise. Since these are permanent cells, skeletal muscle cells get larger rather than divide. Hypertension and a narrowing of the aortic orifice are also common causes of hypertrophy of the left ventricle of the heart (cardiac muscle). These heart cells contain more myofilaments, allowing them to contract more efficiently.

Hyperplasia is an increase in cell number, a phenomenon that can result from physiological or pathological causes. An example in the female of a physiological cause is estrus, a condition in which hormones cause the cells lining the uterus to divide. A similar example can be cited in the male. The prostate gland enlarges as a result of continued testosterone stimulus. Thus, castration and subsequent loss of androgen stimulation lead to atrophy of prostate tissue. An example of a pathological cause is the removal of a piece of liver, which would stimulate the remaining cells to divide and replace the absent tissue.

Agenesis is the absence or imperfect development of a part due to deficient growth. A common example in humans is the congenital absence of the brain's corpus callosum—the massive formation of nerve fibers bridging the two cerebral hemispheres. Renal agenesis is also common and can be unilateral or bilateral. Its etiology is probably multifactorial—both inherited and environmentally caused. It has been observed most commonly in dogs but has also been reported in the alpaca.

Aplasia is a type of agenesis, referring specifically to the total failure of an organ to develop (such as in animals or people born with only one kidney). These conditions occur during development. Reproductive aplasia commonly occurs leaving a genetically programmed male with female gonads. In the absence of testosterone (for one reason or another), the fetal gonad secretes estradiol and thus promotes the female reproductive system.

Hypoplasia also occurs during development, the result of a deficiency of growth and a diminution of size. Hypoplasia can be viral induced (feline panleukopenia—cerebellar folia are stunted) or congenital (hypoplasia of pituitary, thyroid, pancreas, kidney). An example of a common congenital hypoplasia is microhepatia, a condition that's characterized by a small liver. With canine distemper virus, it is also common to see hypoplasia of the young dog's (2 to 6 months of age) dental enamel upon examination of the oral cavity.

Atrophy occurs after full development from physiological or pathological causes; therefore, it's the reduction in size of a fully developed tissue. For example, disuse atrophy causes the withered appearance of an arm or leg (muscle cell down-regulation) after a cast is removed.

Metaplasia refers to the replacement of one fully differentiated cell type into another type where normally it is not found. For example, columnar epithelium in the bronchus can turn into stratified squamous epithelium from cigarette smoking. If the stimulus that caused metaplasia stops, it is generally reversible. If the stimulus is continual, the metaplasia may progress to dysplasia.

Dysplasia is the abnormal differentiation or development of a tissue with disordered architecture. This condition may be the precursor of a neoplasia.

Neoplasia is a pathological formation or growth of an abnormal tissue. The term is sometimes used interchangeably with *cancer* when the neoplasm is malignant. *Neoplasia* comes from the Greek *neos* + *plasia* meaning "new formation." However, it is much more serious in that it represents a pathological uncontrolled and progressive overgrowth of cells or tissue. These tissues can be small, solid masses or highly diffuse and infiltrative lesions.

Pathological Mineralization and Pigmentation

Pathological mineralization, or **calcification**, refers to the abnormal deposition of calcium salts together with small amounts of iron, magnesium, and other mineral salts. It's a common phenomenon occurring in a variety of pathological processes.[1] You should recall that necrosis sometimes results in mineralization. Pathological mineralization is divided artificially into dystrophic, metastatic, and the less common calcinosis calcification.

Dystrophic Calcification

When calcium is deposited in injured, degenerating, or dead tissue, it's termed **dystrophic calcification**. This is the most common form of abnormal calcification. Protein denaturation is an early consequence of cell death, and these denatured proteins bind calcium salts. Usually irreversible, the clinical significance of dystrophic calcification depends on what tissues are affected. For example, calcification of the pericardium, or the sac that surrounds the heart, can make heart contraction more difficult and result in heart failure. Calcification can be seen in lesions of tuberculosis, trichinosis, and caseous lymphadenitis.

Metastatic Calcification

Metastatic calcification is the deposition of calcium in soft tissues that aren't the site of previous damage. The condition is usually associated with **hypercalcemia**, which is an elevation of blood calcium levels. Causes of hypercalcemia include

---

[1]Miller, S. (2001, January). Pathologic mineralization in aquatic animals. *Exotic DVM, 3*(1), 12–13.

excessive parathyroid activity (the parathyroid is the endocrine gland that controls blood calcium levels), neoplastic involvement of the bone, hypervitaminosis D (vitamin D allows calcium to be absorbed from the intestine), bone atrophy, and increased alkalinity in tissues, which lowers the solubility of calcium. The clinical significance, as with dystrophic calcification, depends upon the tissue being affected.

Other Forms of Mineralization

Calcinosis can occur in normal or injured tissue and is usually found in skin or subcutaneous tissue. There are two forms:

- *Calcinosis circumscripta* consists of localized deposits, usually found on the extremities (such as the elbows of large dogs) as a result of the pressure put on these areas.

- *Calcinosis universalis* is widespread calcification.

Heterotopic bone formation is the formation of bone where there's normally cartilage; it occurs with metastatic or dystrophic calcification. Arterial calcification, also known as arteriosclerosis, is the result of high cholesterol levels, which bind calcium (Figure 2-3). Lithiasis refers to the formation of pigment or

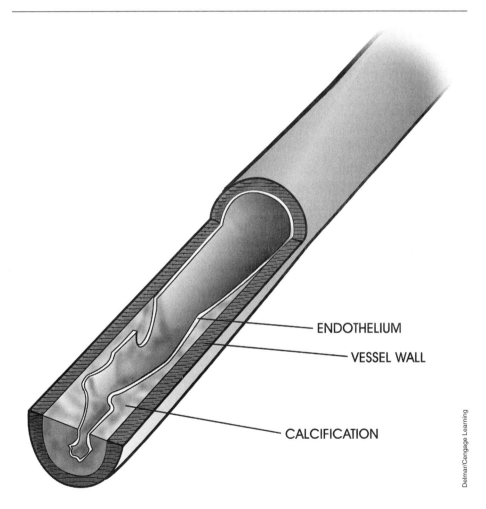

FIGURE 2–3

Arterial calcification can cause blockage within blood vessels.

ENDOTHELIUM

VESSEL WALL

CALCIFICATION

Delmar/Cengage Learning

protein "stones" or a calculus of any kind in hollow organs; examples include gallstones and kidney stones.

Pathological Pigmentation

Pathological pigmentation is an accumulation or deposition of abnormal amounts of pigment in tissue cells or tissue fluids. Colored materials or pigments are normally present (such as in the eye's iris) or deposited in a number of conditions in the skin or in the internal tissues. There are two classes of pigmentation.

The first class, *exogenous pigmentation*, results in conditions that fall under the category of "occupational hazards." Although most of these "environmental hazards" traditionally have not been seen in companion animal veterinary medicine, they are discussed briefly here for a better general overall understanding of pathology and to better prepare the RVT/CVT/LVT working in research institutions. In addition, many companion animals follow their owners into hazardous areas and thus are capable of contracting all of these occupational hazards as listed:

- *Anthracosis:* resulting from the inhalation of carbon (found in coal miners)

- *Silicosis:* resulting from the inhalation of silica dust

- *Siderosis:* resulting from the inhalation of iron dust

- *Asbestosis:* resulting from the inhalation of asbestos

- *Argyria:* an accumulation of silver in the skin or other tissue

- *Plumbism:* an accumulation of lead in the skin or other tissue (especially seen in avian practices)

The second class of pigmentation, *endogenous pigmentation*, refers to the three main types of pigment produced by the body. They are:

- *Melanin.* Made by cells called melanocytes, this is a brown-black pigment normally found in the skin, hair, and pigmented coat of the retina and inconstantly in the adrenal glands, and parts of the brain. Its presence is associated with cancer, such as malignant melanoma, or signified by "moles" and "liver spots." Its purpose in mammals is protection against solar ultraviolet radiation. In lower vertebrates, it is thought to be protective in inflammation.

- *Lipochromes (lipofuscin and ceroid).* These consist of a yellow-brown pigment made from the breakdown of fatty acids and cell membranes. Lipofuscins increase with advancing age and tend to develop in neurons, thyroid epithelium, and muscle cells. One autosomal recessive congenital disorder (neuronal ceroid lipofuscinosis) is seen in cattle, sheep, cats, dogs, and humans.

- *Hemoglobin derivatives.* Hemoglobin is the pigment within red blood cells responsible for the transport of oxygen. When these cells die, hemoglobin is released and generally broken down into amino acids, iron, and bilirubin, which is converted into bile by the liver. Alterations of this breakdown can produce a number of pigments, such as the following:

- Hemosiderin is a golden brown pigment that's found inside cells of the spleen and bone marrow. It is produced by the phagocytic digestion of hematin. The production of hemosiderin is the body's way of storing iron. However, excessive amounts may accumulate with increased red blood cell destruction.

- Hematin is a bluish black, amorphous substance containing iron in the ferric or Fe^{+3} form. Normally heme contains the iron in the ferrous or Fe^{+2} form (see methemoglobin below). Heme gives the red blood corpuscles their oxygen-carrying ability along with their red color.

- Methemoglobin is a bluish chocolate-brown pigmented product of Fe^{+3}. Methemoglobinemia occurs when the iron that is part of the hemoglobin is altered such that it does not carry oxygen well. Certain compounds introduced into the bloodstream can cause this oxidation. The iron in hematin is Fe^{+3} instead of Fe^{+2} (ferrous iron). The Fe^{+3} (ferric iron) does not combine with oxygen, forming methemoglobin instead of hemoglobin. Certain chemicals (nitrates, chlorates) also produce methemoglobinemia. Cattle grazing highly fertilized forage develop nitrate poisoning and cause the formation of methemoglobin. Livestock consuming certain weed killers that contain chlorates do so as well.

- Porphyrins are pigments found in both animal and plant life which can make urine, bones, tissues, and teeth red in appearance. Porphyrins act like chlorophyll. They are light sensitive; thus, people with excess porphyrins in their uncovered skin blister easily when exposed to light. Animals with porphyria have an inability (usually inherited) to produce an enzyme required for normal hemoglobin synthesis. This results in an overproduction of porphyrins and makes the animal more photosensitive. Thus animals tend to be hypersensitive to light for unpigmented areas of the skin. Porphyria has been reported in cattle, swine, cats, and humans. Animals with porphyria excrete abnormal uroporphyrins and exhibit brown discoloration of their teeth.

- Bilirubin is a red pigment formed from hemoglobin during both normal and abnormal destruction of erythrocytes by the reticuloendothelial system. In healthy animals, after 3 to 4 months, red blood cells are destroyed by this system releasing hemoglobin into the blood. Macrophages engulf the cellular debris; iron is recycled to form new red blood cells or sent to the muscles to form myoglobin. The remaining product after the protein and iron are stripped away from the hemoglobin is called biliverdin. Biliverdin eventually is reduced to bilirubin and transported to the liver and then to the gallbladder in the bile. An excess of bilirubin in the blood causes the skin and mucous membranes to appear yellow or jaundiced (icterus). Bilirubin excess is generally a sign of liver disease, occlusion of the bile ducts, or excessive red blood cell destruction.

- Hematoidin is another golden brown pigment that's probably composed of the same chemical as bilirubin, differing only in its site of origin. Hematoidin is frequently found in scars and necrotic debris. It contains no iron.

| CASE STUDY | Rat Poisoning: Rodenticide Toxicity (Cell Damage) |

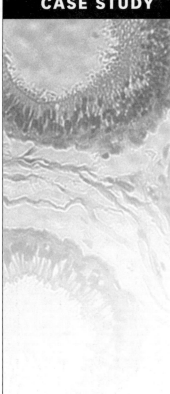

A black Labrador (canine) was presented with vomiting, lethargy, and ataxia. The black Lab also had pale mucous membranes. The owner had noticed a container of rat poison chewed open in her garage, which the dog had access to from the dog run. The owner believed it must have happened the previous night.

Rodenticides of the first generation, such as warfarin (coumarin) or diphacinone, can depress the production of the clotting factors (factors I, II, VII, IX, and X) in the liver for 7 to 10 days. The second generation, such as bromadiolone, can depress the clotting factors for 3 to 4 weeks. The rodenticides inhibit the vitamin K–dependent coagulation factors by blocking the activity of vitamin K in forming carboxylated proteins (clotting factors), which is the conversion of vitamin K to its active form. Clinical signs start to show after the depletion of active clotting factors. This usually happens 24 to 48 hours after the animal has ingested the poison.

Hepatocytes are metabolically active cells that serve many functions. For example, they take up glucose, minerals, and vitamins from portal and systemic blood and store them. In addition, hepatocytes can produce many important substances needed by the body, such as blood clotting factors, transporter proteins, cholesterol, and bile components. These cells are compromised in a poisoning situation.

Diagnosis is based on the history of exposure to the toxin, prolonged bleeding times determined by the PIVKA (Thrombotest), and the response to vitamin K therapy. Response to vitamin K therapy usually happens within 24 to 48 hours after the initial treatment. Two other options for diagnosis are the activated partial thromboplastin time (APTT) and activated coagulation time (ACT) tests.

TREATMENT

Inducing vomiting (if not contraindicated) can be done within a few hours of ingestion. Administering activated charcoal, such as Toxiban, is another option. In severe stages, whole blood transfusions may be necessary to replace the blood clotting factors and red blood cells if the animal is anemic. If the animal is not anemic, fresh frozen plasma can be used to replenish the clotting factors. Vitamin K_1 should be administered every 24 hours until the toxic levels are no longer in the animal.

After blood work was performed, the Labrador presented severely depressed with declining packed cell volume (PCV). He was given fresh frozen plasma, IV upon presentation, plus vitamin K_1 treatment. This provided the needed clotting factors to stop localized areas of bleeding, especially epistaxis (bleeding from the nose).

Summary

Cell injury can be progressive, sublethal, or lethal if leading to cell death. Reactions to cell injury can be observed microscopically. Factors that affect how injured cells respond to injury include both the environment and genetics. Examining the type and location and the extent of cell injury and the exposure of the animal to various environmental toxins and cell irritants can aid the veterinary staff in diagnosing and treating animal disease. Tissue necrosis is classified based on cytoplasmic changes and by cause. Pathological mineralization or calcification occurs when abnormal depositions of calcium salts occur in various body tissues. Pathological exogenous pigmentation can result from environmental hazards.

REVIEW

FILL IN THE BLANKS

1. _____ is a brown-black pigment normally found in the skin, adrenal glands, and parts of the brain.

2. _____ is an accumulation of extracellular, proteinlike substances in tissues and organs.

3. _____ is the pathological death of one or more cells.

4. _____ is a red bile pigment formed from hemoglobin during destruction of red blood cells and found in the blood; an excess causes the skin and mucous membranes to appear yellow or jaundiced.

5. _____ is the reduction in size of a fully developed tissue.

6. _____ is a localized collection of pus in tissue.

7. _____ is endogenous programmed cell death or normal tissue.

8. _____ is a cheeselike necrosis that typically involves the slow, progressive destruction of cells and that prevents a high response of leukocytes.

9. _____ is a condition in which animals tend to be hypersensitive to light for unpigmented areas of the skin and in which the animal has an inability to produce an enzyme required for normal hemoglobin synthesis.

10. _____ is a type of necrosis that involves a deficiency of blood to an area and therefore a lack of oxygen.

DISCUSSION QUESTIONS

1. Name three of the six types of cellular degeneration.

   _____

   _____

   _____

2. What are the five types of cellular necrosis?

   _____

   _____

   _____

3. What are the two basic cellular responses to injury?

   _____

   _____

   _____

4. By what methods can exogenous pigments gain entry into the body?

_____

_____

_____

CHAPTER 3

Inflammation and Healing

OBJECTIVES

Upon completion of this chapter, the reader should be able to:

- List the cardinal signs of inflammation.
- List the main components of acute inflammation.
- Describe the inflammatory response.
- Cite the functions of leukocytes at the sites of inflammation.

KEY TERMS

abscess

adhesions

anastomosis

bacteremia

calor

catarrhous (or catarrhal)
 exudate

cellulitis

chemotaxis

dolor

emigration

erosion

exudation

fibrinopurulent exudates

fibrinous exudate

granulation tissue

granuloma

hemodynamic

hemorrhagic exudate

histamine

incision

inflammation

keloid

leukocytosis

leukopenia

lymphocytosis

lymphokines

mediator

monocytosis

neoplasm

phagocytosis

phlegmon

prostaglandin

pseudomembranous exudate

purulent or suppurative
 exudate

Inflammation

According to *Stedman's Medical Dictionary*,[1] **inflammation** is defined as "a fundamental pathologic process consisting of a dynamic complex of cytologic and histologic reactions." These reactions occur in affected blood vessels and adjacent tissues in response to an injury or abnormal stimulation caused by a physical, chemical, or biological agent, including (1) the local reactions and resulting morphological changes, (2) the destruction or removal of the injurious material, and (3) the responses that lead to repair and healing. Figure 3-1 illustrates the visible elements of inflammation that result from a simple flesh wound.

The four main cardinal signs of inflammation are **rubor**, or redness; **calor**, or heat (warmth); **tumor**, meaning swelling; and **dolor**, or pain. A fifth sign, *functio laesa*, which refers to inhibited or lost function, is sometimes included as well. Redness and warmth result from an increased amount of blood in the affected tissue, which is usually congested. Swelling ordinarily occurs from this congestion or migration of fluid from vessels and from exudation. Pressure on (or the stretching of) nerve endings, as well as changes in osmotic pressure and pH, may lead to pain. Finally, a disturbance in function may result from the discomfort of certain movements or the actual destruction of an anatomical part. All of the preceding signs may be observed in certain instances, but no one of them is necessarily always present.

Inflammation literally means "burning." This symptom is the body's most important and useful host defense mechanism and the most common cause of tissue injury as well. A significant number of animals die from inflammatory disease every year. The processes of inflammation, healing, and repair all occur at the same

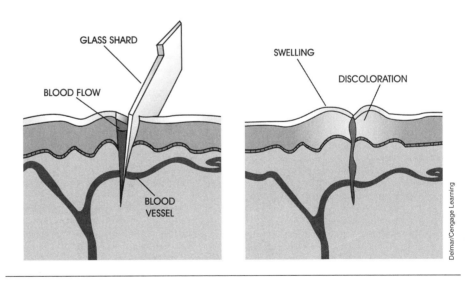

FIGURE 3–1

The right-hand illustration shows the local reactions to an injurious agent (a piece of glass) and the morphological changes that occur at tissue and blood levels.

[1] *Stedman's medical dictionary* (26th ed.). (1995). Baltimore: Williams and Wilkins.

KEY TERMS (continued)

pyrogen

regeneration

repair

resolution

rubor

scarring

septicemia

serotonin

serosanguineous exudate

serous exudate

systemic

toxemia

tumor

ulcer

vascular permeability

vasodilatation

vasodilation

time or lead into one another. This is because inflammation involves a reaction of living tissue to local injury. The reaction involves changes in the following:

- Vascular bed (blood vessels and capillaries)
- Blood (the liquid component as well as the cells)
- Connective tissue

The processes, which occur in these three tissues, are designed to eliminate the irritant and repair the damaged tissue. As stated previously, inflammation is a process involving multiple participants on the cellular, humoral (blood), and tissue level. It occurs only in living tissue, because only living tissue is vascularized. The series of overlapping events that occur during inflammation end in the healing or destruction of the tissue. Even though these events occur in an orderly fashion, the process can restart itself repeatedly. Inflammation is an evoked response, meaning that it's called into action by the body itself. This process is known as cause and effect.

The body's inflammatory response can actually be more harmful than the cause of the inflammation. This response is fundamentally a defense reaction, always following the same steps regardless of its cause. The amount of time that each step is experienced and the severity of the reaction are the only factors that vary among instances. Most of the body's reactive components are in the blood. Certain chemical factors exist that help control inflammatory reaction.

Three terms are used to describe the severity of an inflammatory reaction: mild, moderate, and severe. A mild inflammatory reaction involves little or no tissue destruction and only slight evidence of vascular involvement. There's also little exudation. **Exudation** refers to the presence of fluids and the migration of inflammatory cells into an inflammatory lesion.

A moderate inflammatory reaction involves some tissue damage and a visible host reaction. The host reaction refers to the presence of inflammatory cells and vascular phenomena. A severe inflammatory reaction exhibits considerable tissue damage and abundant exudation.

A number of terms are used to describe the location of lesions within an organ or tissue. Becoming familiar with these terms will make it easier for you to describe the gross appearance of inflammation:

- *Focal:* pertaining to a single abnormality, or focus
- *Multifocal:* meaning there are many scattered *foci* (plural of focal)
- *Locally extensive:* meaning that all of the tissue within a particular zone is affected
- *Diffuse:* involving an entire organ or tissue

Inflammatory processes can last for a short period of time or lag on indefinitely. They can also have different onsets:

- *Peracute:* inflammatory episodes usually caused by a potent stimulus and becoming apparent rapidly
- *Acute:* inflammatory episodes occurring 4 to 6 hours following the stimulus and remaining fairly constant in appearance

- *Subacute:* inflammatory reactions that are characterized by a decrease in the redness of the tissue and fluid accumulation

- *Chronic:* a clinical concept referring to the existence of a persistent inflammatory stimulus, usually accompanied by (1) an immune response, (2) evidence of repair (such as scarring), and (3) certain types of inflammatory cells found within the lesion

The Acute Inflammatory System

An acute inflammatory event is the response of living tissue to a local injury in which the reaction has a fairly rapid onset and a clear and distinct end. It's basically an exudative phenomenon, with blood fluid and cells entering the affected tissue. Remember, exudation refers to blood cells and fluid oozing through the blood vessel wall and into the tissue. An exudate is thus high in protein, cells, or solid materials derived from these cells. The term *exudate* comes from the Latin prefix *ex* meaning "out" and the Latin suffix *sudo* meaning "to sweat."

Vasodilatation (also known as **vasodilation**), or the dilation of blood vessels, results in increased blood flow to the affected area and, in turn, the onset of heat and redness. An increase in **vascular permeability**, or "leakiness" of the blood vessels, occurs as a result of the stretching of cells lining the blood vessels and the formation of gaps between them. This phenomenon allows for the escape of fluid, which causes swelling. Increased sensitivity or pain occurs at the affected area due to pressure or irritation. Along with these signs, there may be a decrease or loss of function.

Vascular Changes in Tissue

During inflammation, vascular changes occur in every organ system. Initially, there's an increase of blood flow, due primarily to the dilation of arterioles (small arteries), capillary beds, and venules (small veins). These are known as **hemodynamic** changes.

Increased vascular permeability follows, primarily in the venules and then the capillaries. Normal low venous pressure is insufficient to return blood back to the heart. As veins pass between skeletal muscle groups, a massaging or pumping action is produced. This and one-way valves within the venous system ensure the blood flows in one direction: toward the heart. However, because of their pumping action, venules are less likely to drain; this allows fluid to escape from the vasculature. The fluid leakage and changes in the amount of dissolved salts in the leaked fluid can cause the blood flow to slow or even stop. Endothelial cells contract, allowing even more leakage.

Cellular events now begin to take place. Blood elements, which include the white blood cells (WBCs) and red blood cells (RBCs), may stop or tumble out of the affected vessels. Certain WBCs marginate, meaning they stick to the inside lining of the blood vessel wall. A clumping of many cells in the vessel can also occur, possibly due to an electrical charge.

Emigration, which refers to cells leaving the blood vessels to enter tissue, may occur as a response to chemical cues. **Chemotaxis** is the chemical process by which cells orient themselves and "home in" on a target; this process directs migration in the tissues. **Phagocytosis**, the ingestion of such things as dead cells, foreign particles, or bacteria by WBCs, occurs when WBCs enter the inflammation site (Figure 3-2). As mentioned earlier, this accumulation of cellular debris, liquefied by proteolytic enzymes and WBCs, is called pus. Of course, WBCs are necessary for removing any damaged or dead cells as well as the agent that caused the inflammation.

FIGURE 3-2

Phagocytosis is said to occur when a white blood cell "consumes" bacterial and foreign particles or dead cells. The white blood cell membrane accomplishes this by enveloping the bacteria within its walls.

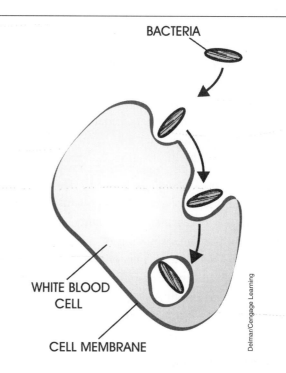

BACTERIA

WHITE BLOOD
CELL

CELL MEMBRANE

Delmar/Cengage Learning

Factors Influencing the Acute Inflammatory Response

The body organizes inflammatory reactions like a well-rehearsed play. There are multiple actors with multiple jobs. The body controls its "actors" in a variety of ways, primarily by chemicals and the lymphatic system. **Mediators** are the components that help control the inflammatory reaction.

Chemical mediators include chemicals such as vasoactive amines, which are combinations of nitrogen and chemicals that can constrict or dilate blood vessels. Among the more common vasoactive amines are histamine and serotonin. **Histamine** dilates blood vessels and increases vascular permeability. When you have a head cold, your nose runs because of the inflammation and release of histamines. Histamine dilates the blood vessels in your nose, increasing their "leakiness"; this dilation results in a runny nose. That's why you would take an antihistamine to counteract the histamine. **Serotonin** is found in high concentrations in some areas of the central nervous system. It is released, however, by blood platelets (thrombocytes) during the inflammatory response and is responsible for stimulating smooth muscles and inhibiting gastric secretion. Serotonin is a vasoconstrictor.

Another category of chemical mediators is the plasma proteases. Members of this group include a group of chemicals called complement and coagulation factors.

Prostaglandins form the third chemical group and are the most important vasodilators. Drugs such as aspirin help inhibit prostaglandins, lessening the severity of headaches by decreasing the inflammation in the head. The last group of chemicals is the **lymphokines**. The most important of these, the interferons, are being developed to treat a number of diseases because of the way they affect cells, particularly WBCs. Lymphokines are hormone-like peptides; they are produced by T lymphocytes and help to regulate and promote the interaction of cells within the immune system.

The lymphatic system is utilized to help drain fluids away from sites of inflammation. The cells that make up the walls of these capillaries get stretched apart, allowing fluid to flow in and away from the inflamed area.

Types of Inflammation Based on the Nature of Exudate

As you learned earlier, an exudate consists of the fluid that collects within tissues that are experiencing inflammation. This fluid comes from leaky blood vessels, which are the result of hemodynamic changes and increased vascular permeability. The nature of exudate can allow us to classify inflammations in a different manner. The following list organizes the types of exudate that are currently known to exist:

- **Serous exudate** is a fluid low in dissolved protein; it has no clotting factor and a low number of cells. Cutaneous blisters are examples of sores that contain serous exudate.

- **Serosanguineous exudate** is a fluid containing both serum and blood. It is a thin, watery exudate.

- **Purulent or suppurative exudate** is a cloudy yellow, white, or green fluid containing many leukocytes; it's often referred to as "pus."

- **Fibrinopurulent exudate** is a fluid containing the clotting protein fibrin and pus.

- **Fibrinous exudate** refers to a substance containing fibrin but lacking pus because of a severe necrotizing infection. Fibrinous exudate is seen in "hardware disease" in cattle.

- **Hemorrhagic exudate** contains many RBCs—essentially, blood.

- **Catarrhous (or catarrhal) exudate** contains mucus.

- **Pseudomembranous exudate** is found in the respiratory and gastrointestinal tract; it consists of areas of tissue that mat together (due to fibrin) and appear membranous.

Types of Inflammation Based on Location

In addition to "naming" inflammatory processes by duration and exudate, we can also name them by location:

- **Abscesses** are localized collections of pus in tissue. The pus can thicken or inspissate.

- **Cellulitis** is an inflammation of cellular or connective tissue.

- **Phlegmon** is an obsolete term that describes a suppurative inflammation of subcutaneous connective tissue. A common example is pharyngeal phlegmon.

- **Ulcers** constitute the loss of a superficial layer of an organ or tissue, with acute inflammation at the base of the lesion.

- **Erosion** is a shallow ulcer usually limited to the mucosa with no penetration of the muscularis mucosa.

Systemic Reactions to Inflammation

Because of the variety of tissues that make up a living creature as well as the number of different cells and chemicals involved in an inflammatory reaction, a number of reactions can occur that are systemic in nature. **Systemic** means that they occur throughout the entire body. Fever is a common systemic reaction. Substances called **pyrogens**, released by WBCs, cause the body temperature to increase.

Leukocytosis is an increase in the total leukocyte count of the blood and is one reason veterinarians perform complete blood counts (CBCs) when they suspect infections. **Leukopenia** is a decrease in the total number of circulating leukocytes. This can occur when certain inflammatory processes "consume" more leukocytes than the body can manufacture. If the body can keep up with or surpass the demand for leukocytes, **lymphocytosis** (an increase in lymphocytes) and **monocytosis** (an increase in mononuclear cells, or monocytes) are common, as these cells help to influence the inflammatory response.

Bacteremia is a condition by which the body is unable to keep bacteria from circulating into the blood. Some bacteria make toxins (poisons). If the bacteria and their poisons get into the blood, a condition known as **septicemia** develops. If just the toxin circulates, this is known as a **toxemia**.

Chronic and Granulomatous Inflammation

The persistence of an injurious agent for weeks to years provides a continual stimulus for an inflammatory response. Chronic inflammation can be the result. The chronic inflammatory response is primarily caused by an increase in the number of reactive cells, as opposed to the fluid generated in an acute inflammatory response. A chronic response may last anywhere from weeks to years. The three factors that induce the reaction are (1) the extent of the inflammation, (2) the intensity, and (3) the age of induction.

Again, the swelling in chronic inflammation is due to large numbers of cells, not fluid. A buildup of fibrous connective tissue from this cellular response can prevent an abnormally functioning organ from scarring. **Scarring** is characterized by contraction and distortion of the tissue or organ. The location of the inflammation can affect organ function as well. For example, a narrowing of the colon can result in the formation of **adhesions** (a result of chronic inflammation), which in turn affects the animal's ability to have a bowel movement. Adhesions are simply the union of apposing inflammatory bands of tissue.

The initial cause of chronic inflammation may be difficult to identify due to time lag. Repeated or continuous trauma is frequently involved. Some common examples of these are bladder or kidney stones, a sore tooth, bacterial or viral infections, anatomical abnormalities, injuries, and rheumatoid arthritis. Figure 3-3 shows the severe intestinal inflammation brought about by Johne's disease in calves.

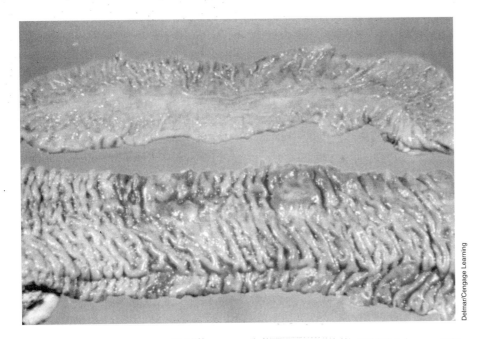

FIGURE 3-3

A normal intestine (top) and an intestine from a calf with Johne's disease (bottom). Johne's disease is the result of a bacterial infection that causes severe inflammation and a widespread thickening of the intestine itself.

One type of chronic inflammation, granulomatous inflammation, consists of focal inflammatory lesions in which the tissue reaction is primarily of chronic inflammatory cells. Chronic inflammatory cells include macrophages, giant cells, lymphocytes, plasma cells, and fibroblasts. Vascular leakage of fluid and blood cells is usually minimal or absent.

In general, **granulomas** consist of a sometimes necrotic central core containing a foreign body, which in turn is surrounded by zones of chronic inflammatory cells. Older granulomas tend to contain more collagen, a type of connective tissue, and may eventually be represented by fibrous nodules in which there's little or no evidence of the original cause of inflammation.

A common cause of granulomas in cattle is the inhalation of feed particles that are contaminated with bacteria. These particles are surrounded by neutrophils (a type of leukocyte and the body's first line of defense against infection), monocytes (which become macrophages), foreign-body giant cells (from the fusion of macrophages), epithelioid cells, lymphocytes, plasma cells (B-cell lymphocytes that are "turned on" and produce antibodies), a proliferation of fibroblasts, and eosinophils (a type of leukocyte often associated with parasites and inflammation).

The gross morphology of a chronic inflammatory lesion is a discrete white-gray, roughly spherical mass that's well defined. These lesions may be difficult to detect in the lung. They're often focal and small—sometimes only one millimeter in diameter. If the inflammation occurs over a large area, individual foci are no longer discrete, and the lesion may come to resemble a **neoplasm**, or tumor.

Throughout a normal course of events, the reaction subsides, collagen in the periphery contracts, and the central area may calcify. In our example of a cow inhaling feed particles, the bacteria are destroyed, the dead cells in the center of the lesion are removed, and the core solidifies with calcium. When the surrounding collagen contracts, the result is a tiny ball of scar tissue surrounding a solid nugget of "stone"—the calcium.

Healing of Injured Tissue

You have learned that inflammation is a continuum that, ideally, leads to the healing of injured tissue. Healing relies on the ability of different tissues to either regenerate or repair. The regeneration or repair of an injured tissue is often crucial to the survival of the mammal itself. But these two processes differ from one another in their final results.

During **regeneration**, the tissue in question undergoes restructuring by the cells that belong there. An example of this is the lizard that breaks off its tail to escape an enemy. The tail eventually regrows and resembles the original tail.

Repair is the replacement of destroyed tissue by connective tissue. An example of repair is a puppy that has its tail docked at birth; the stump can't regenerate. So instead, the end heals with the formation of a scar.

Two other terms you should become familiar with at this point are resolution and granulation tissue. **Resolution** is the absorption or breaking down and removal of the products of inflammation. **Granulation tissue** is another name for vascular connective tissue.

The length of time it takes for healing to occur is an important factor in determining if repair or regeneration will occur. The longer the inflammatory reaction goes on, the more collagen and scar tissue is produced. Scar tissue implies a process of repair, because it takes up the space where cells that belong in the tissue would normally be found.

Many factors stimulate cell growth and influence healing. These include the following:

- Epidermal growth factor

- Platelet-derived growth factor

- Fibroblast growth factor

- Macrophage-derived growth factor

- The presence of fibrin and fibrin-degradation products

- The loss of contact inhibition

- Inhibitory substance loss (chalones, which are mitosis-inhibiting factors)

The action of steroids is often assumed to stimulate cell growth; however, steroids don't stimulate cell growth! They decrease the inflammatory response through a variety of mechanisms and contribute to faster healing of an inflammatory lesion.

In the course of their work, veterinarians often find it necessary to operate on animals. This requires making deliberate wounds in an animal known as **incisions**. With a clean incision the edges are apposed together and skin granulation can start in 24 to 48 hours. A healthy immune system, antibiotics, and a sterile surgery field help to ensure that no infection results from the wound (incision). Because these wounds are kept clean and sutured to ensure that no infection develops, they heal faster than other injuries, such as untreated cuts from broken glass.

A surgical incision that's sutured forms a primary union. The healing of such a wound is termed first intention. Healing by second intention occurs when there's an extensive dermal wound with complete loss of the epithelium. The healing process is identical in both, but faster in the first intention process and without much scarring because of the small size of the epithelial destruction.

Let's examine the healing process of a clean surgical wound over a period of 5 days:

- *Day 1:* The defect fills with fibrin. A blood clot, or scab, forms.

- *Day 2:* Neutrophils release lysosome contents to break down the clot. Neutrophils phagocytize bacteria, dead cells, and debris.

- *Day 3:* Macrophage infiltration occurs in order to remove the fibrin and other cellular debris (Figure 3-4). With contaminated wound sites this phase can last for several weeks.

- *Day 4:* Capillary bud ingrowth begins at the ends of the damaged vessels. They grow from about 0.1 to 0.3 mm per day.

- *Day 5:* Activated fibroblasts produce collagen after 5 days, forming fibrovascular tissue (Figure 3-5 A and B). This is known as epithelial regeneration.

The initial injury causes a breakdown of collagen during the first 4 to 5 days due to enzyme release. Collagen is then produced over a 2-month period in the wound, a plateau is reached, and the wound regains its maximum strength. However, only 80 to 85 percent of the original structure now remains.

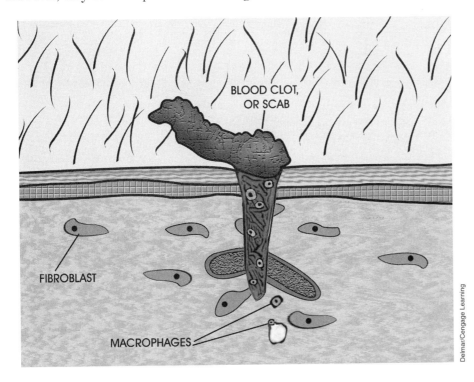

FIGURE 3–4

Macrophages infiltrate the wound to "clean it out" in preparation for repair.

Delmar/Cengage Learning

FIGURE 3–5

The collagen deposited by activated fibroblasts re-forms the wounded tissue (top). As time passes, the tissue undergoes further repair and strengthening until the wound is completely healed (bottom).

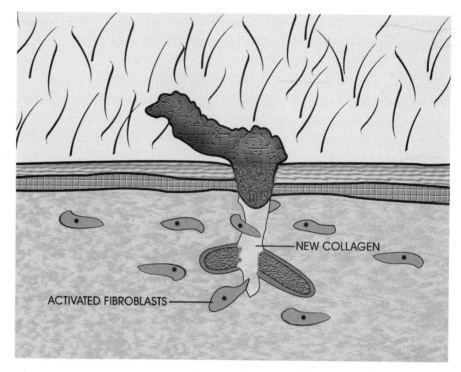

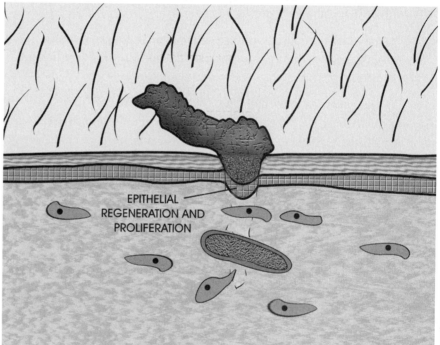

Delmar/Cengage Learning

Sometimes a **keloid**, an abnormally large scar, forms due to excess collagen production. Horses, for instance, will frequently exhibit a tumorlike mass of collagen and granulation tissue. Nerves show little regeneration, although there may be some innervation to vessels. Epithelial cell proliferation allows any gaps to be bridged after the scab is removed by the leukocytes. Structures such as hairs or sweat glands are re-formed if remnants were originally present. Decreased cellularity follows, as well as resolution of any edema. **Anastomosis**, or a uniting of the capillaries, follows. The lesion becomes smaller as exudate lessens and the fibroblasts contract. Breakdown of collagen and restructuring occurs, with a scar being the final result.

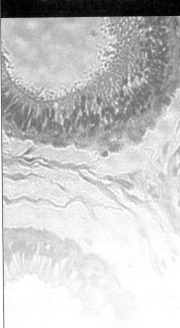

CASE STUDY

Disseminated Intravascular Coagulation (Mammary Tumor Neoplasia)

A 9-year-old mixed-breed Labrador female dog was presented for evaluation of mammary tumor, hematuria, and gingivae that bleed easily. The CBC results showed a decrease in platelets and a high count of WBCs. The cytology showed (see Figure 3–6) a numerous quantity of schistocytes (RBC fragments), and neutrophils that look toxic (Döhle bodies) and have basophilic vacuolated cytoplasm. Döhle bodies (toxic neutrophils) in the blood smear suggested bacterial infection, poisoning, or endotoxin. The mixed-breed Labrador had a prolonged thromboplastin time and prolonged activated coagulation time. The owner made the decision to euthanize his dog since she was suffering from neoplasia (cancer) and disseminated intravascular coagulation (DIC).

DIAGNOSIS

The veterinarian initially made a clinical diagnosis as regenerative anemia. Platelet decrease with schistocytes suggesting DIC. The mammary tumor neoplasia was primary and the DIC was secondary to the cause. DIC is not a primary disease, but a disorder secondary to several triggering events such as bacterial, viral, protozoal, or parasitic diseases; heat stroke; burns; neoplasia; or severe trauma (shock). To confirm DIC, coagulation screening tests such as prothrombin time, activated partial thromboplastin time, and an activated coagulation time are performed. In the case of DIC, death is generally caused by extensive microthrombosis or circulatory failure, which leads to single or multiple organ failure. Treatment is usually directed toward the primary cause (underlying problem). Supportive care is necessary. Administration of electrolyte solutions is imperative to maintain effective circulating volume within the body. Prognosis is usually poor, especially if DIC is not detected early.

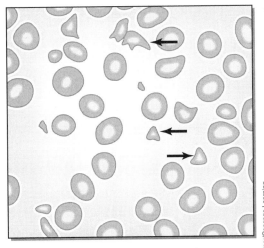

Schistocytes (RBC fragments)

Delmar/Cengage Learning

FIGURE 3–6

Arrows point to numerous schistocytes (RBC fragments).

CASE STUDY | Johne's Disease (Paratuberculosis)

A rancher notices that four of his calves have had thick diarrhea for a few weeks. The four calves are bright and alert with great appetites, but they are losing weight. The diarrhea has no blood or mucus, but there is an offensive odor. The feedlot rancher has poor management skills. The pen where the calves are being housed is contaminated with feces. The calves are being infected by nursing on their mother's udders, which are completely contaminated with feces. The calves have ingested the causative organism known as *Mycobacterium avium paratuberculosis (Mycobacterium paratuberculosis)*.

DIAGNOSIS

Because there is not one single test for Johne's disease, a combination of a few tests are performed to diagnose the disease. A microscopic exam of a fecal sample using Ziehl-Neelsen stain can be used to diagnose mycobacteria in heavy shedders of the disease. Cultures, serological test, and Enzyme-linked Immunosorbent Assay (ELISA) test are other forms that can be used for diagnosis. The intradermal Johnin test (cell-mediated immunity) detects infected animals before an outbreak occurs. Mycobacteria lesions can be found in the intestines with a histopathological examination during a necropsy. An acid-fast stain, Ziehl-Neelsen, can help to reveal these lesions. The ileum is thickened by granulomatous inflammation. This leads to the clinical signs of diarrhea and weight loss. A condition known as a protein-losing enteropathy also occurs (i.e., protein absorption is impaired and excess protein is lost in the feces). The result can be measured as subnormal levels of total serum proteins.

Johne's disease is incurable, but it can be controlled with good sanitation and management skills. Herds should be tested that show signs of this disease. The positive animals are slaughtered. Vaccines are available but are usually not recommended due to their efficacy and the uncertainty of the antigens of this organism.

Summary

Inflammation consists of a complex of both cytological and histological reactions resulting in morphological changes that can be identified by the veterinary technician. There are many mediators that help animals control the inflammatory reaction. These chemical compounds include histamines, prostaglandins, lymphokines, and serotonin. Types of inflammation are classified by their location and by their nature of exudate. The healing process follows a structured order varying in length of time based on the insult to the tissue, contamination of wound site with microorganisms, the animal's immune system, and several other mitigating factors.

REVIEW

FILL IN THE BLANKS

1. _____ is a nodular inflammatory lesion containing a necrotic core with a foreign body surrounded by chronic inflammatory cells.

2. _____ is an obsolete term meaning discharge pus.

3. _____ is an inflammation of cellular or connective tissue.

4. _____ is a term that describes fluid that collects within tissues that are experiencing inflammation.

5. _____ are one type of chemical produced by the body as a vasodilator. The effects of this particular endogenous chemical are mediated by the use of aspirin.

6. _____ is a vasodilator released by thrombocytes which stimulates smooth muscle and inhibits gastric secretion.

7. _____ dilates the blood vessels, increasing their permeability.

8. _____ is the ingestion of such things as dead cells or bacteria by white blood cells.

9. _____ are localized collections of pus in tissue.

10. _____ is a condition by which the body is unable to keep bacteria from circulating into the blood.

11. _____ are inflammatory bands that connect opposing serous tissues.

DISCUSSION QUESTIONS

1. What are the three levels on which an inflammatory response occurs?

   _____

   _____

   _____

2. What important task do white blood cells perform at sites of inflammation?

   _____

   _____

   _____

(continued)

REVIEW *(continued)*

3. What are the main signs of inflammation? The main causes?

_____

_____

_____

4. Explain the difference between repair and regeneration.

_____

_____

_____

5. What is the relationship of histamines to the inflammatory response?

_____

_____

_____

6. Are there any beneficial effects of inflammation on the body?

_____

_____

_____

7. What are the main harmful effects of inflammation on the body?

_____

_____

_____

8. What is keloid formation and what causes it?

_____

_____

_____

CHAPTER 4

Fluid and Circulatory Disturbances

OBJECTIVES

Upon completion of this chapter, the reader should be able to:

- Trace the flow pattern of blood through the heart and body.

- List the blood components and explain their basic functions.

- Describe the distribution of fluid between the intracellular, extracellular, intravascular, and interstitial compartments.

- Define hemodynamic disturbances and give examples both of a physiological and pathological nature.

- Define and list clinical symptoms of hemorrhage.

- Describe the steps of hemostasis.

- Define and explain the pathogenesis of shock, emboli, and edema.

KEY TERMS

basophil

blushing

complete blood count (CBC)

congestion

dehydration

ecchymosis

edema

embolism

embolus

endothelium

eosinophil

epistaxis

erythrocyte

fibrin

hematemesis

hematoma

hematuria

hemopericardium

hemoperitoneum

hemoptysis

hemorrhage

hemostasis

hemothorax

hyperemia

hypotension

hypovolemic shock

infarction

interstitial

intracellular

intraintestinal

intravascular

leukocyte

lymphocyte

melena

monocyte

myocardial infarction

Review of the Circulatory System

The circulatory system's function includes transportation of cells, antibodies, nutrients, waste products, hormones, and gases throughout the body. There is a normal range for each of these products that can be measured and assessed in a clinical setting. The heart serves as a pump for this system. There are three major types of blood vessels. Arteries take blood away from the heart, veins bring blood back toward the heart, and capillaries allow the exchange of these products between the extracellular fluid and the blood.

Understanding the normal flow pattern of blood through the heart assists the veterinary technician in detecting abnormalities which can occur and which may be monitored by various examination procedures (Figure 4–1).

Blood enters the heart on its right side through two large veins: the cranial and caudal venae cavae. Blood is returned to the heart after picking up carbon dioxide and exchanging it with oxygen. Thus this oxygen-poor blood must be reoxygenated before returning to the body. The vena cava brings blood into the right atrium. As the right atrium contracts, it forces blood through a one-way valve known as the tricuspid or right atrioventricular valve (RAV), and into the right ventricle. When the right ventricle contracts, the RAV closes—forcing blood through another one-way valve, the pulmonary semilunar valve, and into the pulmonary artery. This oxygen-poor blood eventually reaches the pulmonary capillaries for exchange of carbon dioxide with oxygen.

Oxygen-rich blood returns to the heart from the lungs via pulmonary veins entering the heart at the left atrium. The left atrium contracts, forcing blood through the bicuspid valve, also known as the left atrioventricular valve (LAV) or mitral valve. Blood leaves the heart when the left ventricle contracts. The LAV closes, forcing the newly oxygenated blood through a one-way valve, the aortic semilunar valve, and into the largest of the body's arteries—the aorta.

Each heartbeat starts without any outside nerve stimulus. The heart's pacemaker, also known as the sinoatrial node (SA), located within the right atrium and close to the vena cava, controls the heart's rhythmic contractions. In order for the heart to properly function, the two atria must contract before the two ventricles. There is a second node located at the junction of the atrium and ventricle, the atrioventricular (AV) node. It receives input from the two atria. Its function is to slow the signals and allow the ventricles to fill before they contract. The AV node prevents rapid conduction to the ventricles in cases of rapid atrial rhythms. The contraction impulse then continues to the bundle of His–Purkinje conduction system which spreads across the two ventricular walls.

The stethoscope allows the veterinary staff to assess rate and rhythm of the heart. More specifically trained staff may pick up abnormal sounds that might indicate problems with the heart valves, skipping heartbeats, or the addition of extra heartbeats. As part of the physical exam the chest should be auscultated to determine if a heart murmur, arrhythmia, muffled heart, and lung sounds (pulmonary crackles and wheezes) are present.

KEY TERMS (*continued*)

neutrophil

normovolemic shock

petechia

plasma

platelet

platelet plug

purpura

shock

thrombocytopathy

thrombocytopenia

thrombus

vasoconstriction

water intoxication

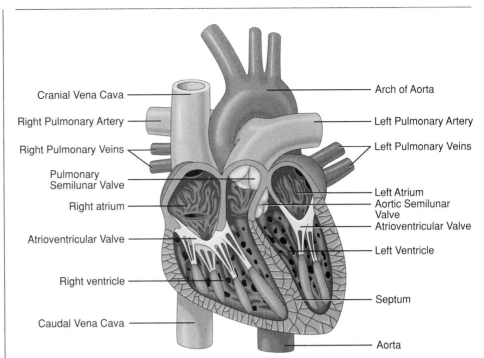

A.

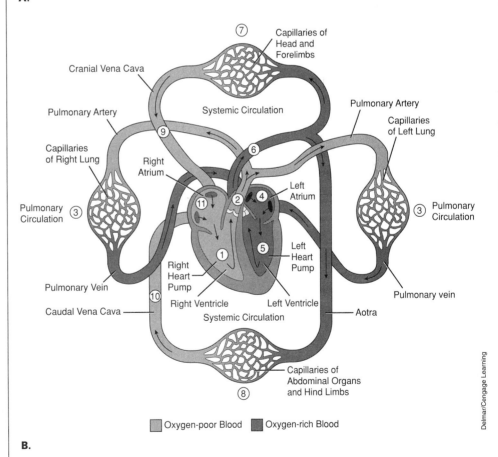

B.

FIGURE 4–1

A. The internal structure of the heart. B. The flow of blood through the body. (Diagram courtesy of Lawhead and Baker, Introduction to Veterinary Science. 2005. Delmar Publishers.)

Blood

Blood consists of a fluid portion called plasma and the cellular portion composed of red blood cells, white blood cells, and platelets. The **plasma** constitutes greater than one-half of the blood volume and is 90 percent water. The remaining 10 percent is composed of large protein molecules which help keep the fluid within the vascular system, electrolytes, antibodies, nutrients, waste products, and clotting factors. Red blood cells account for 99 percent of the cellular portion. Only 1 percent is composed of white blood cells and **platelets**.

Blood can be separated into its major components by centrifugation. The heavier portion (red blood cells) goes to the bottom, and the lighter portion (plasma) rises to the top of the hematocrit tube. A very thin middle layer is composed of white blood cells and platelets.

The portion of whole blood volume that is occupied by red blood cells is the hematocrit (Hct), also known as the packed cell volume (PCV). Low PCV values may indicate anemia whereas elevated PCV values can be caused by stress, dehydration, splenic contraction, as well as other factors.

The red blood cells **(erythrocytes)** are the most numerous of the cells. They carry a molecule of hemoglobin which increases the carrying capacity of oxygen. In most mammals, red blood cells are shaped like biconcave disks and contain no nuclei. In birds, reptiles, and fish they are normally nucleated and egg shaped. Members of the camelid family (camels, llamas, and alpacas) generally reside in arid environments. Their red blood cells are nonnucleated but are shaped more elliptically. The thinner cell walls and elongation are theorized to assist in passing through small capillaries, minimizing the likelihood of sludging when the viscosity of the blood increases during dehydration. The life span of the red blood cell varies with the species and is outlined in Table 4–1.

TABLE 4–1   Life Span of Red Blood Cells in Various Species

Species	Life Span of Red Blood Cells (in days)
Horses	140–145
Cattle	130
Humans	120
Dogs	100–115
Cats	73
Mice	43

Dietary factors, endoparasites, ectoparasites, genetics, and other factors may shorten the life span of the red blood cell within species.

White blood cells (WBCs), **leukocytes**, serve to fight infection. There are five major white blood cell types; each has its own functions. Identifying and quantifying these five types of cells is an important portion of the complete blood count (CBC). Each type of cell has an independent role in fighting infection. Three of the WBC types have granules visible within their cytoplasm. They are thus called granulocytes. They include the **neutrophils**, acidophils **(eosinophils)**, and **basophils**. The agranulocytes (those WBCs without granules in their cytoplasm) include the **lymphocytes** and **monocytes**. Identifying abnormal numbers of the individual types assists the veterinary staff in their diagnosis.

The most numerous WBC is the neutrophil. Neutrophils are so named because their granules do not pick up either eosin, a red acidic stain, or methylene blue, a basic or alkaline stain. They are the body's chief defense against bacterial infection. Neutrophils have the capacity of leaving the vasculature and migrating to areas of infection or tissue damage to phagocytize microorganisms and other foreign material. Band neutrophils are immature neutrophils. Both neutrophils and band neutrophils increase with acute bacterial infections. Viral and some bacterial infections (i.e., brucellosis) can actually lower the neutrophil count, causing neutropenia.

Eosinophils increase in numbers in response to parasitic infections, allergies, pregnancy, and sometimes with neoplasia. Basophils are much more rare. Basophil granules contain both histamine and heparin, have receptors for Immunoglobulin E (IgE) (responsible for the immune response of allergies), and participate in the degranulation of WBCs that occurs during allergic reactions. Although they are capable of ingesting foreign particles, they are most commonly associated with acute allergies and with mast cells. Both heparin and histamine heighten the movement of WBCs to inflammatory sites. Identifying an absolute basophilia is important in that it may indicate the presence of a hematologic malignancy or a change in bone marrow function.

Lymphocytes constitute approximately 25 percent of WBCs. Their main function is to react with an antigen and, together with monocytes, mediate the immune response. They include B and T cells. The B cells produce antibodies that attack both bacteria and toxins. The T cells attack body cells that have been taken over by viruses, intracellular parasites, or cells that have become cancerous. T cells are important in antigen recognition.

Monocytes are the largest (in size) of the WBCs. The monocyte is capable of leaving the blood and entering the tissues, where it matures into a macrophage. Macrophages and monocytes both function as phagocytes. They ingest microbes and cellular debris. The monocyte-macrophage system plays a major role in the body's immune response.

Thrombocytes or platelets are fragments of megakaryocytes. They are an essential component of the blood's ability to clot in mammals. An abnormality or disease of the platelets is called a **thrombocytopathy**. It is characterized by dysfunctional platelets resulting in prolonged bleeding time, defective clot formation, and increased hemorrhaging. The causes are varied and include nutritional factors, complications with viruses, and inherited thrombocytopathies. One such genetic disorder is von Willebrand's disease, which is most commonly reported in dogs but also observed in swine, horses, cattle, and cats.

Complete Blood Count and Differential WBC Count

The **complete blood count (CBC)** evaluates the red blood cell (RBC) count, platelet count, hemoglobin concentration, and WBC count. It also provides a breakdown of the types of cells present, their percentages, and total numbers (differential

WBC count). The CBC is important from a clinical standpoint as identifying changes in numbers and percentages to aid in diagnosing anemia and certain cancers of the blood, and to monitor blood loss, infection, and inflammatory diseases. The platelet count is important in the diagnosis and/or monitoring of bleeding and clotting disorders.

Chemistry Profile

Blood chemistry panels are an essential part of any thorough examination. Most veterinary clinics have access to in-house testing or a local lab that analyzes the serum portion of the patient's blood. The concentration of the various chemical elements of the blood is maintained within strict limits. These normal parameters can be compared to the patient's test results in helping to diagnose subclinical findings. The most valuable chemistry panel is one combined with a urinalysis and a complete blood count. The chemistry panel should evaluate levels of glucose, electrolytes, protein, liver and pancreatic enzymes, bilirubin, creatinine, and blood urea nitrogen. Liver, pancreatic, kidney, adrenal, and endocrine function; hydration status; and muscle damage are better evaluated with these data. The urinalysis includes examination of color, pH level, specific gravity, chemical analysis for blood proteins, glucose and microscopic examination of RBC and WBC, and bacteria. This additional testing aids in diagnosing kidney and bladder infections.

Fluid Homeostasis

The concentrations and volumes of body fluids are maintained within narrow limits. These fluids are compartmentalized, or restricted to certain spaces within the tissues, organs, or the cells themselves. The three compartments are (1) **intravascular**, meaning within blood vessels; (2) **interstitial**, or between the cells; and (3) **intracellular**, meaning within the cells. Sometimes, the fluid found within the digestive system forms a fourth compartment, called **intraintestinal**.

In disease, significant alterations of the fluid volumes may occur. There are no barriers to the movement of water among the various compartments of the body. Therefore, it's the quantity of solute, or materials dissolved in the fluid, rather than the amount of solvent, or body of fluid, that defines the size of each compartment. Each body compartment contains at least one solute that, because of its basic restriction to that compartment, determines the size of the space. Table 4-2 shows the divided amounts that make up each compartment.

TABLE 4–2   Compartmentalization of Total Body Water

Intracellular	Two-thirds of total
Extracellular	One-third of total
Intravascular	25–33% of compartment
Interstitial	67–75% of compartment

The cells and tissues of the body are dependent on a normal fluid environment and an adequate blood supply. Fluid imbalances (which include edema, dehydration, and water intoxication) and hemodynamic disturbances (which include active hyperemia, passive congestion, hemorrhage, thrombosis, embolism, and infarction) are common and may cause death in certain pathological states.

All mammalian hearts, including human hearts, are basically the same, so the same terminology should apply across species. The various cardiac pathologies are also basically the same. There are some conditions, however, that will usually occur in only one species, such as "hardware disease" in cattle. Any animal (including humans) can suffer from cardiac puncture but the terms used to describe these signs or disorders may differ from one species to another. The terminology used to describe the same condition will also vary depending on the practitioner's experiences, background, and prejudices. Dogs are the usual victim of heartworm, but cats may also contract the parasite. There are also breed predispositions for various conditions.

An **infarction** occurs when the blood supply is cut off from any portion of any organ or tissue by any etiology long enough for the tissue to die. It usually appears as a pale triangle, since the blood comes in at a single vessel and then branches out several times to the capillaries. If this happens in the heart muscle, it is a myocardial infarction. If enough heart muscle dies, the animal dies.

Animals experience the same clinical signs as humans with similar heart diseases. Depending on the practitioner, different terms will be used to describe the same illness and/or clinical signs (heart attack, cardiac arrest, cardiac failure). The only thing different about veterinary medicine from human medicine is that humans have "symptoms" they can describe themselves. Animals have "clinical signs" because they can't talk, even though these signs are the same as those found in humans.

Hemodynamic Disturbances

Alterations of the fluid flow are known as hemodynamic disturbances; two specific types of hemodynamic disturbances are hyperemia and congestion. **Hyperemia** is an active process in which the arterioles fill with blood. **Congestion** is a passive process in which venules engorge with blood. Arterioles are under direct pressure from the pumping of the heart; venules are not.

The abnormalities that cause hyperemia and congestion can be physiological or pathological. **Blushing** is an example of physiological hyperemia—a red face is the result of an increase in blood flow through the vessels. Inflammation, as we learned earlier, involves the increased blood flow of a vascular component. The redness of an inflammatory lesion is the result of pathological hyperemia.

Hyperemia and congestion are also described according to their duration, extent, and mechanism. Durations can be acute (a quick onset) or chronic (developing slowly). "Extent" refers to the amount of tissue affected; it is localized (restricted to a specific area) or generalized (affecting the entire body). "Mechanism" refers to the process itself. Hyperemia, as stated previously, is active, whereas congestion is passive. We use this information to describe the process by way of an anatomical-mechanistic classification system.

First, let's examine hyperemia in more detail. Hyperemia is an excess of blood within vessels, and particularly tissues. During hyperemia, the arterioles are like a river that's swollen with floodwaters from an increased inflow. Thus, active hyperemia is an increase in arterial blood in the microvasculature, caused by physical, chemical, or biological injury.

You'll recall that many chemicals, called mediators, are released during inflammation. Unfortunately, these chemicals can dilate the arterioles. (Inflammation, as you'll also recall, can be local, but not generalized or diffuse.) Hyperemic tissues appear bright red and are warm from the increased blood flow. The clinical terminology for this condition is *acute local active hyperemia.*

Congestion is passive because it involves the venules. While hyperemia is comparable to a swollen river, congestion resembles the buildup of water behind a dam (Figure 4–2). Congestion may be associated with an obstruction of venous outflow, also called a thrombus. **Thrombus** is the proper term for a blood clot within a vessel.

FIGURE 4–2

The effect of congestion on a network of venules. The blood clot, or thrombus, causes outflowing blood to "dam up" and engorge the blocked vessel with blood.

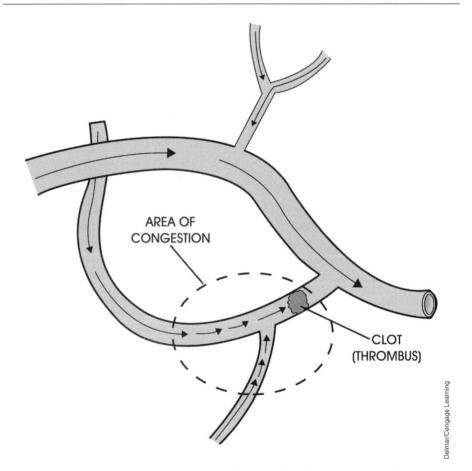

AREA OF
CONGESTION

CLOT
(THROMBUS)

Delmar/Cengage Learning

The heart, which is the pumping mechanism for the entire vascular system, can also cause congestion if it fails to maintain adequate cardiac output. A problem with outflow is called local congestion. When the heart is diseased and blood builds up throughout multiple organs or the entire body, you have what is called *generalized* congestion.

Acute, subacute, and *chronic* describe the duration of congestion. Acute local passive congestion, for example, occurs in a tissue as the result of a clogged vein. Chronic generalized passive congestion characterizes an animal suffering from a condition such as coronary artery disease, which slowly kills the heart cells and makes the heart beat less effectively. Although traditionally this is seen more commonly in humans, with an increase in obesity in our pets and exposure of our pets to similar environmental hazards, toxins, and microbes, veterinarians are seeing an increase in similar circulatory disorders in dogs and cats.

Because all blood in the body must flow through the heart and lungs, generalized passive congestion always involves disease in these organs. The congestion often leads to escape of fluid from the circulatory system.

A **hemorrhage** is the escape of blood from the cardiovascular system. Any part of the cardiovascular system can be affected—the heart, arteries, arterioles, capillaries, venules, or veins. Hemorrhaging can also occur within the skin, mucous membranes, serous cavities, or any tissue. The causes of hemorrhage are many and include trauma, poisoning, vitamin K deficiency, liver disease (because the liver makes clotting factors, chemicals which allow clotting to occur), and **thrombocytopenia**, which is a deficiency of platelets in the blood.

The amount of blood lost is a crucial factor in determining the severity of the hemorrhage. Up to one-fourth the total volume of blood can be lost with no sign of illness. A loss of one-third of the total blood can cause shock, and losing one-half of the total blood supply can result in death. Quick loss of blood can cause hypovolemic shock, whereas a slow loss may allow the body to replace some of the blood at the same time.

The site of a hemorrhage is important as well. Bleeding out of the body or into the abdominal cavity, although bad, doesn't cause a major problem. This is because the blood that is lost isn't compressing any vital tissues or becoming trapped in an organ. On the other hand, bleeding even a small amount of blood into an organ such as the brain could be fatal. The following medical terms refer to the different sites in which bleeding can originate:

- **Epistaxis**: bleeding from the nose

- **Hemoptysis**: blood coughed up from the lungs

- **Melena**: blood in the feces

- **Hematemesis**: vomited blood

- **Hematuria**: blood in the urine

- **Hemothorax, hemopericardium, hemoperitoneum**: an effusion of blood in the abdominal or peritoneal cavity

There are also definitions regarding what the blood looks like in the tissues:

- **Petechiae**: small, pinpoint hemorrhages (Figure 4-3)

- **Ecchymoses**: larger, more spread-out areas (Figure 4-4)

- **Hematoma**: essentially a blood blister or localized collection of extravasated blood confined within an organ, tissue, or space

- **Purpura**: results when ecchymoses combine with each other

FIGURE 4–3

The kidneys of this dog are speckled with numerous pinpoint hemorrhages, or petechiae, which in this instance were brought about by canine herpes virus infection.

FIGURE 4–4

Ecchymosis is very similar to petechiae, only larger in size.

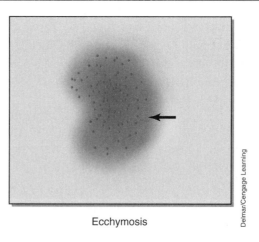

Ecchymosis

A discussion of hemorrhages wouldn't be complete without a discussion of **hemostasis**, or clotting. Clotting disorders can result in the hemodynamic disturbances of thrombosis, embolism, and infarction.

Hemostasis is the blood coagulation that occurs to prevent blood loss. The process occurs step by step, starting with **vasoconstriction** to reduce blood flow. A **platelet plug** then forms to block the damaged blood vessel. Next, a cascade of coagulation chemicals, termed *chemical hemostasis*, occurs. Chemical hemostasis is a complex mechanism because of the large numbers of chemicals involved and the way in which they react with each other.

The ultimate result of the coagulation cascade is the formation of **fibrin**, a chemical that solidifies and sticks to platelets and tissues, forming a clot. The formation of fibrin occurs in what we refer to as the common pathway of the coagulation cascade. This pathway can, in turn, be triggered by the extrinsic or intrinsic pathway of the coagulation cascade.

The extrinsic pathway is triggered by the release of "tissue juices." Suppose that blood vessel cells are damaged and leak their contents. These juices trigger the extrinsic pathway, leading eventually to the formation of fibrin in the common pathway.

The intrinsic pathway is activated by the exposure of collagen. If an injury scrapes cells lining the blood vessel free from the underlying basement membrane, collagen is exposed. This exposed collagen attracts platelets and chemicals, which activate the intrinsic pathway and eventually the common pathway. The result is a blood clot, which the body will eventually remove by resorption or replace with connective tissue.

A thrombus forms as a result of injury to the **endothelium** (the cells that line blood vessels), alterations in normal blood flow, or hypercoagulability of the blood. A thrombus can also break loose from the blood vessel in which it is formed to circulate throughout the bloodstream. This transported material is called an **embolus**. Most emboli are solid, but fat globules and gaseous emboli also exist. An **embolism** is an impaction too large to complete the circulatory route of the embolus. If an embolism blocks off the blood supply or venous drainage of a tissue, the result is an infarction. The sudden reduction of oxygen to a tissue as the result of an infarction causes coagulation necrosis.

Infarctions (or infarcts) are classified according to their color and the presence or absence of bacterial contamination:

- White infarcts (little blood is present in the tissue)

- Red infarcts (there is hemorrhage in the tissue)

- Bland infarcts (no bacteria are present)

- Septic infarcts (bacteria are present)

Infarcts heal by a process known as organization. The clot is converted into fibrous tissue. The tissue that died is also replaced by scar tissue; healing occurs via repair. If an infarct should take place in an organ such as the heart or the brain, the animal may die before healing can occur.

Shock is a syndrome resulting from **hypotension**, or decreased blood pressure, which is caused by acute generalized circulatory failure of the capillary bed and/or pooling of blood. We know that "acute" signifies a quick onset, "generalized" refers to the whole body, and capillaries are the networks of small blood vessels in tissues where oxygen exchange takes place. If all of the capillaries in the body should suddenly allow blood to flow through them at once, blood pressure would drop severely, resulting in shock. The body's total blood volume is just too small to fill the entire vascular system.

Several types of shock can occur. **Hypovolemic shock** is due to a loss of blood volume. Lack of oxygen to the tissues will cause the capillaries to dilate, decreasing blood pressure and inducing shock. **Normovolemic shock** can occur with a normal quantity of blood. A heart attack, or **myocardial infarction**, can damage the heart muscle, causing a decrease in blood output. This decrease in blood supply to the tissue causes a lack of oxygen, which also causes the capillaries to dilate, and blood pressure to go down.

Normovolemic shock may also occur another way. Overwhelming bacterial infections can cause blood to collect in the infected area, effectively removing the blood from circulation and mimicking hypovolemic shock. This phenomenon is known as *peripheral pooling*.

Fluid Imbalances

Edema is an excess of fluid in the interstitial fluid compartment of the body. The tissues that experience this excess fluid between the cells can be localized or generalized throughout the body.

Fluid is kept in the blood vessels by three mechanisms: (1) an intact functioning circulatory system, (2) a normal functioning lymphatic system, and (3) albumin (a protein made by the liver) in the serum. The Starling equilibrium is a mathematical equation that explains how these three mechanisms interact to keep most of the fluid in blood vessels. All edema results from an imbalance in one or more of these three mechanisms. Fluid then leaks into the tissue space by one of four ways:

- A decrease of albumin in the blood allows more fluid to leak out of arterioles because of osmotic pressure changes.

- An increase in blood pressure can force fluid from the blood vessels.

- Lymph vessels drain the small amount of fluid that normally leaks into the tissues back into the circulatory system. If the vessels become blocked (a condition known as *lymphatic obstruction*) the fluid in the tissues increases.

- Increased vascular permeability can occur, brought about by inflammation. When this happens, the cells lining the blood vessels form gaps between each other and allow more fluid to escape from the circulatory system.

Dehydration and water intoxication are the two remaining types of fluid imbalances. Water is one of the largest constituents of the body. It varies in amount from 50 percent in a finished steer to 80 percent in a newborn lamb. Younger animals have a greater percentage of water in the body. In addition, fatter animals have a lower percentage of water. **Dehydration** is due to the loss of water from the body or from a decreased intake of water. Large deficits of water (20 percent of the body weight or as little as 10 to 12 percent of the water in the body) may lead to death. Problems such as diarrhea and kidney failure can cause the body to lose more water than can be taken in. An animal that is dehydrated will have a dry mouth, dry-looking eyes, reduced urine output, reduced feed consumption, and skin that appears excessively wrinkled. Calculating the water loss in dehydrated animals is essential to their recuperation. A 100-lb calf, for example, with mild diarrhea can lose 10 percent of its body weight in water. This would require 10 lb or more in fluid replenishment depending on how long the diarrhea continues.

Skin turgor is used to obtain an estimate of an animal's hydration status. The skin is picked up and released (Figure 4-5). It normally will return within 1 second. As an animal becomes dehydrated the skin loses some of its elasticity, causing it to return more slowly (Table 4-3). This test is less accurate in obese, emaciated, or very old animals.

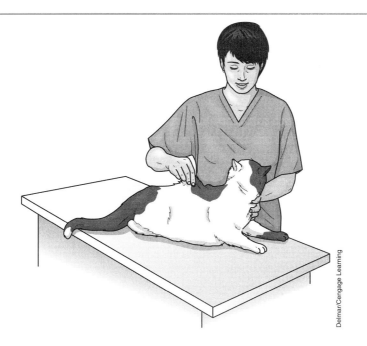

FIGURE 4–5

A registered veterinary technician conducting the skin turgor test.

TABLE 4–3 Skin Turgor Test

Degree of Dehydration	Clinical Signs
Less than 5%	Cannot be detected clinically.
5–6%	Doughy skin and tacky mucous membranes.
6–8%	Slightly sunken eyes, dry mucous membranes, and inelastic skin.
10–12%	Increased skin turgor (no elasticity; the skin will stay up in the air after it is pulled up). Very sunken eyes and dry mucous membranes. Status of this animal is poor.
12–15%	Shock, low temperature, and prolonged capillary refill time. Status of this animal is grave.

Water intoxication usually occurs with salt loss or when animals are on limited salt intake and are allowed unlimited access to water. If an animal that lacks salt drinks an excessive amount of water, the animal develops tremors, incoordinated movement, and convulsions. The animal's brain cells can swell, along with its RBCs, and kill the animal. This is common in both humans and in calves that drink large amounts of water after a short period of deprivation.

CASE STUDY | Toxicity

A 2-year-old tabby cat was found playing with an opened bottle filled with pills. The owner found a few pills on the floor but did not think that her cat ate any of the pills. The cat seemed to be acting normally. Several hours later, the cat became depressed. His face was a little puffy and he was starting to have trouble breathing. The owner took her cat to the veterinarian.

The diagnosis was acetaminophen (Tylenol) toxicity. Acetaminophen acts as an oxidant of hemoglobin, resulting in the formation of Heinz bodies (Figure 4-6). The Heinz bodies cause hemolytic anemia. Apparently, the cat must have eaten a few of the pills. Emesis could not be performed to eliminate the toxin, since the cat ingested the pills several hours earlier. An emetic can be administered within 2 to 4 hours of ingestion. The mucous membranes were a muddy brown color, which is a clinical sign of toxicity. The cat was placed on oxygen therapy for dyspnea. A blood sample was taken for cytology. The blood was a brown color. The cytology revealed Heinz bodies (Figure 4-7). At least 80 percent of the erythrocytes were affected. To confirm the Heinz bodies, new methylene blue stain was used, after the regular Wright's stain was used on the slide.

Acetaminophen (Tylenol) is contraindicated in cats. Severe hematuria, methemoglobinemia, and icterus can be seen. When injury or toxicity occurs in the body, agents convert a large amount of hemoglobin into methemoglobin, which does not function as an oxygen carrier. Methemoglobin is a compound made from hemoglobin by oxidation of the iron atom from the ferrous to the ferric state. Ascorbic acid was administered to treat the methemoglobinemia. Acetylcysteine (Mucomyst) was also administered to provide building blocks for glutathione, which serves as a reducing agent in most biochemical reactions. The erythrocytes are protected from oxidation and hemolysis or deficiency when glutathione is reduced. The 2-year-old tabby's prognosis was grave. It died soon thereafter. Generally the prognosis is poor, but it would be improved depending upon the amount ingested (50 to 100 mg/kg, 1 to 2 tablets, can be fatal) and how quickly emesis can be performed.

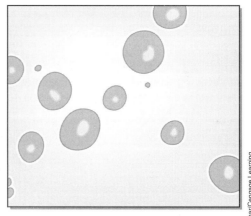

Heinz bodies – Oxidized hemoglobin

Delmar/Cengage Learning

FIGURE 4–6

Heinz bodies, oxidized hemoglobin, using new methylene blue stain.

CASE STUDY	Toxicity *(continued)*
	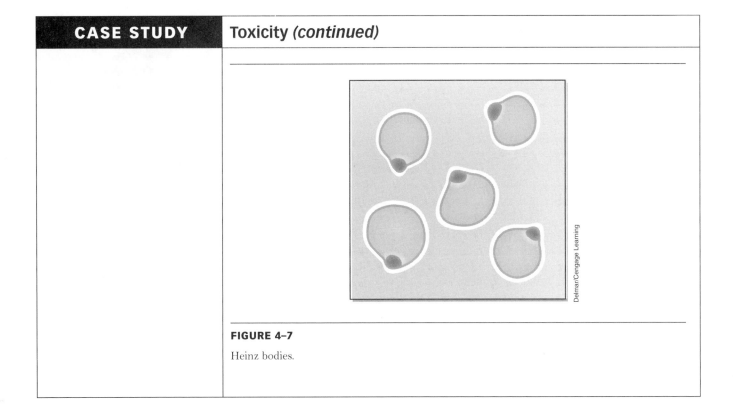
	FIGURE 4–7 Heinz bodies.

Summary

Understanding the normal flow pattern of blood through the heart assists the veterinary technician in detecting abnormalities which can occur and which may be monitored by various examination procedures. Homeostasis of body fluids must be maintained within narrow limits to preserve normal health of the animal. Various blood and urine tests are routinely performed to help diagnose and assess various pathologies. The same disorders that affect human hearts can affect the hearts of animals living longer lives and experiencing similar environmental stressors that we humans experience. Modern veterinary medicine allows us to observe or detect changes within the cardiovascular system, warning us of abnormalities and treating these clinical signs thus extending the lives of our companion animals. Hemodynamic disturbances include hyperemia, hemorrhage, and congestion. Through proper nutrition and animal management, our companion and livestock species can live longer and more productive lives similar to their human counterparts.

REVIEW

FILL IN THE BLANKS

1. _____ is the excess of fluid in the interstitial fluid compartment and/or the body cavities.

2. _____ is a term meaning blood in the urine.

3. _____ is a thrombus that breaks loose from the blood vessel in which it's formed in order to circulate throughout the bloodstream.

4. _____ is a term meaning blood in the feces.

5. _____ is the result of an embolism blocking off the blood supply or venous drainage of a tissue.

6. _____ is a general term describing low blood pressure.

7. _____ is an active process in which the arterioles fill with blood.

8. _____ is the dilation of the blood vessels.

9. _____ is the process by which overwhelming bacterial infections cause blood to collect in an infected area, effectively removing the blood from circulation and mimicking hypovolemic shock.

10. _____ is the proper term for a blood clot within a vessel.

11. _____ is a condition in which there is a deficiency of platelets in the blood.

12. _____ is a condition in which there is bleeding from the nose.

13. _____ is a localized collection of extravasated blood that is from a vessel and is confined within an organ, tissue, or space.

14. _____ is a term meaning vomited blood.

15. _____ is one way to estimate the state of hydration (fluids) or degree of fluid loss or dehydration.

16. A/an _____ occurs when the blood supply is cut off from any portion of any organ or tissue by any etiology long enough for the tissue to die.

17. _____ (general term for type of blood vessel) carry blood toward the heart.

18. Another name for the pacemaker is _____ .

19. Another name for hematocrit is _____ .

DISCUSSION QUESTIONS

1. What are the three fluid compartments of the body?

   _____

   _____

   _____

2. With what condition will you always observe hyperemia of the tissues?

   _____

   _____

   _____

3. Describe fibrin.

   _____

   _____

   _____

4. What three mechanisms keep fluid in the blood vessels?

   _____

   _____

   _____

5. Trace the flow of blood through the heart starting at the vena cava.

   _____

   _____

   _____

6. What is the difference between neutrophils and band neutrophils?

   _____

   _____

   _____

7. What additional information might a practitioner discover from a blood chemistry panel over a CBC?

   _____

   _____

   _____

(continued)

REVIEW   *(continued)*

8. What might an increased basophil count indicate?

_____

_____

_____

9. List several contributing factors that might shorten the normal life span of a red blood cell within a given species.

_____

_____

_____

CHAPTER 5

Neoplasia: An Introduction to Tumors

OBJECTIVES

Upon completion of this chapter, the reader should be able to:

- Define neoplasia, tumor, cancer, metastasis, and neoplasm.

- Cite differences between benign and malignant tumors.

- Describe abnormal histological differences in a tumor cell compared to normal tissue.

KEY TERMS

anaplasia

benign

cancer

carcinogen

carcinoma

descriptive classification
  system

karyotypic

malignant

melanoma

metastasis

monoclonal

neoplasia

neoplasm

osteoma

parenchyma

practical classification
  system

regional classification system

stroma

tissue classification system

TNM classification system

tumor

Classification of Tumors

The term **tumor** originally referred to any swollen tissue. Currently, tumor is commonly used as a synonym for neoplasm. Tumors are caused by abnormal regulation of cell division. Tumors are classified as benign (slow growing and usually harmless) or **malignant** (faster growing and usually able to spread to other parts of the body). A **neoplasm** is an abnormal tissue that grows by cellular proliferation more rapidly than normal and continues to grow after the stimuli that initiated the new growth cease (Figure 5-1). **Neoplasia** is the pathological process that results in the formation and growth of a neoplasm.

Cancer is the term used to indicate the various types of malignant neoplasms. It is characterized by the uncontrolled growth of abnormal cells. Malignant refers to the ability of a neoplasm to invade and spread (metastasize) to remote sites in the body.

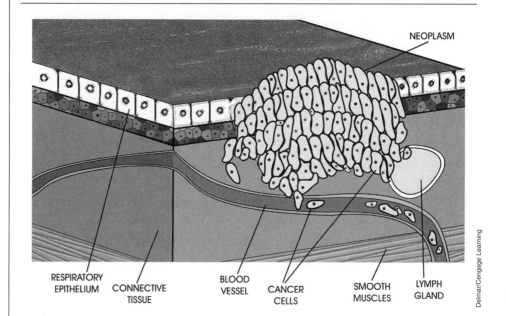

FIGURE 5–1

A close-up view of a malignant neoplasm on the wall of a lung. Malignant neoplasms are largely characterized by their nonresponsiveness to normal growth regulation.

Behavior of Tumors

To understand the behavior of tumors, you must first learn the ways in which tumor cells behave differently from normal cells. The differences between normal and neoplastic cells in regard to their growth properties are fairly simple: Tumor cells have unregulated proliferation. In other words, the body isn't able to "turn off" the division of tumor cells.

Most tumors are clonal, resulting from the progeny of a single somatic cell, by the time they are diagnosed. Tumor growth is a balance between proliferation, differentiation, apoptosis, and necrosis. Tumors are classified by their histogenesis and their behavior.

Tumor growth is regulated by the host immune response, adequacy of blood supply, and nutrients and hormonal and growth factor availability. Unless they are removed or are killed by the body's defenses, chemotherapy, radiation therapy, hormone therapy, biological therapy (using the body's immune system to fight the cancer), or photodynamic therapy, the cells will go on dividing forever until the animal dies. The most difficult aspect of tumor behavior to understand is loss of *contact inhibition*. Normal cells divide until they come in contact with cells completely surrounding them. This signals to the cells that there's no more room to expand and that they should stop dividing. Tumor cells go right on dividing no matter how crowded conditions become.

Morphological Differences

Neoplastic cells may appear different from normal cells under microscopic examination. These changes are called *morphological differences*. For example, normal cells are programmed to be of certain types—bone, skin, kidney, and so on. Tumor cells, because they divide so rapidly, never fully mature in the normal sense and they lack the ability to differentiate among different parts of the body. This failure within a cell to differentiate is called **anaplasia**. Anaplasia is seen in most, but not all, malignant tumors.

As previously mentioned, tumor cells also have a **monoclonal** origin. This means that the entire tumor, no matter how large, arose from a change in a single cell. **Karyotypic** changes are changes in chromosome number or shape. Abnormal cells will often rearrange the genetic material in their nucleus, making the chromosomes appear different.

Metastasis

Metastasis is the growth away from the primary site in which a tumor originates. It can occur by direct seeding in a cavity, which happens when cells break off and grow where they lodge. Tumor cells get their nutrients to grow from the host's vasculature.

There are several pathways for metastasis:

- Lymphatic spread results when there's a tumor cell invasion into the lymphatics (Figure 5-2).

- Hematogenous spread is a tumor's use of the bloodstream as a route.

- Transplantation is known to occur in dogs. An example of this type of tumor is a venereal tumor that dogs transmit to each other when they copulate. Warts can also be transplanted among many animals in many different species.

- Direct seeding of body cavities or surfaces occurs whenever a malignant tumor penetrates a naturally open cavity such as the peritoneal or pleural cavities.

Benign tumors grow expansile masses that remain localized to their site of origin. They normally do not have the capacity to infiltrate, invade, or metastasize to distant sites. Most malignant tumors are invasive and can be expected to penetrate the wall of the colon, or uterus, for example, or fungate through the surface of the skin. They recognize no normal anatomical boundaries. Some

FIGURE 5–2

Squamous cell carcinoma, a common malignant tumor of the feline oral cavity, is an example of a neoplasm that often spreads to the lymphatic system.

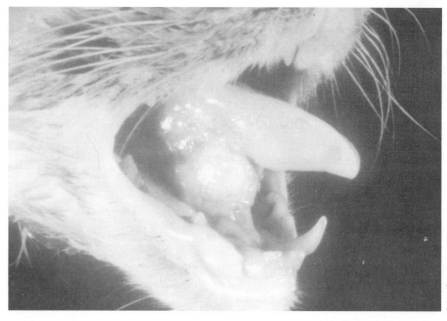

tumor cells are more mobile, produce more enzymes, or are less adhesive than others and easily grow into normal tissue. Individual cells or small clumps may embolize, or be moved by the circulatory system to new sites. The cells may prefer localization, growing in certain new areas, which can be explained by the route of blood flow or as a result of gradual selection by the cells. In any event, once the cells stop, or are arrested at the new site, they can leave the blood vessel and grow into a new tumor at the new location.

Chemical Carcinogenesis

Carcinogens are physical agents, viruses, or chemicals that can cause mutation, resulting in tumor formation. Tumor cells are different from normal cells and tissues (Figure 5-3).

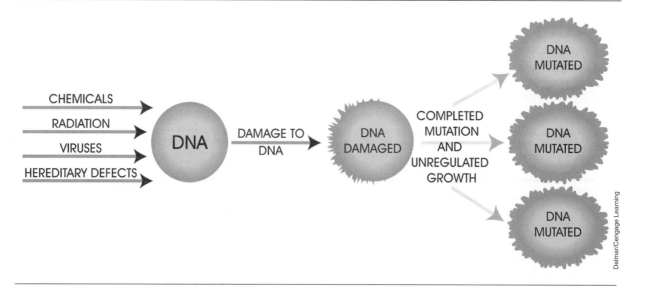

FIGURE 5–3

This drawing illustrates the ultimate effect of carcinogens. Once mutated, the stricken cells are unable to control their growth.

Tumors are composed of living cells and have the ability to create new blood vessels (angiogenesis) to maintain growth and metastasize. Tumors don't have their own lymphatics, but a tumor may be drained by local lymphatics. A tumor is able to subvert host tissue stroma and reprogram the stroma's activity to its own advantage. A stroma is a normal body tissue that acts as a framework and grows the blood vessels, which nourish the tumor.

A well-defined, discrete, fibrous tissue layer known as a capsule surrounds some benign tumors. Capsules are formed by pressure atrophy of the existing connective tissue and new connective tissue. **Benign** tumors usually lack the ability to invade and grow at sites away from the original tumor.

A tumor cell is a mutated cell that is characterized by a lack of growth and differentiation control. Tumors are known from an almost infinite variety of cell types, including fibroblasts, renal tubular epithelium, chondroblasts, islet cells, hepatic cells, and astrocytes. Essentially, any cell capable of dividing can become a tumor.

Some tumors may resemble normal tissue but with some important differences. For instance, abnormalities exist in a tumor cell's:

- Size

- Shape

- Staining properties

- Spatial arrangements

Tumors often exhibit a decrease or total loss of specialized functions or a development of new functions. A tumor cell may perform the same functions as normal cells of the same type. Sometimes, though, these functions don't occur if the cells are reproducing rapidly.

As previously mentioned, every tumor starts out as a single, normal cell. It's only when the cell "goes haywire" and no longer responds to controls over its division that it is considered neoplastic. The evolution of a tumor represents the development of a colony of permanently altered cells. Most neoplasia have irreversible genetic alterations so that growth becomes "autonomous." Sometimes, however, cellular proliferations that are induced by viruses may be reversible after the viral infection is eliminated.

Cellular proliferation, or growth, is the most conspicuous feature of cancer. A tumor cell divides into two; further division continues to occur until a large colony is formed, each cell possessing the properties inherited from the parent cell. Neoplasia differs from normal cell division and hyperplasia in that the tumor's growth never attains normal equilibrium. When normal cells reach their adult size, equilibrium between cell birth and cell death occurs. With neoplasia, cells continue to divide until such time as their growth is inhibited by limits in blood supply and/or nutrients. The growth can be slow or rapid and is almost always progressive. Tumors rarely undergo regression, so their continued growth is very destructive.

Malignant versus Benign

Malignant tumors display several features not observed in benign tumors. Foremost and unique of these is invasion and spread (metastasis), or the ability to infiltrate and destroy normal tissue. Benign tumors contain well-differentiated cells that are clustered together in a single mass. Benign tumors are often contained in a capsule composed of compressed connective tissue. Cells of a malignant tumor, on the other hand, are not restrained by a capsule. If a capsule were present, it might help to contain the growth of the tumor.[1] Normally, benign tumor cells do not cause death unless their location interferes with normal vital functions.

Metastasis formation, as discussed earlier, is the development of secondary centers of tumor growth at a distance from the primary site. The malignant tumor can spread by breaking off pieces that are able to grow in a new location, often in a completely different organ. For example, **melanoma** is a type of cancer that's frequently triggered by exposure to the sun. It can originate in the skin (a good reason for sunscreen) and then spread, or metastasize, to the brain and liver by way of blood vessels.

Classification and Appearance of Tumors

Classifying tumors makes it easier to discuss the presence of disease within an animal. Tumors are composed of two types of tissue: **parenchyma** and **stroma**. The parenchyma represents the functional tissues of an organ, whereas the stroma is composed of the supporting structures of lymph, blood vessels, and connective tissues. The stroma provides support for growth and survival of the tumor. The parenchyma determines the behavior of the tumor and is the component for which the tumor is most often named. Tumors are usually given the suffix *-oma* to the parenchyma tissue type from which the growth began. Thus a benign tumor of the bone would be given the name **osteoma**.

We classify tumors to understand them and their behaviors more clearly. The **Tissue Classification System** is one method for categorizing tumors. The principal basis of this system is the type of tissue from which the tumor arises. Some examples of these tumors and the tissues they attack include the following:

- Squamous cell carcinoma or papilloma (epithelium tissue)

- Fibroma (connective and muscular tissue)

- Leukemia (hemopoietic tissue)

- Astrocytoma (nervous tissue)

- Tumors of more than one type of tissue (such as a teratoma or mixed mammary tumor) and miscellaneous types of tumors that don't fit into any of the preceding four groups (such as melanoma)

The **Practical Classification System** is based on a number of the tumor's criteria (manner and rate of growth, cell characteristics, etc.). It utilizes some behaviors we've

---

[1]Barr, L. C., Carter, R. L., & Davies, A. J. S. (1988, July 16). Encapsulation of tumors as a modified wound healing response. *Lancet*, 135–137.

already discussed, including encapsulation, invasion, and metastasis. There are two basic categories in the practical classification system: benign and malignant.

The **Regional Classification System** classifies tumors according to where they occur within or on the body. However, because different kinds of tumors may arise in the same part of the body, this system can be awkward to use. The significance of this system is related to the regional incidence of tumors in certain sites of the body. For instance, white cats tend to get a malignant tumor of the skin, squamous cell carcinoma, on the tips of their ears. This tumor is found there more frequently than other types of tumors. Thus, whenever a tumor is located in this area, there's every reason to be concerned.

The **Descriptive Classification System** is based on the gross or microscopic appearance of a tumor, often utilizing eponyms. Eponyms constitute an ineffective way to classify tumors, because they name a disease based on the name of the person who discovered or described it. An example of this is Hodgkin's disease. From the name alone, you would never know it was a tumor of lymphoid tissue.

The **TNM Classification System** was created by the International Union Against Cancer (IUAC) and the American Joint Committee on Cancer Staging and End Stage Reporting (AJCCS). This system classifies neoplasia based on the extent of the primary tumor (T), the involvement of the regional lymph nodes (N), and the extent of metastatic involvement (M). This system is used by many cancer facilities in providing information regarding the prognosis of the disease, as well as the progress and success of treatment.

The Terminology of Tumors

As previously mentioned, most tumors end in the suffix -*oma*, an ending that usually, but not always, implies a benign neoplasm. A benign tumor of epithelial tissue is a *papilloma* or *adenoma*, whereas a malignant tumor is called a **carcinoma**. Benign tumors of connective tissue or muscle receive their names by joining Greek terms for tissue to -*oma*; malignant tumors are named the same way, except with the ending *sarcoma*. For example, a benign tumor of cartilage (Greek = *chondros*) is a *chondroma* whereas a malignant tumor of cartilage is a *chondrosarcoma*. In addition, certain carcinomas have special names. A liver carcinoma is often called a *hepatoma*.

The Morphology of Tumors

If you examine a tumor removed from an animal, it appears as a lump or mass of tissue that consists of tumor cells, materials that the cells produce, and the stroma. The principal function of the stroma is to provide nutrients to the tumor by way of a blood supply and remove metabolic waste. Without a stroma, the tumor would die. The stroma is provided by the normal tissues of the body. Connective tissues near the tumor form new vessels and new collagenous tissue for the proliferating tumor cells. The size of the stroma varies with the type of tumor. And yet the structure of the stroma determines the texture and consistency of the tumor.

Tumor cells may secrete a variable quantity of material specific to the type of cells that they are. For example, connective tissue tumors are identified by the material they produce.

The change in tumor morphology helps practitioners diagnose the stage and development of the cancer. The arrangements of tumor cells help determine the gross and microscopic appearance of a tumor. Cells tend to be arranged differently in different kinds of tumors (Figure 5-4). In epithelial tumors, cells tend to form clumps, or groups. Tumor trauma or massage tends to increase total tumor cells and clumps released into the effluent.[2] Within any clump, each cell is fastened to the next. Connective and muscular tissue tumors are characterized by cells that lie singly. Hematopoietic tumors originate from plasma cells or lymphocytes. Lymphosarcoma is a neoplasm of lymphocytes arising as solid

FIGURE 5-4

Cell arrangement in different types of tumors.

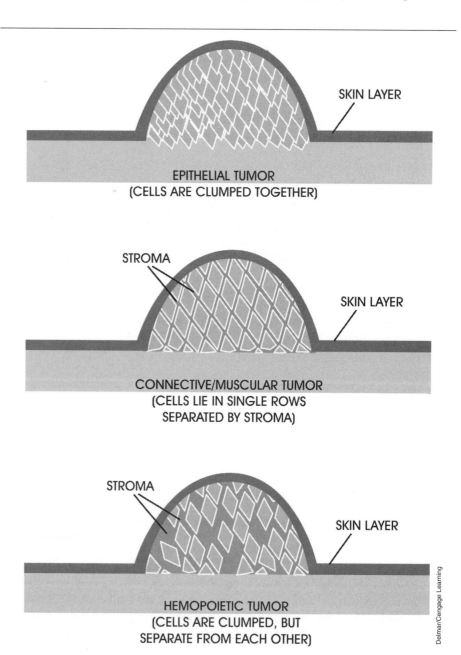

SKIN LAYER

EPITHELIAL TUMOR
(CELLS ARE CLUMPED TOGETHER)

STROMA

SKIN LAYER

CONNECTIVE/MUSCULAR TUMOR
(CELLS LIE IN SINGLE ROWS
SEPARATED BY STROMA)

STROMA

SKIN LAYER

HEMOPOIETIC TUMOR
(CELLS ARE CLUMPED, BUT
SEPARATE FROM EACH OTHER)

Delmar/Cengage Learning

[2]Liotta, L.A., et al. (1976). The significance of hematogenous tumor cell clumps in the metastatic process. *Cancer Research, 36*(3), 889–894.

tissue masses in lymphoid organs other than bone marrow. Leukemias are also neoplasms of hematopoietic cells but originate in the bone marrow.

Tumors may be of many shapes. Surface-type tumors tend to be flat plaques, indurations (hardened areas), or bodies that stand above the surface. Of the latter group, stalked tumors are known as polyps; those with fronds, or leaflike extensions, are called papillary (papillomas are warts); and those with a broad attachment are said to be fungating. Malignant tumors may form an ulcer or ulcerate, which is the development of a sore open to the body surface or a body cavity. Some benign tumors (i.e., histiocytomas in dogs) may also ulcerate.

Cancer Research

Animals suffer from many types of neoplasms including bone, skin, and breast (mammary) cancer and leukemia. In particular, cattle suffer from cancer eye and lymphosarcoma; cattle, horses, dogs, and cats suffer from squamous cell carcinoma; and dogs and cats suffer from malignant melanoma.[3] Cancer is responsible for almost half of the deaths in pets older than age 10. Animal medicine has become much more sophisticated allowing veterinarians to use radiation therapy, chemotherapy, and other cancer-treating regimens similar to what is offered to humans. Cancer research on animals for the benefit of humans has returned full circle, enabling tens of thousands of animals to live and to have a more normal life after treatment.

The nude mouse, a mouse without hair and without a thymus gland, is used in cancer research. Because the mouse has the ability to maintain human tissue (does not have functioning T cells that would cause the animal to reject cancerous grafts) it has been used successfully in this regard. This research, in turn, has provided many tools used by veterinarians today in treating similar cancers in domestic animals. The study of tumor growth in companion and lab animals has helped veterinarians in their ability to diagnose, prevent, and treat cancer. The treatment of cancer in companion animals utilizes many of the same methods available in human medicine. Many animal facilities are using a combination of chemotherapy, radiation therapy, hypothermia, immunotherapy, and photodynamic therapy as part of the arsenal assembled to fight cancer in the pet animal.

Most clinical research trials investigating new methods of cancer diagnosis or treatment are conducted at colleges of veterinary medicine using laboratory and companion animals. Once these new treatments have proven safe and effective in these clinical trials, the therapy can then be considered a potential benefit to other animal patients. Many naturally occurring cancers in pet animals closely resemble human cancer and provide meaningful systems for cancer research to benefit both humans and animals. Many times specimens obtained from these clinical trials (blood, urine, and/or tumor samples) are provided to basic researchers to gain additional information that may benefit present and future human and animal cancer patients.

---

[3]Verhaert, L. (2001). *Oral proliferative lesions in dogs and cats.* Presented at the 26th World Small Animal Veterinary Association World Congress, Vancouver.

CASE STUDY — Squamous Cell Carcinoma

A cat with white hair and a thin coat had a lump on the tongue. The owner of this white cat noticed that her cat was losing weight. She had also noticed that her cat had bad breath.

The clinical signs of squamous cell carcinoma are excessive salivation, halitosis, dysphagia, and weight loss. Some of the common areas that squamous cell carcinoma presents itself are on the tongue, gingiva, eyelids, and the tips of the ears (especially in white cats). A squamous cell carcinoma lesion/mass is pink in color with irregular margins. Squamous cell carcinomas are the most common intraoral tumor in felines. The prognosis for a skin tumor depends on the degree of cellular anaplasia, whereas intraoral tumors have poor prognosis due to location and uncontrolled recurrence. A fine-needle aspiration is the method used to obtain a sample for cytology. The cytological signs are cells in epithelial clusters. The epithelial cell may be single, too. The nuclei can be small and condensed or large with anisokaryosis (inequality in the size of the nuclei from what is normal for the tissue) and prominent nuclei. Sunlight can be a probable influence regarding the pathogenesis.

Surgical excision is the treatment of choice, but the 1-year survival rate is usually less than 10 percent in felines.[4,5] The tumors are controlled locally in only 25 percent of the animals 1 year after surgical resection or radiation therapy. The prognosis is poor. The cancer takes a rapid course. The cat with white hair lived for 15 months after he had surgical excision of the tumor. He had difficulty eating toward the last month of life.

Summary

A neoplasm is an abnormal tissue that grows by cellular proliferation more rapidly than normal and continues to grow after the stimuli that initiated the new growth cease. Benign neoplasms grow expansile masses that remain localized to their site of origin. The term *cancer* is used to describe any of a variety of malignant neoplasms. Due to animal research that was searching for preventions and treatments of human cancers, many animal cancers are now often preventable and treatable. The key to a successful cancer treatment is in identifying it before it has a chance to metastasize. Veterinarians are increasingly able to classify tumors with the same classification system used by physicians. Improved opportunities in cancer prevention and treatment should continue to provide our animals with prolonged life with greater normalcy and less pain.

[4]Fossum, T.W. (1997). *Small animal surgery* (p. 226). St. Louis: Mosby.

[5]Aiello, S.E. (1998). *The Merck veterinary manual* (8th ed., p. 690). Whitehouse Station: Merck & Co.

REVIEW

FILL IN THE BLANKS

1. _____ is a tumor that does not invade surrounding tissue or spread to other parts of the body.

2. _____ is a general term meaning difficulty in swallowing.

3. _____ is a type of laboratory animal used in cancer research that is hairless, lacks a normal thymus gland, and has a defective immune system.

4. _____ when describing tumors means that they were produced by or are composed of cells derived from a single cell.

5. _____ is the chromosomal characteristics of a somatic cell.

6. _____ is the failure within a cell to differentiate parts of the body.

7. _____ is a nerve-tissue tumor.

8. _____ is a type of cancer that is frequently triggered by exposure to the sun.

9. _____ is a name of a disease based on or derived from the name of a person.

DISCUSSION QUESTIONS

1. What is the difference between chondroma and chondrosarcoma?

   _____
   _____
   _____

2. What is metastasis?

   _____
   _____
   _____

3. What are carcinogens?

   _____
   _____
   _____

(continued)

REVIEW   *(continued)*

4. Tumors are composed of what three tissues or tissue types?

_____

_____

_____

5. Name the six tumor types using the Tissue Classification System.

_____

_____

_____

6. Define neoplasm, polyps, stroma, and sarcoma.

_____

_____

_____

7. Why are tumors sometimes considered to be "immortal"?

_____

_____

_____

8. How do neoplasms metastasize?

_____

_____

_____

9. What is an example of an eponymic tumor?

_____

_____

_____

10. Is it ethical to use information on cancer research from a mouse to save a dog's life? A human's life? Why or why not?

_____

_____

_____

11. How would you explain (from an ethical viewpoint) to a client the use of this research that could save a pet's life?

_____

_____

_____

CHAPTER 6

Immunological Diseases

OBJECTIVES

Upon completion of this chapter, the reader should be able to:

- Define and differentiate between acquired and passive immunity.

- Define antigen, antibody, and retrovirus.

- List an autoimmune disease in animals and describe its pathogenesis.

- Define the function of mast cells.

KEY TERMS

acquired immunity

active immunity

antigen

autoimmune reaction

colostrum

heparin

histamine

hypersensitivity reaction

mast cell

passive immunity

opsonization

retrovirus

Immunity

More than 4,600 years ago, the Chinese recognized that an individual who recovered from smallpox was resistant to any future attacks of the disease. Many centuries later, the English physician Edward Jenner created a smallpox vaccine from cowpox lesions. Today we recognize this immune response as active or acquired immunity. Thus, exposing an animal to a foreign antigen (either by vaccination or direct exposure to the disease) is called **acquired immunity** or **active immunity**. The body's immune response is to produce specific antibodies to the foreign antigen.

When a newborn animal consumes **colostrum** (first milk) rich in antibodies, its consumption provides passive or temporary immunity to the young. Likewise, if a horse that was vaccinated against tetanus has its partially purified blood serum injected into another animal (i.e., tetanus antitoxin) the antibodies in the serum provide passive or temporary immunity to the animal. **Passive immunity** does not stimulate antibody production by the recipient animal since only the antibodies have been injected or consumed and not the antigens. The protection is thus short lived. Antigen stimulus takes an actively working immune system (which newborns are lacking) and time (2 to 3 weeks in most cases for the start of antibody production and 12 to 16 weeks for most breeds of dogs and cats to develop an active working immune system). Thus passive immunity is a temporary means of providing protection to specific antigens for newborns and unimmunized animals. In addition, passive immunity interferes with the development of active immunity during the first few weeks of life. Thus, vaccination of a newborn (even if the newborn had a mature immune system) would not stimulate production of antibodies.

Autoimmunity

Coating the outside of an animal's cells are different types of chemicals with all types of necessary functions. If we could strip these chemicals off, place them in a syringe, and inject them into a different species of animal, the chemicals would be recognized as foreign. The second animal's body would mount an immune response against them, the same as if the animal were injected with bacteria. However, if the syringe were injected into the animal that we purified them from, there would be no immune reaction.

Because these chemicals have the ability to cause an immune reaction, they are called **antigens**. Figure 6-1 illustrates several types of foreign-body antigens. All living cells produce self-antigens by simply performing their normal functions. It was once believed that the body didn't react to these self-antigens because of self-tolerance, or the body's ability to recognize its own chemical agents. However, the body isn't always able to make this distinction.

Some tumors are perceived as foreign when they change the chemicals described previously; the tumors are then recognized as nonself cells. These nonself cells can provoke the formation of antibodies against themselves, which is good. But under certain circumstances, even normal cells can provoke the formation of autoantibodies, or autoreactive T cells, which results in the destruction of normal tissues. This is known as **autoimmune reaction**.

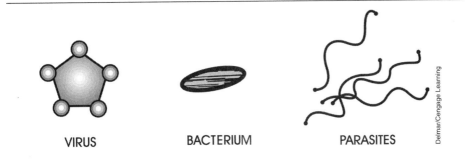

VIRUS   BACTERIUM   PARASITES

FIGURE 6-1

Each of these antigens is a chemical body that provokes an immune response in animals.

With the appropriate stimulation, autoantibodies are produced relatively easily. Certain lymphocytes, the immune cells that make antibodies, are reactive to the normal tissue antigens, or self-antigens, that are always present. Common self-antigens include DNA, IgG (an antibody), phospholipids, erythrocytes (red blood cells), and lymphocytes.

The antibodies made can react with normal tissue and have no effect. This is because the antibodies are regulated. It's when this regulation breaks down that immune diseases can result. Regulation is affected by a variety of problems. For instance, normal cells can often appear indistinguishable from infected cells, enough so that they're considered abnormal too. This is known as *cross-reactivity* with microorganisms.

Most autoimmune disorders likely occur as a result of immune cells that had previously been suppressed by normal control mechanisms of the body. When these cells are no longer suppressed, they attack normal cells. Viruses, particularly those that infect lymphoid tissue, are sometimes capable of interfering with the immune system's control of recognizing "self." The result is autoimmune disease.

Mechanisms of Tissue Damage in Autoimmune Disease

Tissue damage results from a **hypersensitivity reaction**, which is an inflammatory response of immunological origin. There are four types of hypersensitivity reactions.

In type I hypersensitivity, or immediate hypersensitivity, the antibodies IgE and IgG attach to mast cells, which then bind antigen and release histamine (Figure 6-2). **Mast cells** are found in loose connective tissue surrounding blood vessels. They produce and release both histamine and heparin. **Histamine** causes local acute inflammation by dilating blood vessels. **Heparin** is an anticoagulant. If all mast cells release histamine at the same time, acute systemic anaphylaxis, or shock, is the result.

An example of autoimmune disease with type I hypersensitivity is milk allergy. Imagine a cow with a milk-swollen udder. Calves will butt the udder to start the milk flowing. But the head-butting can cause milk to leak into blood vessels. There are many fats and proteins in cow's milk that are not generally found in the circulation. Thus, the cow's body thinks it's being exposed to something foreign, and antibodies are created against some of these substances. The next time the cow gets head-butted and milk leaks into the blood, the antibodies can bind with the mast cells and cause the hypersensitivity reaction, possibly killing the cow.

FIGURE 6–2

Type I hypersensitivity involves the release of histamine from mast cells.

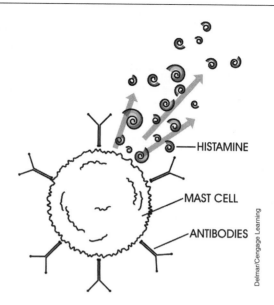

HISTAMINE

MAST CELL

ANTIBODIES

Delmar/Cengage Learning

The second kind is called type II hypersensitivity, or cytotoxic hypersensitivity. Cells that are destroyed by antibodies or by other chemically assisted means that result in their lysis may initiate an acute inflammatory response. This response is due to the release of biologically active cell-breakdown products. Let's examine this phenomenon in more detail.

Autoimmune hemolytic anemia, which is the destruction of RBCs, occurs when there are autoantibodies directed against RBC surface antigens, causing their disintegration. For some reason, the body causes antibodies to be produced against RBCs, to which they stick. A special type of white blood cell (WBC) can bind to the antibodies stuck to the RBCs and can, in turn, cause the RBCs to be destroyed. When this occurs, the RBCs release chemicals that trigger inflammation. And as you have learned, inflammation can be harmful. In this example, the inflammation becomes so bad that the destruction of RBCs continues and the animal can die.

The third category of hypersensitivity is type III hypersensitivity, or immune complex hypersensitivity. Immune complexes are clumps of antibodies (or another chemical called *complement*) that are bound to an antigen. The immune system doesn't "like" immune complexes and attempts to destroy them. Defense and attack immune cells called neutrophils are attracted because of the activation of the complement. Neutrophils release proteolytic enzymes (chemicals that break down proteins), resulting in tissue destruction. This in turn activates more complement. The total effect is acute inflammation and complete tissue destruction.

Rheumatoid arthritis is an example of type III hypersensitivity. Within joints are found antigens that resemble the antibodies attracted to joint cartilage. The antigens cause the formation of antibodies, which stick to them. The result is an immune complex within the joint capsule and close to or attached to the cartilage. Neutrophils are sent to destroy the immune complexes and, in the process, release their proteolytic enzymes, which also damage the joint. Painful, swollen joints are the result.

Type IV hypersensitivity, or delayed hypersensitivity, is called such because of the long time it takes to develop. Type IV reactions are immune cell-assisted reactions that participate in acute inflammation and generally take 24 hours to reach maximum intensity. They are caused by the release of chemicals (such as histamine), which allow inflammatory cells to enter the tissue. Once there, these inflammatory cells begin to slowly kill normal cells in the tissue. Ulcerative colitis is a common example of this. WBCs within the lining of the colon begin to kill the cells around them. The remaining cells don't divide rapidly enough to replace the killed cells, and ulcers result, causing painful defecation and bloody stool.

Disturbances in Immune Function

Disturbances in immune function include deficiencies that arise as a result of inherited defects in the development of the immune system, defects secondary to other causes, or tumors of the immune system. Defects are usually made apparent by an increased susceptibility to infectious diseases. But it's important to be able to distinguish between a defect and a deficiency. Suppose you are buying a car, and the one you are interested in has the wrong carburetor on its engine. As a result, it runs poorly. This is an example of a defect. However, if the car were actually missing the carburetor, it wouldn't operate at all. This is an example of a deficiency.

Inherited Defects in Antigen Processing

The diseases that occur due to defects fall into two major classes: failure of opsonization and failure of intracellular killing. **Opsonization** is a way the immune system has of "coating" antigens with chemicals, such as complement or antibody, in order to make it easier for immune cells to phagocytize them. Problems of this sort are unknown in animals. However, there are cases of antigen-processing defects in animals that fall into the category of failure of intracellular killing. Under this type of defect, cells that have engulfed antigens are unable to destroy them.

Inherited Deficiencies in the Immune System

With deficiencies of this type, actual members of the immune system are missing. Deficiencies in the immune system have helped confirm the "arrangement" of the immune system and allow us to understand how it works. Most immunodeficiencies involve the lack of certain immune cells—either one type or several types—or the lack of certain antibodies.

Secondary Immunological Defects and Deficiencies

The immune system may become suppressed or defective because of an outside influence. Four of these influences can have a negative effect on the body's ability to fight disease.

Malnutrition

In general, severe nutritional deficiencies reduce immune cell function and impair immune cell responses. This can allow increased severity of bacterial infections. Because viruses require healthy cells in which to replicate, malnutrition can cause increased resistance to viruses. This is why overfed dogs are more susceptible to viruses.[1] Several research studies on obesity have demonstrated significant increases in circulatory, locomotion, skin, reproductive, neoplastic, and hepatic diseases.[2,3,4]

Virus-Induced Immune Disorders

Viruses are divided into two groups depending on how they attack the immune system: those that affect primary lymphoid tissue (thymus, skin, bursa of Fabricius, Peyer's patches), and those that attack secondary lymphoid tissue (lymph nodes, spleen, bone marrow, tonsils). Some viruses, such as canine distemper, feline panleukopenia, simian (monkey) AIDS, infectious bursal disease of chickens, mouse thymus herpes virus, and Newcastle disease of birds, can destroy lymphoid tissue. Other viruses stimulate lymphoid tissue to an unusual extent, such as the Aleutian disease of mink virus. Still other viruses cause lymphoid neoplasia, such as Marek's disease of chickens, feline leukemia, bovine leukemia, lymphosarcoma, and mouse leukemia.

Toxin-Induced Immunosuppression

Environmental toxins such as polychlorinated biphenyls, polybrominated biphenyls, iodine, lead, cadmium, methyl mercury, and DDT have a suppressive effect on the immune system. Mycotoxins are chemicals produced by fungi. Being biological in origin, they can also suppress the immune system.

Other Secondary Immunodeficiencies

Protein loss can be a secondary cause to immunodeficiency. Protein loss makes the formation of antibodies difficult. Other secondary causes of immunodeficiency include chronic diarrhea, the chilling of newborn puppies (hypothermia), rapid weaning, prolonged transportation, crowding, trauma (and destruction of lymphoid tissue), and certain endocrine disorders.

Neoplasms of Lymphoid Cells

The replication of immune cells is closely regulated. This regulation is accomplished through rigorous surveillance and destruction of abnormal cells, a process that keeps "bad" cells from replicating. When replication isn't regulated, neoplasia of the immune system is the result. Since the immune system regulates cell replication, any failure or suppression of the immune system can have a bad effect on an animal. Immunosuppressed mammals have an increased rate of cancer as a result of the immune system's failure to regulate cells.

---

[1]Lewis, L.D., et al. (1994). *Small animal clinical nutrition III.* Mark Morris Institute.

[2]Chandra, R.K. (1980). Cell mediated immunity in genetically obese mice. *American Journal of Clinical Nutrition, 33,* 13–16.

[3]Newberne, P.M. (1966). Overnutrition and resistance of dogs to distemper virus. *Federation Proceedings, 25,* 1701.

[4]Williams, G.D., & Newberne, P.M. (1971). Decreased resistance to *Salmonella* infection in obese dogs. *Federation Proceedings, 30,* 572.

Canine lymphosarcoma, a neoplasm of lymphocytes, is responsible for 57 percent of all canine malignancies (Figure 6-3). Classified by site of origin and by cell type, this is a disease of the immune cells and differs from lymphoid leukemia only if there are circulating malignant cells.

Delmar/Cengage Learning

FIGURE 6–3

Cutaneous lymphosarcoma is a type of canine malignancy that almost always spreads to internal organs.

Feline leukemia is caused by an oncogenic (cancer causing) **retrovirus** causing proliferative and degenerative conditions. Retroviruses are potent disease agents. Included in this family of viruses is the human immunodeficiency virus (HIV) that causes human AIDS. Some retroviruses cause neoplasms, as in the case of feline leukemia. Proliferative conditions include lymphosarcoma and leukemia. Degenerative conditions include immunosuppressive diseases, anemia, and pancytopenia (a decrease in all circulating blood platelets, erythrocytes, and leukocytes). Another immune system neoplasm, Marek's disease, is a virus-induced tumor found in chickens.

CASE STUDY   Feline Infectious Peritonitis (FIP)

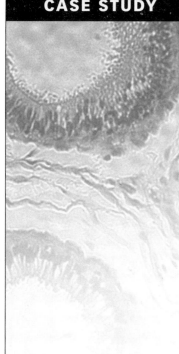

A 5-month-old kitten was adopted from the shelter. The new owner noticed that the kitten wouldn't eat. The kitten's abdomen was distended, doughy, and painful when the owner tried to hug her. The kitten had a fever of 104.5°F and was depressed. The kitten was placed on antibiotics to help with the infection and fever. Unfortunately, the fever did not respond to the antibiotics. The kitten died overnight.

Diagnosis was based on the history, clinical signs, and the laboratory results. A complete blood count (CBC) and chemistry was performed with results indicating leukocytosis with neutrophilia and lymphopenia. A necropsy was performed to confirm wet FIP. The fluid was a thick yellow pyogranulomatous exudate. The specific gravity and the total protein were high. Granulomatous lumps were found on the surface of the kidneys, and the liver had some focal pale lesions. Kittens raised in infected colonies can contract the virus from their mother. These kittens are usually placed in their new homes before the clinical signs start showing, weeks or months later.

Transmission of FIP is by ingestion of the virus or aerosol exposure. The virus replicates in the tonsils or intestinal epithelium. Macrophages and monocytes transport the virus to the primary organs, such as spleen, liver, and visceral lymph nodes. The intrinsic immune responses of the cat determine the form of the disease. The two forms are effusive or noneffusive. The clinical signs of noneffusive FIP are weight loss, ocular and central nervous system (CNS) signs, nonresponsive fever, and serum protein abnormalities.

There is no treatment for FIP, once the clinical signs start to show. Most felines die several weeks or months after diagnosis. Supportive care can make the kitten more comfortable, or possibly extend its life by several months. Supportive treatment, including anti-inflammatory and immunosuppressive drugs, can be given to felines that are in good physical condition, show good appetites, have no neurological signs, and do not have Feline leukemia virus **(FeLV)**-induced malignancy or bone marrow suppression.

A modified live virus intranasal vaccine can help prevent FIP. Efficacy is 40 to 80 percent in kittens over 16 weeks of age that are not already infected with feline enteric corona virus (seronegative). However, vaccination alone should not be relied upon to control FIP. Isolation of felines that are tested positive, weaning kittens earlier, good sanitation, and use of a virucidal disinfectant all help to reduce exposure to this disease.

Summary

Active immunity is generated by an animal's own immune system following exposure to foreign proteins or antigens. Passive immunity is conferred by the administration of preformed antibodies in either serum or colostrum. For most animals, active immunity is not possible for several weeks after birth. The animal's immune system can be adversely affected congenitally, nutritionally, or by exposure to environmental toxins or microbes. Autoimmune disorders are diseases caused by an immune response that is directed against the animal's own tissues. They most commonly occur as a result of immune cells previously suppressed by normal control mechanisms of the body. Disturbances in immune function include deficiencies that arise as a result of inherited defects, tumors of the immune system, and effects that are secondary to other causes.

REVIEW

FILL IN THE BLANKS

1. _____ is a neoplasm of lymphocytes that is responsible for 57 percent of all canine malignancies.

2. _____ are toxic chemicals produced by fungi.

3. _____ are cells found in loose connective tissue along blood vessels that respond to injury by producing and releasing both heparin and histamine.

4. _____is a histopathological term meaning inflammation process in which there is infiltration of polymorphonuclear white blood cells into a more chronic area of inflammation characterized by mononuclear cells, lymphocytes, and macrophages.

5. _____ is a chemical or substance that has the ability to cause an immune reaction within the body.

6. _____ causes local acute inflammation by dilating blood vessels.

7. Mycotoxins are produced by _____.

8. _____ is a process by which the immune system coats antigens with chemicals, such as complement or antibody, in order to make it easier for immune cells to phagocytize them.

9. A/an _____ is a virus that contains RNA and reverse transcriptase. The virus transcribes its RNA into a complementary DNA (cDNA) provirus that is then incorporated into the host cell.

DISCUSSION QUESTIONS

1. What is an autoimmune reaction?

   _____
   _____
   _____

2. Name the three causes of disturbance in immune system function.

   _____
   _____
   _____

3. Explain the phenomenon of cross-reactivity among cells.

   _____
   _____
   _____

4. Name four outside "influences" that can prevent the body from fighting disease.

_____

_____

_____

5. Give an example of an animal retrovirus.

_____

_____

_____

6. List three lymphoid tissues of the body.

_____

_____

_____

7. List three examples of animal diseases that destroy lymphoid tissue.

_____

_____

_____

8. How can obesity in dogs (overfeeding your dog) lead to higher incidence of viral infections in the animal?

_____

_____

_____

9. Why can't you vaccinate a newborn and expect immediate protection from a disease?

_____

_____

_____

10. What is the difference between passive and active immunity?

_____

_____

_____

11. What is colostrum and its importance in immunity?

_____

_____

_____

CHAPTER 7

Mechanisms of Infections and Causes of Infectious Disease

OBJECTIVES

Upon completion of this chapter, the reader should be able to:

- Describe methods by which parasites are capable of causing disease.

- Describe means by which an organism can be affected by the animal itself.

- Describe how a virulent organism can be attenuated to stimulate antigenicity and not disease.

- List several common sources of infection in animals.

- Explain why an infectious disease doesn't have to be contagious but a contagious disease must be infectious.

KEY TERMS

asymptomatic carrier

attenuation

bacteremia

bacteriophage

carrier

contact carrier

contagious

convalescent carrier

endotoxin

entrapment

exotoxin

fomite

iatrogenic

immunosuppression

immune carrier

inert

infection

infectious disease

nosocomial

parasite

pathogenicity

phytotoxin

septicemia

superinfection

synergism

thermolabile

toxemia

toxoid

vector

virulence

zootoxin

Infection

When living agents enter an animal's body and disturb the function of any part, **infection** is said to have occurred. An **infectious disease**, therefore, is one caused by the presence in or on an animal body of a foreign living organism, which creates a disturbance that leads to signs of illness. Most infections are caused by organisms that have escaped from the same species of animal.

Most infecting organisms are destroyed by the host tissue. This is because the host-parasite relationship isn't natural. But in some instances, the animal's resistance may not be sufficient to prevent growth. When this occurs, infection is the result. A chronic infection is from the persistent presence of an organism and the body's persistent attempt to destroy or contain it. An acute or peracute infection occurs when resistance is overwhelmed.

The Fate of the Infecting Organism

Infecting organisms are eliminated in secretions or excretions of the host. These include pus, respiratory secretions, feces, and urine. Chronic infections can result in large numbers of these organisms being eliminated from the body. The escape route from the body may be peculiar and specific to certain organisms, such as saliva for the rabies virus.

If a disease is fatal, many infecting organisms are destroyed with the carcass. However, improper disposal of the body can result in outbreaks. In some instances, the organism is arrested, or inactivated; in other words, although the organism is unable to cause serious damage, the host is unable to eliminate it.

Animals who exhibit flare-ups of disease or who appear altogether normal may also discharge organisms in excretions. These animals are known as **carriers**. A **convalescent carrier** is one who has had a recognized disease but has not rid itself of the infecting agent. **Immune carriers**, or **asymptomatic carriers**, are animals who have eliminated a virulent infection, though they've never actually suffered from the disease. **Contact carriers** are animals that may harbor and eliminate dangerous organisms that were picked up from contact with other animals.

Sources of Infection

The sources of infection are often indirect and complicated. Here are 10 examples:

1. *Direct or immediate contact with a diseased animal.* This involves actual contact between a diseased and normal animal surface. Some specific examples of direct contact infections are ringworm, venereal infections, and bites.

2. *Contact with fomites.* **Fomites** (the plural of *fomite*, also known as *fomes*) are inanimate objects that carry infections from one animal to another. Vaccinating an animal that has a bloodborne disease and using the same needle on another could transfer the disease. The needle would serve as a fomite. This used to be a common occurrence in immunizing cattle on

the range and still occurs with some frequency. Hairbrushes for grooming dogs, beds, water troughs, and blankets for horses, cats, and dogs can all act as fomites in the transmission of infection.

3. *Contact with disease carriers.* Remember, carriers may appear normal even though they're shedding an organism. This can result in direct or indirect infection (Figure 7-1).

FIGURE 7–1

As normal as these dogs appear, one of them could easily be contaminating the others with an infectious organism.

Delmar/Cengage Learning

4. *Infection from soil.* Certain spore-bearing organisms that live in the soil are able to produce disease if they are carried into tissues, usually through wounds. *Clostridium tetani* (the organism that causes tetanus) enters the body when an animal steps on a rusty nail or cuts itself on a dirty can.

5. *Infection from food or water.* Although this is a factor to keep in mind, contaminated food and water are more likely to cause disease in humans than animals. Dogs and cats are commonly infected with *Giardia* from drinking water out of ponds, streams, puddles, and public fountains.

6. *Airborne infections.* Infectious agents use a variety of ways to enter the body when airborne. These include droplets of moisture (saliva and other excretions) that are sneezed or coughed, dust particles, and spores.

7. *Infections from bloodsucking arthropods.* Insects, mites, and ticks are examples of arthropods. Arthropods can be biological vectors, in which they are necessary to part of the infective organism's development, or they may simply carry the organism from one host to another, in which case they are known as mechanical vectors. **Vectors** are simply organisms that carry pathogens from one host to another.

8. *Infections from organisms that are normally carried.* Sometimes an organism that's already found in the body can cause disease, possibly mediated by immune depression. These infections can be diagnosed by serological surveys. The host's immune status, antibody response, and other factors may cause these "inapparent infections" to not be seen. Epidemiologically, however, these infections are still quite important because they serve as an unrecognizable source for the spread of viruses that may interfere with the animal's overall immune system.

9. *Infections acquired in the laboratory or hospital* (**nosocomial** *infections*). Isolation wards, quarantine times and areas, foot baths, rodent control, control of arthropods, and general aseptic techniques minimize the spread of disease in the laboratory and veterinary clinic. It is estimated that 5 to 10 percent of hospitalized patients become ill due to nosocomial infections. There are many factors that predispose the hospital or clinic patient to nosocomial infection (burn patients, congenital deficiencies, glucocorticoid therapy, and antimicrobial therapy). These and other factors need to be considered when treating, handling, or housing patients (see Chapter 10 for zoonotic diseases).

10. *Iatrogenic responses.* **Iatrogenic** refers to negative responses to therapy or medicine. For instance, administering penicillin to someone who's allergic to it will cause a severe, even life-threatening reaction.

Infection and Contagion

An infectious organism that can be transmitted from one body to another by direct or indirect contact is said to be **contagious**. All contagious diseases are infectious. But an infectious disease isn't necessarily contagious. How contagious something is depends on the method of elimination from the host and the organism's opportunity of reaching another body. A highly contagious disease is likely to infect another animal, a slightly contagious disease poses some risk of infection, and a disease that isn't contagious offers no risk of further infection. For example, a bird that gets bitten by a mosquito carrying malaria will probably get the disease, because it is highly infectious. Other birds kept in the same cage probably won't get malaria from the infected bird, though, because it is not contagious. A bite from an infected mosquito is required to spread the disease.

A **superinfection** is a fresh infection in addition to one already present. The term is also used for a reinfection. The infection may have been passed from the same species of organism or from a different species. If more than one organism is present, we have a **mixed infection**. A secondary infection is a type of mixed infection in which the first organism makes a favorable environment for a second organism. For example, puppies that get Parvovirus have a suppression of their immune system. Since the immune cells can no longer fight off foreign invaders, bacteria are able to infect the puppy as well. The bacteria are secondary infectors.

Synergism occurs when the action of two or more agents produces a result that neither could bring about alone. For example, an organism called *Treponema* (anaerobic bacteria) isn't very dangerous by itself; but when it interacts with normal gut microbes in pigs, the result is swine dysentery, or inflammation of the intestine.

Properties of Pathogenic Organisms

Virulence refers to the disease providing the power or malignancy of an infectious organism. We call this pathogenic power (the ability to cause disease in susceptible animals). Virulence can be altered. **Attenuation** is the process by which virulence is diminished. Methods of attenuation include:

- Cultivating an organism at unfavorable temperatures.

- Heating an organism to below the thermal death point.

- Cultivating an organism on an unfavorable medium.

- Selecting a nonvirulent strain.

- Injecting the organism into a resistant species. This is most commonly used when developing vaccines for viruses. Rinderpest virus was first attenuated by growth in goats and later in rabbits to develop a working vaccine for cattle. The development of a rinderpest vaccine devoid of residual virulence was thus developed.

- Repeated passage and adaptation. (In some cases, however, repeated passage can also increase virulence.)

Another method of attenuation of viruses is the growth and repeated passage of the virus through eggs. Rabies vaccine for dogs and cats is developed by this method. This high egg passage vaccine (HEP) has 178 passages in eggs and thus is completely safe for both dogs and cats.

Entrapment refers to a malignant growth that causes a pathological situation in which an organism, organ, or tissue (i.e., nerve) is trapped in an abnormally produced anatomical or physiological site. The characteristic clinical features of most metastatic disease is the unrestricted growth which leads to local infiltrates and entrapment of structures. The growth can in turn "influence" the second structure or organism. **Bacteriophages**, viruses that infect bacteria, are able to transfer genes to bacteria in order to change their virulence.

The **pathogenicity** of an organism can also be affected by the animals themselves. Many diseases are not as pathogenic as they once were. This is due to factors such as herd immunity, better nutrition, better hygiene, and genetic factors. Our ability to alter the pathogenicity of organisms has allowed us to develop vaccines, which we can then use to provide protection against deadly forms of the same disease. If we can create bacteria that lack the ability to cause disease, we can inject these bacteria to cause an immune reaction (Figure 7-2). That way, if an animal ever gets infected with the actual disease-causing organism, it will be protected.

Mechanisms of Disease Production by Pathogenic Organisms

The ability of an organism to produce disease depends on two properties of the organism:

1. *The ability to multiply in tissues or on body surfaces.* Once established in the tissues, a localized or generalized infection can result. Generalized infections,

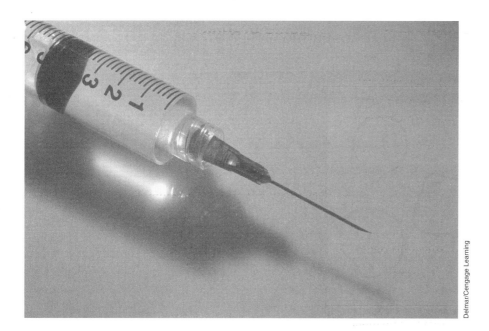

FIGURE 7–2

Vaccines play a large part in warding off infectious diseases in humans and animals.

also called systemic infections, are the result of metastasis and spread via the circulation. **Bacteremia** means that bacteria are circulating within the blood. **Septicemia** means that bacteria and the toxins they produce are circulating in the blood. A **toxemia** results when a bacterial toxin enters and circulates in the blood.

2. *The ability to form toxins.* Toxins come in a variety of types. **Phytotoxins** are produced by plants, an example being mushroom poison. **Zootoxins** include poisons such as snake venom (Figure 7-3). Then there are the bacterial toxins, chemicals produced by bacteria, which can cause disease or even death. These toxins usually exhibit specific tissue preferences; as such, they can act upon nerve tissue, blood, or WBCs. **Endotoxins** are made within the bacteria and may be structural components of the organism. Endotoxins are sometimes released when the bacteria die and break open or are extracted.

Exotoxins are secreted or excreted. Exotoxins are primarily proteins that are produced within some gram-positive bacteria and secreted into the surrounding media. Clostridial toxins are examples of exotoxins. Diseases they produce would include botulism, tetanus, malignant edema, blackleg, and overeating disease. Exotoxins are very antigenic and potent.

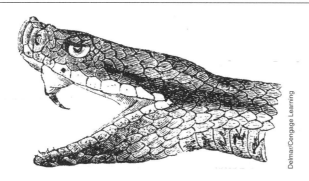

FIGURE 7–3

Toxins formed by bacteria on the cellular level are comparable to such common zootoxins as snake venom.

Toxins have a number of important properties. Certain toxins have the following qualities:

- Are **thermolabile**, or destroyed by heat

- Consist of large molecules

- Require a period of incubation (Because they're big, toxins often need time to diffuse through tissues.)

- Deteriorate with age

Being large in size, a toxin consists of several parts. As it ages, the poisonous part of the toxin disappears first, leaving behind a toxoid. Although a **toxoid** isn't poisonous, it is antigenic. Because the toxoid is antigenic, injection of one will cause an antibody reaction. (Toxoids are used, and sometimes specifically made, as vaccines.) And since it isn't poisonous, injection of a toxoid won't cause disease. For instance, every time you get a tetanus shot, you are being given an injection of tetanus toxoid, the antigenic portion of the tetanus poison, to protect you in the event that you become infected by tetanus bacteria.

In addition to toxins, there are other active constituents produced by pathogenic bacteria. These include coagulase, which accelerates clotting of plasma; fibrinolysin, which breaks down fibrin and allows bacteria to move more readily through tissue; and hyaluronidase, which is a spreading or diffusing factor.

Pathogenesis of Parasitic Infections

Parasites are organisms that live on or in another organism, getting their nourishment from that other organism. The animal being parasitized is called the host. The pathogenic effects of a parasitic infection may be so subtle as to be unrecognizable, or they may be very obvious. Healthy animals can be host to hundreds of parasitic worms with no indication of stress or illness. Or they may be in danger of death from anemia, unthriftiness, or retarded development. Pathogenic effects fall into one of four categories.

The first category, trauma, involves the destruction of cells, tissues, or organs. Trauma can be mechanical or chemical. Examples of mechanical trauma include roundworm or hookworm larvae burrowing through lung tissue, adult hookworms performing mucosal grazing in the intestine, *Capillaria* (parasitic nematode) causing hepatitis and gastroenteritis, *Trichuris* (whipworms) destroying the liver and preventing liver regeneration with eggs, and sparganum-type tapeworm larvae tunneling through the dermis, creating thousands of cavities. The digestion of the intestinal mucosa by *Entamoeba* (an amoeba) is an example of chemical trauma.

The diversion of host nutritive substances constitutes the second category. Tapeworms and Acanthocephala (thorny-headed worms) lack digestive systems (Figure 7-4). They can absorb significant amounts of food from the host intestine. *Diphyllobothrium*, a tapeworm of fish-eating mammals (including dogs, cats, and humans), absorbs vitamin B_{12}, which is necessary for RBC production. A lack of this vitamin can cause anemia. *Ascaris* also consumes a good deal of the food ingested by the host. *Giardia*, a protozoan, covers the intestinal cells and

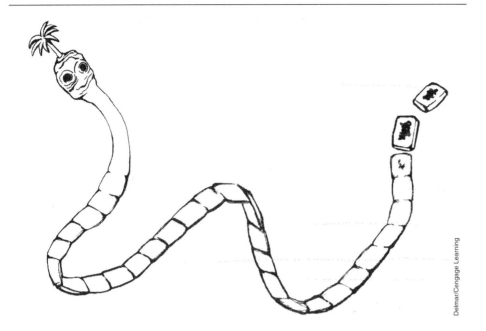

Delmar/Cengage Learning

FIGURE 7–4

Once the tapeworm attaches itself to the host's intestine, it can steal away important nutrients for itself.

prevents normal absorption. This interferes with normal intestinal absorption, and the unused nutrients are passed uselessly.

Toxin production is the third category. Reactions in the host can be directly due to toxin release or toxin release coupled with an immune reaction. *Trypanosoma cruzi*, a protozoan, causes cells in cardiac and smooth muscle to rupture and release a neurotoxic substance. This substance in turn attacks autonomic ganglion cells, ruining nervous control of peristalsis and heart contraction. *Trypanosoma brucei gambiense* produces a neurotoxin that causes severe brain damage. *Plasmodium*, the parasite that causes malaria, proliferates in RBCs, which can rupture simultaneously and release the parasites' waste products. This often induces a characteristic malarial fever. Verminous intoxication, caused by the waste products of tapeworms, can cause dizziness and nausea. *Trichuris* can cause rectal prolapse by producing a toxin that affects the nervous control of the intestinal muscles. Reactions due to toxin release combined with an immune reaction include examples such as *Onchocerca* larvae, which wander into the retina of the eye, where they die and elicit a powerful immune response. This combination of foreign protein and WBC invasion destroys the retina, causing blindness. *Wuchereria* (a filarial worm), which normally lives in the lymphatics, may elicit an immune reaction by releasing larvae in aberrant sites, resulting in elephantiasis.

The fourth category consists of serving as vectors for other disease agents. As you learned earlier, vectors are organisms that carry a disease pathogen from one animal to another. The salmon-poisoning fluke carries a rickettsia (*Neorickettsia helminthoeca*), which produces salmon poisoning in dogs that eat infected fish (salmon, trout, or Pacific giant salamanders). *Heterakis*, a nematode worm of birds, transmits the protozoan *Histomonas* within its eggs, which can be highly pathogenic in birds.

Tolerance in the Host-Parasite Relationship

Successful parasites evolve to avoid the protective immune response of the host. Adaptation tolerance, or mutual tolerance, occurs when the parasite becomes immunologically inert. The term **inert** means that the parasites are essentially invisible to the host. One method of becoming inert involves finding immune "blind spots." When this happens, the parasite is able to mimic the host in some way—essentially masking itself—so as to avoid a defensive response (i.e., the host's creation of antibodies).

A nonpathogenic parasite has adaptive value. For one thing, it doesn't destroy the host and can continue to reproduce. As previously stated, the parasite is able to manufacture antigens that don't cause an immune reaction. In fact, some parasites have the ability of antigenic variation. These parasites are able to continually change their surfaces to keep the immune system from mounting an effective immune response.

Immunosuppression is another method by which a parasite can cause the host to tolerate its presence. During immunosuppression, parasites can kill immune system cells, stimulate cells until the supply is exhausted, or even release immunosuppressive chemicals to inhibit the immune system from mounting a defense.

Case Study	**Parvovirus**
	A 10-week-old male Rottweiler puppy is presented for anorexia, vomiting, and diarrhea. The puppy was obtained from a backyard breeder about 5 days previously. According to this owner the puppy has not had any vaccinations and did not show any signs of illness for the first 4 days. The dog is being fed generic dog food that is being fed to the other two dogs in the household. Vomiting and anorexia started one day ago. The owner describes one bout of diarrhea the morning of presentation when it was noticed that the diarrhea had a foul smell. This can be attributed to an increase of protein in the feces.

On physical examination the dog appears depressed, mucous membranes are pale, yet capillary refill time (CRT) is delayed by more than 2 seconds (normal is 1 to 2 seconds). The heart and lungs sound normal. The abdomen is tender upon palpation. The pup's temperature is 101.8°F, but there is observed blood on the thermometer. Abdominal radiographs reveal intussusception—a consequence of Parvo infection.

The primary differential diagnosis for this case, based on the history and the physical examination, is Parvovirus. Other differentials that should be considered are foreign body ingestion, high worm burden (whipworms, hookworms), and intussusception.

Diagnostic tests that are run on this pup include a Parvo Cite test (in house) by obtaining some feces followed by an ELISA test that is sent to an outside laboratory for testing of antigens. Another test that is helpful is a CBC. Results indicate a + Cite test and a leukopenia. A diagnosis of Parvovirus can be made.

The treatment for Parvo is supportive. The animal is placed on appropriate IV fluids to support the severe diarrhea caused by the denuded and necrotic intestinal lining. (See Animal Disease 6 and 7 from Appendix III.) For support the animal is also kept NPO (nothing per os—which means nothing by mouth) to allow the intestinal lining to heal. Antiemetics and broad-spectrum antibiotics are used to treat the vomiting and secondary septicemia respectively. Most dogs can recover with treatment if caught early enough—otherwise prognosis can be poor.

Summary

Infectious disease is caused by the presence of one or more foreign living organisms that create a disturbance leading to signs of illness. The fate of the infecting organism, the sources, and the method of contamination and infection all determine its identification, prevention, and treatment. Sources of infection can include fomites and vectors. Infections can be acquired in a hospital setting, on the farm, by consumption of contaminated food and water, and by airborne sources. *Virulence* refers to the power or strength of infectious capability. Organisms can produce disease by multiplying in animal tissues or on body surfaces and by producing and releasing toxins.

REVIEW

FILL IN THE BLANKS

1. _____ is a term that describes a malignant growth in the presence of another organism.

2. _____ are viruses that infect bacteria.

3. _____ is also known as a fresh infection or reinfection.

4. _____ means that bacteria and their toxins are circulating in the blood.

5. _____ are toxins produced by plants.

6. _____ are substances that resemble bacterial toxins in antigenic properties but are actually found in an animal's fluids such as snake venom.

7. _____ is a toxin that has been treated or aged to destroy its toxic properties but retain its antigenicity.

8. _____ are organisms that live on or in another organism, getting their nourishment from that other organism.

9. _____ tolerance occurs when the parasite becomes immunologically inert.

10. _____ are nonliving objects that possibly harbor or are capable of transmitting disease.

11. _____ is the condition or ability to cause disease.

12. _____ is a disturbance in the function of any part of an animal's body caused by the entering of living agents into that body—the presence of endoparasites in a host.

13. _____ are organisms that carry pathogens or infective agents from one host to another.

DISCUSSION QUESTIONS

1. What is the relationship between a mixed infection and a secondary infection?

_____

_____

_____

2. What is a fomite (fomes)?

_____

_____

_____

3. What is meant by a convalescent carrier?

_____

_____

_____

4. Name three of the four methods by which parasites are capable of causing disease.

_____

_____

_____

5. What is the difference between an infection and being contagious?

_____

_____

_____

6. Define and give an example of an iatrogenic infection.

_____

_____

_____

7. Describe the differences between endotoxins and exotoxins.

_____

_____

_____

CHAPTER 8

Genetic Disorders

OBJECTIVES

Upon completion of this chapter, the reader should be able to:

- Describe the fundamental differences between genetic, nutritional, and microbial disorders.

- Discuss dominant versus recessive traits and their relationship to genetic disorders.

- Describe common genetic disorders in horses, dogs, cats, and livestock.

KEY TERMS

congenital

cytogenetic

dominant

gene tracking

genetics

genome

hereditary

hermaphrodite

inheritance

mutation

pseudohermaphrodite

recessive

recombinant DNA

reverse genetics

transgenics

Genetic Technology

Today's current progress in medical **genetics** is directly due to progress in the field of molecular biology, involving recombinant DNA technology. **Recombinant DNA** is DNA that's made in the laboratory when new DNA sequences are inserted via chemical or biological means (Figure 8-1).

All diseases have a molecular basis. Molecular biology has helped to solve genetic diseases using several techniques. The classical approach to genetic disorders was to identify an abnormal gene product so that the normal gene could be found and the changes within it determined. Today, scientists have developed a method known as **reverse genetics**, in which marker genes are placed within an animal to localize mutant genes. **Transgenics** is the exciting new technology in which this recombinant DNA is placed into an animal. Once there, it can trigger the production of unique proteins or "knock out" other proteins that the animal would normally produce. Knockouts are animals that have been "engineered" to lack a gene they normally have.

As our discussion turns to genetic disorders, several definitions become important. The term **hereditary** refers to a trait derived from one's parents and transmitted in the genes through the generations. Since the trait is passed down through families, it's also called a familial trait. Similarly, the term **congenital** means "born with." Thus, we can say that certain individuals possess congenital hereditary traits. *Homozygous* means having identical genes at the same location on paired chromosomes. *Heterozygous* means having nonidentical genes at the same location on paired chromosomes.

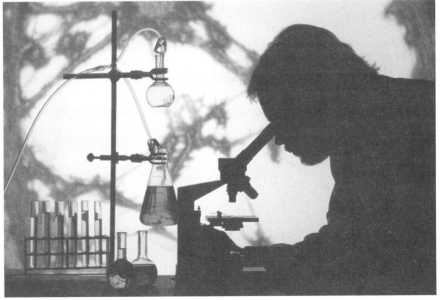

Delmar/Cengage Learning

FIGURE 8–1

Today, scientists are able to engineer recombinant DNA in the laboratory.

Mutations are permanent changes in DNA. **Genome** mutations involve the loss or gain of an entire chromosome. Chromosome mutations are the result of a rearrangement of genetic material and give rise to the visible structural changes in chromosomes. Gene mutations are partial or complete deletions of genes on the chromosomes. Mutations interfere with protein synthesis. When proteins are not made properly, disease is the result.

Categories of Genetic Disorders

There are four major categories of genetic disorders:

- Mendelian, in which mutant genes have a large effect
- Multifactorial, in which multiple genes and the environment influence the disorder
- **Cytogenetic**, which result from chromosomal disorders
- Nonclassical **inheritance**, in which a single-gene disorder with a nonclassical pattern of inheritance exists

Let's examine each of these in more detail.

Mendelian Disorders

There are over 4,500 Mendelian disorders, all the result of an expressed mutation in a single gene that has a large effect. There are several transmission patterns for inheriting this type of single-gene disorder.

With autosomal dominant disorders, only one copy of the gene is necessary, and the disorder becomes relevant because it's the dominant gene. **Dominant** genes always express themselves in the first generation. **Recessive** genes have their traits hidden or masked in the heterozygous condition. Autosomal recessive disorders are those in which the disease does not usually affect the parents; however, one in four offspring will be affected when he or she inherits two copies of the gene, one from each parent. X-linked disorders involve daughters as carriers of a disease, while sons suffer the disease's effects (Figure 8-2).

The inheritance of one of the many colors of cats is a classical example of sex-linked inheritance in domestic animals. The heterozygous offspring of a pure-bred black bred to a purebred orange cat will exhibit a tortoiseshell coloring (mixing of black and orange hair) in female offspring. The male offspring will be black (if black is dominant). The expression of this trait is thus linked to the sex of the offspring.

The biochemical and molecular basis of single-gene disorders is a result of either defective enzymes, defects in receptors and transport systems, or alterations of nonenzyme proteins. Defective enzymes can cause an accumulation of material that's normally broken down, a decrease in the amount of a product that's supposed to be produced, or the failure to deactivate tissue-damaging substances. Several diseases due to defects in enzymes are seen in animals and

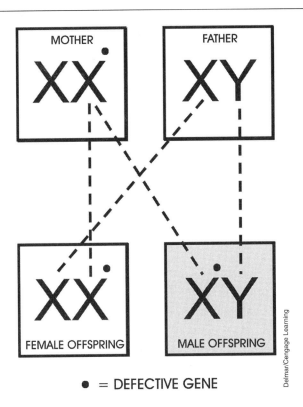

= DEFECTIVE GENE

Delmar/Cengage Learning

FIGURE 8–2

In this illustration, a gene that carries disease is passed from the mother to both the female and male offspring. Because the disease is an X-linked disorder, only the male offspring exhibits traits of the disease; the female offspring is merely a carrier.

include lysosomal storage diseases, mucopolysaccharidoses, and glycogen storage disease (glycogenosis). Any disorder associated with defects in proteins that regulate cell growth could result in the growth of tumors.

Multifactorial Disorders

Multifactorial disorders are due to the combined action of the environment and two or more mutant genes having an additive effect. An example was previously discussed when describing renal agenesis. Heart disorder, diabetes, and cancer are additional examples of multifactorial disorders. Currently, these disorders and the interactions of the environment and genetics are not well understood. Animals differ genetically in their nutritional requirements. Animals also differ in their abilities to withstand nutrient deficiencies and toxicities. The White Leghorn chicken, for example, thrives on a diet that contains only 30 ppm (parts per million) manganese. Many varieties of the Rhode Island Red chicken need as much as 50 ppm to prevent the disorder perosis. Zebu cattle are able to perform well at 46°C, whereas Holstein cattle have elevated respiration rates, increased body temperature, and lowered performance at 36°C.

Cytogenetic Disorders

These disorders are due to chromosome mutations in which there may be an abnormal number of chromosomes or alterations in the structure of one or more chromosomes. An abnormal number of chromosomes can be expressed as euploid, where the number present is an exact multiple of the normal number. For example, humans normally have 23 chromosomes; therefore, if someone had 46 (2 × 23), he or she would be termed euploid. An aneuploid individual possesses a number of chromosomes that's not an exact multiple of 23, such as 24.

Along with abnormal amounts, cytogenetic disorders involve abnormal alterations in the structure of one or more chromosomes. For example, a deletion is the loss of a portion of a chromosome. A translocation occurs when a segment of one chromosome is transferred to another. Isochromosome formation results when one chromosome arm is lost and the remaining arm duplicates itself to make up for the missing piece. Ring chromosomes result when the ends of a chromosome fuse with each other. And inversion occurs when the chromosome breaks into three pieces and one piece reverses its position before joining back with the remaining pieces.

The most obvious results of a cytogenetic disorder are those that involve the sex chromosomes. **Hermaphrodites**, animals that have both ovarian and testicular tissue, are the result of cytogenetic disorders, as are **pseudohermaphrodites**. A female pseudohermaphrodite has ovaries but has the external appearance of a male, whereas a male pseudohermaphrodite has testicular tissue but outwardly resembles a female.

Nonclassical Inheritance

Disorders of this sort include triplet repeat mutations (or fragile X disease), mutations in mitochondrial genes, and genomic imprinting. Fragile X syndrome is the most common inherited cause of mental retardation in humans. The full mutation occurs 1 in 3,600 males and 1 in 5,000 females. The mutation occurs in the FMR1 gene on the X chromosome. Fragile X is not the result of a single change in a base. It is a trinucleotide repeat disorder in which there is a multiplication of part of the genetic information. Researchers have developed a fragile X syndrome animal model that has been associated with cognitive deficits. The studies involve the brains of wild-type and FMR1 knockout mice. Fortunately, none of these disorders are common in animals.

Molecular Diagnosis

Advances in molecular biology have aided greatly in the diagnosis of genetic disease. Genetic diseases and their effects can be determined only if you understand gene structure, and molecular biology has allowed us to do that. **Gene tracking** is a process that helps determine whether members of the same family have inherited a defective gene. Animals identified in this manner can then be prevented from breeding. Other diagnostic applications of recombinant DNA techniques help aid in the diagnosis and understanding of cancer and how it affects cells, forensics, and infectious disease diagnosis. Polymerase chain reaction (PCR) analysis can determine the actual size of repeat sections in individuals with trinucleotide repeat disorders and compare them to the stable version or with the permutation. Cytogenetic tests have been around since the 1980s to help in diagnosing carriers of various genetic disorders.

Severe combined immunodeficiency (SCID, formerly known as just CID) in Arabian horses and in basset hounds is an autosomal recessive disease. The carriers of the disorder are completely free of any clinical signs of the disease. If two carriers are mated, there is a 25 percent chance that the resulting offspring will be affected and a 50 percent chance the resulting offspring will be a carrier. Affected foals with SCID usually die before 2 months of age to secondary infections. Puppies with SCID appear normal until 6 to 12 weeks of age. The most common cause of death in SCID puppies is from canine distemper due to routine immunization with modified live virus distemper vaccine. Until recently, molecular diagnosis of a carrier of SCID was very difficult. Today testing is highly accurate and easy to perform. A simple cheek or lip swab or a blood sample is used. For ethical reasons no carriers should be used as breeding animals.

CASE STUDY — Canine Panosteitis

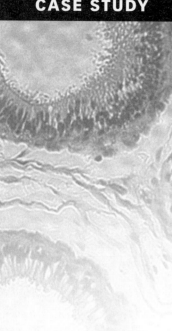

An 8-month-old male German shepherd shows lameness that started in the left foreleg and migrated to the right foreleg. The puppy's legs were painful on palpation. The veterinarian requested radiographs and routine blood work (CBC/biochemistry) to rule out any other possible diseases due to the prolonged fever. Radiographs showed a mottled density in the medullary cavity of the diaphysis (long bone) of the affected bone. The puppy had a low-grade fever for a few days. The blood work presented normal. The puppy was given a nonsteroidal anti-inflammatory drug (NSAID) to minimize pain and decrease inflammation. The lameness lasted for 3 weeks and the puppy made a full recovery.

CLINICAL SIGNS AND DIAGNOSIS

Pathogenesis of panosteitis is unknown. There is no proven genetic transmission, but German shepherds are predominantly affected by panosteitis, which strongly suggests an inheritable basis for this disease. Panosteitis is a self-limiting, painful condition that can affect one or more of the long bones in young, medium- to large-breed dogs. This disease is characterized clinically by lameness and radiographically by high density of the marrow cavity. Depending on the duration of the disease, radiographic lesions fall into three phases. In the early phase the ends of the diaphysis become more prominent, but may appear blurred and granulated. The middle phase has patchy sclerotic opacities, which appear around the nutrient foramen and later throughout the diaphysis. In the late phase the overall opacity of the medullary canal diminishes toward normal pattern, but some granular opacity can remain. Prognosis for recovery is good.

CASE STUDY — Hip Dysplasia

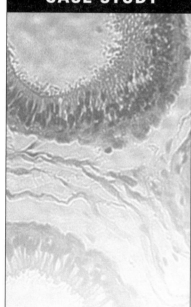

A 12-month-old male Golden Labrador had normal hips at birth. As the dog grew, his hips developed abnormally. The dog had a dorsal displacement of the head of the femur in the hip joint. This caused the cartilaginous dorsal rim of the acetabulum to become distorted. During the later months the acetabulum ossified into a distorted shape (shallow). The dog started to have difficulty getting up and lying down. He did not play ball with the owner as much as he used to. The owner noticed that the dog was reluctant to climb the stairs. The veterinarian recommended conservative medical therapy first, followed by surgery in the future, due to the financial considerations of the owner. Since the owner had a pool, the veterinarian recommended that his dog go swimming to maintain joint mobility while minimizing the weight bearing on the dog's hips.

An orthopedic exam and radiographs were performed. Ortolani's sign was elicited as a guide to estimate the degree of rotation. Orthopedic Foundation of America (OFA) radiographs were performed to determine hip dysplasia and its degree of

severity. Hip dysplasia is the malformation and degeneration of the coxofemoral joints. This disease is a developmental defect initiated by genetic predisposition to subluxation of the immature hip joint. Weight control, NSAIDs, and surgery are the recommended treatments. Rapid weight gain can cause the disease to progress. The dog should not be used for breeding due to the hereditary condition of this disorder. The surgical considerations depending on the severity of joint laxity and the presence or absence of degenerative joint disease are triple pelvic osteotomy, total hip replacement, and excision arthroplasty. Most dogs can lead a normal life with proper medical or surgical management.

CASE STUDY	Hyperkalemic Periodic Paralysis (HYPP)

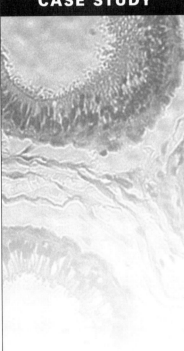

A 3-year-old well-muscled male quarter horse is suffering from muscle tremors shortly after exercising in the arena. A few minutes later the horse is lying down on his side. The owner mentions that her horse has done this a few times before—each time after exercising.

Diagnosis can be made based on the clinical signs presented, along with the gene probe test for the HPP type sodium channel DNA. An electromyogram, which records the intrinsic electrical properties of skeletal muscles, can be helpful in diagnosing the disorder. All affected horses can trace their ancestry to the American Quarter Horse sire Impressive.

Hyperkalemic periodic paralysis is due to an inherited mutation of the sodium channel gene. The disorder is caused by a failure of the sodium ion transport across the skeletal muscle cell membrane. The frequency of the attacks can be controlled with medication such as acetazolamide (a potassium-wasting diuretic). Emergency treatment includes intravenous administration of 5 percent dextrose with sodium bicarbonate or 23 percent calcium gluconate. The dextrose and bicarbonate solution will help move potassium back into the cells, while the Ca-gluconate helps to counteract the effects for hyperkalemia.

Summary

The common use of recombinant DNA, transgenics, and other subdivisions of molecular biology requires a basic understanding of genetics by the veterinary technician. Molecular biology is helping to solve many genetic disorders of animals and humans alike. Multifactorial disorders are due to the combined action of the environment and two or more mutant genes having an additive effect. It is important for the veterinary staff to understand and to help educate animal owners of the risks of genetic disorders in breeding animals, and the care, prevention, and treatment of the same. New advances in detection of inherited or defective genes permit a greater opportunity for prevention.

REVIEW

FILL IN THE BLANKS

1. _____ is an animal that has both ovarian and testicular tissue.

2. _____ are disorders caused by chromosome mutations in which an abnormal number of chromosomes or alterations in the structure of one or more chromosomes have occurred.

3. _____ is an individual that possesses a number of chromosomes that is not normal for that particular species.

4. _____ are permanent changes to the DNA that are perpetuated in subsequent divisions of the cell.

5. A/an _____ disease is one the animal was born with.

6. _____ is a common genetic immunodeficiency disorder in the Arabian horse.

7. _____ is a gene that hides or masks the effect of another gene in the same allelic series.

8. _____ is the transmission and reception of genetics from one generation to the next.

9. _____ occurs when there are permanent changes in the DNA altering the genetic message carried by that gene.

10. _____ gene is one that expresses a hidden characteristic when in the homozygous state.

11. _____ is the biology of heredity and variation in organisms.

12. _____ involves placing recombinant DNA, or cloned genetic material, from one species or breed into another.

13. _____ are animals that have been engineered to lack a gene they normally have.

14. _____ means having nonidentical genes at the same location on paired chromosomes.

15. _____ is a chromosome number that is a multiple of the haploid number of chromosomes.

DISCUSSION QUESTIONS

1. Name three factors that can result in a single-gene disorder.

_____

_____

_____

2. What is gene tracking?

_____

_____

_____

3. Explain the difference between gene mutations and genome mutations.

_____

_____

_____

4. Explain the difference between translocation and deletion.

_____

_____

_____

5. Why would you expect to find more cases of perosis in Rhode Island Red poultry when compared with the White Leghorn breed? How is this problem related to genetics?

_____

_____

_____

CHAPTER 9

Environmental and Nutritional Diseases

OBJECTIVES

Upon completion of this chapter, the reader should be able to:

- List several common types of physical injuries of animals.

- List the fat-soluble and common water-soluble vitamins.

- Describe how secondhand smoke can be dangerous to a domestic cat.

- Discuss global warming and its effects on animal welfare, the environment, and humankind.

- Discuss animal obesity and the consequences on animal health.

KEY TERMS

adverse drug reaction

antimicrobial agent

antineoplastic agent

carcinogens

global warming

hyperthermia

hypothermia

immunosuppressive agent

obesity

undernutrition

Environmental Impact on Health

When disease is not genetic, it's usually the result of environmental factors. These factors can include infectious agents, nutrition, trauma, or pollutants that contaminate the animal's environment. The World Health Organization (WHO) believes that climate change is one of the largest health challenges facing us today. The effects are wide-ranging and have mostly adverse impacts on health. One of the most frightening possibilities is that with, **global warming**, diseases of animals that were once considered tropical may be able to get a foothold in the southern United States (Figure 9-1).

FIGURE 9–1

Global warming may prove to have a negative effect on the disease susceptibility of many animals.

Because most environmental changes can have an impact on nutrition, and because human beings are the ultimate consumers, most discussions have been concentrated largely on the environment's effects upon people. It's safe to say that if people are suffering from these effects, animals are surely suffering as well.

Air Pollution

In many animal environments, indoor air is becoming more of a problem than outdoor air. This is because indoor air is not regulated, there may be poor circulation of air, air is often contaminated with particulates, and many indoor pollutants are not measured to determine the degree of pollution.

Tobacco smoking is a major contribution to disease. Smoke contains **carcinogens**, cell irritants and toxins, carbon monoxide, and nicotine. Coronary heart disease is the number one cause of death in humans in the United States, followed by lung and other cancer and strokes. In addition, involuntary smoke exposure (secondhand smoking) can retard physical and mental maturation, increase the rate of myocardial infarctions, and increase the number of respiratory infections an individual will get.[1] As a result, we are certain to see more of these smoking-related conditions begin to affect animals as well. Dogs and cats kept indoors with smokers can suffer from smoke exposure the same way humans can. Squamous cell carcinoma is the most common malignant cell type in both cats and humans. Research shows a substantial increase in respiratory infections, allergies, and related illnesses in animals and in humans kept indoors with smokers. Tobacco-related pollutants contaminate the fur of household cats, putting them at a greater risk for oral squamous cell carcinoma and lymphoma.[2]

Cats exposed to 5 or more years of secondhand smoke have five times the risk for oral squamous cell carcinoma. The survival rate for these diagnosed felines is less than 10 percent at one year. In another study, cats living in homes where humans smoked at least a pack of cigarettes a day had more than three times the risk of developing lymphoma than cats in nonsmoking houses, cats living with one household smoker had almost twice the risk, and cats living with two or more smokers in a household had nearly four times the risk of developing lymphoma cancer.[3]

Chemical and Drug Injury

Industrial exposure and accidental and, in humans, self-administered drug use can all lead to injury. An **adverse drug reaction** is a serious, undesirable, and possibly life-threatening response to a drug, sometimes occurring at doses normally used for prevention, diagnosis, or treatment (Figure 9-2). Adverse reactions require intervention from a veterinarian. Side effects are defined as secondary reactions to a drug that are usually predictable and not life threatening. Therefore, two types of reaction can occur: a generally predicted side effect or an unpredictable

---

[1]Dunn, A., et al. (1997, September). *Health effects of exposure to environmental tobacco smoke* (Final Report). California Environmental Protection Agency.

[2]Villalobos, A. (2003, May). Animal Oncology Consultation Service. Personal communication.

[3]Bertone, E.R., et al. (2002). Environmental tobacco smoke and risk of malignant lymphoma in pet cats. *American Journal of Epidemiology, 156,* 268–273.

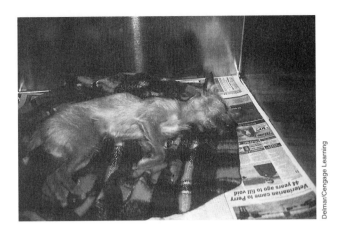

Delmar/Cengage Learning

FIGURE 9–2

Veterinary professionals must always be prepared for adverse drug reactions in an animal.

adverse reaction to the drug. According to the U.S. Food and Drug Administration (FDA) an adverse reaction is unexpected in intensity or in kind—distinct from expected side effects. As previously mentioned, side effects are largely predictable reactions and thus easier to deal with. Adverse reactions often occur when the drug invokes an immune reaction or has an unanticipated toxicity.

Therapeutic agents, such as **antineoplastic agents**, **immunosuppressive agents**, and **antimicrobial agents**, can all cause problems. Antineoplastic agents act on dividing cells and can cause bone marrow suppression, lead to immunosuppression, or even induce cancer. Immunosuppressive agents cause immune system problems, while antimicrobials can provoke hypersensitivity or even lead to resistant infections of organisms. Nontherapeutic agents will not be of great concern to you as a veterinary technician, as few animals are known to be ethyl alcohol abusers. However, heavy metal poisoning from lead or zinc is common in pet birds and can be seen in other pets, such as dogs, when exposed to old paint.

Environmental and occupational carcinogens include arsenic, asbestos, benzene, cadmium, chromium, nicotine, nickel, nitrites, uranium, and vinyl chloride. Street drugs, such as marijuana, cocaine, crack, and heroin, can be ingested by pets when made available by drug-abusing owners.

Physical Injury

There are five types of physical injuries:

- *Injuries induced by mechanical force.* These include fractures, abrasions, lacerations (which, unlike incisions, are irregular tears), contusions due to blunt force, and gunshot wounds (Figure 9-3).

- *Injuries related to changes in temperature.* Also called hyperthermic injuries, these include first-, second-, and third-degree burns and systemic **hyperthermia**, or heat stroke. Abnormally low temperatures (**hypothermia**) can also cause local reactions, such as frostbite, or indirect reactions, such as circulatory changes, and may be life threatening.

- *Injuries related to changes in atmospheric pressure.* These include high-altitude illness, blast injury from close exposure to explosives, air or gas embolism, and decompression disease.

FIGURE 9-3

Veterinarians are called on to treat all types of physical injury, including fractures and lacerations sustained through mechanical force.

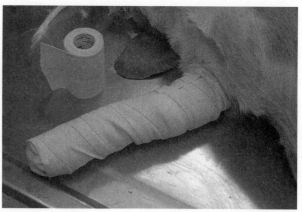

Delmar/Cengage Learning

- *Electrical injuries.* Animals that bite electrical cords can receive severe burns or sustain injuries to the nervous system that may affect other body systems, including normal heart function and death (electrocution).

- *Radiation injuries.* In these cases, the mechanism of action can be a direct, or target, effect, in which the exposed tissue is intentionally killed, or an indirect effect, which can cause cell water to boil and release free radicals, allowing them to attack the nucleus of the cell. When this happens neoplasia may develop. It is thought that an improper balance between the formation and destruction of free radicals may play a role in degenerative disease and in aging.

Nutritional Disease

Undernutrition was once the most prevalent of nutritional diseases. For example, in addition to starvation, protein deficiency results in calorie undernutrition due to a depressed appetite. Appetite depression may then result in inadequate intake of carbohydrates, fats, vitamins, and minerals. We still see seasonal nutritional differences in livestock species related to drought, overcrowding of animals on limited resources, and other factors. With the traditional companion animals (dogs and cats) it is quite rare to see nutritional deficiencies in healthy pets. Where pet owners try to concoct their own animal diets, or feed old or generic feeds to older, younger, or ill pets we still see several deficiency disorders. In addition, many exotic animals (reptiles, birds, rodents) being kept as companion pets are suffering nutritional deficiencies due to owners not understanding what and how to feed them. However, our main topic here will be the different types of vitamin deficiency. Vitamins are divided into two groups: those that are fat soluble and those that are water soluble.

Fat-Soluble Vitamins

Among fat-soluble vitamins, problems involving vitamins A and D are most common. Vitamin A deficiency causes vision problems and alters the mucus secretion of cells. Excessive vitamin A can cause headaches, nausea, vomiting, and diarrhea in humans. Animals may also exhibit some of these signs. Vitamin D deficiency causes rickets, softening of the bones, and calcium-deficiency tremors. Excessive vitamin D can cause metastatic calcification and kidney stones.

Vitamin E and K problems are less common. Vitamin E deficiency causes spinocerebellar degeneration, skeletal muscle changes, white muscle disease of ruminants, and anemia (Figure 9-4). Excessive vitamin E can cause gastrointestinal disturbance. Vitamin K deficiency leads to clotting disorders and bleeding in various tissues. Excessive vitamin K, though rare, induces hemolytic anemia.

FIGURE 9-4

The leg of this calf reveals a disorder known as white muscle disease, which is the result of a deficiency of vitamin E.

Delmar/Cengage Learning

Water-Soluble Vitamins

Water-soluble vitamin imbalances in humans primarily involve thiamine (vitamin B$_1$) and vitamin C. Thiamine deficiency can cause beriberi in humans and avian polyneuritis in poultry. There are no serious consequences of excessive thiamine. In sheep, polio (polioencephalomalacia) is a disease caused by a deficiency of thiamine and is seen most commonly in lambs on high-grain diets. The first sign of the disease is blindness. Lambs will bump into things they ordinarily would see and avoid. If the first symptom of blindness is missed, the lamb will begin to appear wobbly in the rear end. This may be misdiagnosed as lameness. A day later, the lamb may be unable to get up but can still sit up and will be found on his side and unable to hold his head up. Finally, the lamb will die in convulsions. The course of the disease from the first sign of blindness to death will take approximately 5 days if not treated. The further in the disease course the treatment occurs, the longer it will take the lamb to recover. An injection of 3 cc of thiamine in the muscle will correct the condition. Generally, it will take 10 days before the lamb will begin to eat again.

Vitamin C deficiency can cause scurvy, whereas excessive vitamin C may increase iron levels, induce low blood pH, and cause RBC lysis in infants. There is more recent evidence that permanent neurological damage can result from continued excess consumption of vitamin C.[4] Dietary vitamin C is not required for livestock

---

[4]Shapiro, L.S. (2002). *Animal Nutrition, Ag 505* (4th Rev. ed.).

or most companion animals (exceptions being other primates, guinea pigs, and perhaps hamsters). Vitamin B_{12} deficiency can cause anemia in both animals and humans. It is common to see these signs in livestock grazing cobalt-deficient land. Cobalt is a component of vitamin B_{12}. Ruminants can normally produce all of their required vitamin B_{12} so long as they have sufficient cobalt in their diet. Without this requirement they develop "wasting disease," characterized by severe emaciation.

Cats are true carnivores. People who try to feed their cats a vegetarian diet cause irreversible damage to the retina (leading to blindness) and to the heart (causing death) due to a taurine deficiency. Cats require nutrients that cannot be obtained from plant sources (taurine and vitamins A and D, for example). Supplementing vegetarian diets with chemically made taurine and vitamin A and vitamin D adds to what some perceive as an already excessive consumption or exposure of chemicals to our pets. Many companion animals are thus fed "organic" diets. There is a high association between chemical consumption/exposure and neoplasms in animals.

Nutritional Excesses and Imbalances

Obesity is a greater veterinary problem among companion animals than nutritional deficiency. An animal is considered obese when it is 15 percent more than optimum. Obesity can have severe consequences, including diabetes, hypertension, and heart and liver disease. Diet can also encourage systemic diseases, such as arteriosclerosis from a diet rich in cholesterol and/or saturated fats. Obesity can lower resistance to infectious diseases, including viral and bacterial disorders such as canine distemper and salmonella.[5, 6] Finally, diet and cancer are frequently linked. The ingestion of carcinogens, the synthesis of carcinogens from ingested chemicals, and the lack of protective factors can all lead to neoplasia.

Summary

The environment, including nutritional aspects, greatly affects the overall health of our animals. The same environmental pollutants that negatively affect our health (cigarette smoke, heavy metals, and carcinogens) also cause havoc on the overall health of our livestock, exotic, and companion animals. Nutritional excesses and deficiencies as well as nutritional imbalances are common in animal diets prepared by inexperienced and undereducated animal caretakers. The veterinary technician can be a major positive influence in the well-being of animals by first learning sound nutritional guidelines and then, under the direction of the veterinarian, disseminating this information to clients. Understanding how environmental pollutants, including cigarette smoke, affect our companion animals and being able to educate the veterinary client will likely reduce the incidence of several disorders, including certain types of cancer.

---

[5]Williams, G.D., & Newberne, P.M. (1971). Decreased resistance to *Salmonella* infection in obese dogs. *Federation Proceedings, 30,* 572.

[6]Newberne, P.M. (1966). Overnutrition and resistance of dogs to distemper virus. *Federation Proceedings, 25,* 1701.

REVIEW

FILL IN THE BLANKS

1. According to the WWO, one of the largest health challenges facing us today is _____ .

2. Vitamin _____ deficiency causes scurvy in humans and other primates.

3. Vitamin _____ deficiency causes rickets in young animals.

4. The _____ is an example of a companion animal that is also a true carnivore.

5. Wasting disease in cattle and sheep is caused by a deficiency of the mineral _____ .

6. _____ is a greater veterinary problem among companion animals than nutritional deficiency.

7. The mineral _____ is a component of vitamin B_{12}.

8. Feeding cats a vegetarian diet may cause irreversible damage to the retina (leading to blindness) and to the heart (causing death) due to a/an _____ deficiency.

9. _____ is the gradual increase in the earth's surface temperature theorized to be caused by either an increase in atmospheric (CO_2) levels or by the sun getting hotter.

10. Cats exposed to 5 or more years of secondhand smoke have five times the risk for _____ .

11. _____ is a response to a drug that is serious, undesirable, and possibly life threatening. These reactions occur at dosages normally used for prevention, diagnosis, or treatment.

12. _____ agents are a group of specialized drugs used primarily to treat cancer. However, they can also cause bone marrow suppression or immune suppression, or even induce cancer in some cases.

13. _____ is a condition of abnormally high body temperatures caused by factors such as heat stroke and burns.

14. An animal is considered obese when it is _____ percent more than its optimum weight.

15. Increased cases of polioencephalomalacia are seen in young lambs consuming a diet high in _____ .

16. White muscle disease is the result of a deficiency of _____ .

17. Vitamin _____ deficiency leads to clotting disorders and bleeding in various tissues.

18. Avian polyneuritis can be caused by a deficiency of dietary _____ .

(continued)

REVIEW *(continued)*

DISCUSSION QUESTIONS

1. What are the five physical injury agents?

_____

_____

_____

2. Name three adverse conditions that antineoplastic agents can cause.

_____

_____

_____

CHAPTER 10

Zoonoses and Safety on the Job

OBJECTIVES

Upon completion of this chapter, the reader should be able to:

- **Define and give examples of common zoonotic disorders.**
- **Describe the major functions of OSHA.**
- **List ways of decreasing risk of injury in the veterinary workplace.**

KEY TERMS

Baylisascaris procyonis

brucellosis

bubonic plague

chlamydiosis

leptospirosis

ornithosis

ringworm

salmonellosis

tetanus

zoonoses

Zoonoses

As a veterinary technician, it's important for you to become familiar with those diseases that you are in danger of contracting while on the job and methods used to prevent their spread. **Zoonoses** are diseases that animals can pass on to people and vice versa. Over 150 zoonoses are known to exist. Examples of these diseases include rabies, tuberculosis, the plague, hantavirus pulmonary syndrome, ringworm, giardiasis, tetanus, West Nile virus, cat scratch disease, and Lyme disease. Ringworm, for example, can be easily transmitted from animal to human by direct contact with infected dogs, cats, cattle, and rodents or by indirect contact with fomites. Zoonoses are easily transmitted to humans via animal bites or simple contact with infected areas of skin. Zoonotic disorders may also be spread by consumption of contaminated food or water, by insect bites (serving as vectors), or by aerosol means.

Obviously, you will be working closely with animals as a veterinary technician. Thus, it's important for you to remain aware of the different types of zoonotic diseases and always be on guard for them throughout your day-to-day duties. In addition, it's equally important for your employer to provide a safe and sterile working environment.

The Occupational Safety and Health Act was passed specifically to ensure that employers "furnish…employment and a place of employment which are free from recognized hazards that are causing or are likely to cause death or serious physical harm." The Occupational Safety and Health Administration (OSHA) is a division of the U.S. Department of Labor. Since your job will entail consistent exposure to infected animals, it's crucial that the veterinarian you work for abide by standard safety practices. This should include going over with you the most common zoonotic diseases that pass through the office and how you can most effectively protect yourself from them.

The most common injury among workers at veterinary clinics/hospitals occurs when attempting to capture, restrain, treat, feed, exercise, or move animals. Animal bites can not only cause pain and disfigurement but also be instrumental in transmitting zoonotic diseases that are sometimes life threatening. Every employee at the veterinary establishment should be trained in the procedures, policies, and protective equipment needed for restraint and capture of any species of animal that he/she is expected to handle. The American Animal Hospital Association (AAHA) has training videos and workbooks that may help in this regard.[1] OSHA requires each workplace to conduct a hazard assessment for each job to determine the exact protective equipment and training required.

As a minimum, exam gloves, masks, and restraint equipment for teeth and claws are needed to protect employees when handling stray, wild, or unvaccinated

---

[1]AAHA – (800) 252-2242. American Animal Hospital Association, 12575 W. Bayaud Ave., Lakewood, Colorado, USA 80228. http://www.healthypet.com/library_view.aspx?ID=81.

animals. In addition to direct contact with the animal, zoonotic diseases can be spread by contact with blood, feces, urine, and other laboratory cultures. Each practice should have a written guideline for the employees on animal handling, equipment, and protection against zoonotic disorders. Veterinary professionals are required by law to report any suspected diseases that may affect human public health. The Animal and Plant Health Inspection Service (APHIS), a branch of the U.S. Department of Agriculture (USDA), regulates which diseases are reportable on a federal level. Individual state guidelines are usually monitored and regulated by the state's Department of Health or Department of Food and Agriculture.

Brucellosis is a major concern among zoonotic disorders in a veterinary practice.[2] The Centers for Disease Control and Prevention (CDC) and the National Institutes of Health (NIH) publish guidelines to help prevent the transmission of viral and bacterial diseases that are proven zoonotic.[3] Proper means of waste control, along with animal restraint, are concerns of these organizations.

OSHA has also listed anesthetic gases, hazardous chemicals, medical waste, and compressed gas cylinders as common occupational hazards for veterinary employees. To prevent these and other injuries, the following clinic/hospital safety procedures are essential:

- Provide written safety plans for all employees. These should include fire prevention; emergency procedures for fire, earthquakes, floods, and storms; and chemical safety.

- Provide training materials. These should include hazardous materials and conditions at the workplace, warning labels, how and when to use protective equipment, how to lift and carry heavy items (including animals), and employees' rights and responsibilities.

- Conduct routine safety inspections. Inspections should include safety drills for fire; eyewash condition; electrical safety; emergency exit procedures; storage, labeling, and handling of chemicals; and general housekeeping.

- Ensure that protective equipment is used when appropriate. This includes goggles, safety glasses, face shields, aprons, vinyl sleeves, work shoes or boots, eyewash station, earmuffs (in some situations), work gloves, lead shielding (gowns, gloves, thyroid shields, glasses), and nitrile gloves (protection from chemicals).

---

[2]Centers for Disease Control and Prevention (CDC), AAHA, AVMA. Ettinger, Stephen J.; Feldman, Edward C. (1995). Textbook of Veterinary Internal Medicine (4th ed.). W.B. Saunders Company; http://www.avma.org/reference/zoonosis/znbrucel.asp; http://www.cdc.gov/ncidod/EID/vol10no4/03-0805.htm; Compendium of Veterinary Standard Precautions for Zoonotic Disease Prevention in Veterinary Personnel National Association of State Public Health Veterinarians Veterinary Infection Control Committee 2008.

[3]*Biosafety in microbiological and biomedical laboratories (BMBL)* (4th ed.). (1999, May). U.S. Government Printing Office.

- Provide Material Safety Data Sheets (MSDSs) for all hazardous materials that animals or employees might come in contact with at the workplace. These will include drugs, pesticides, anesthetic gases, disinfectants, and all other cleaning materials.

- Prepare a written pregnancy policy including special safety measures to protect the fetus and the pregnant employee. Safety procedures should include, but not be limited to, the following:

- Exposure to chemicals or hormones that may cause birth defects

- Exposure to radiation

- Movements and postures that would cause prolonged periods of physical and/or mental fatigue

- Contact with blood and other body fluids, infected animals, laboratory cultures, feed, and water that may prove to be injurious to the fetus or the employee

- Exposure to various microbes (bacteria, viruses, protozoa, molds, fungi)

Responsibility of the Veterinary Staff and Public Health

The veterinarian and his or her staff are required to report diseases that affect the public's health. Thus the veterinary staff must be able to recognize the transmission of zoonotic disorders, recognize means of prevention, and help with the education of these concerns to the public.

Transmission of zoonotic disorders may occur by several means. The following is a partial list of some of the more common disorders and means of transmission:

- *Bites and scratches*—rabies, cat scratch fever

- *Fecal contact*—*Campylobacter*, salmonellosis, **Baylisascaris procyonis** (raccoon roundworm), cryptosporidiosis (protozoa)

- *Arthropod transmission*—Lyme disease, ehrlichiosis, Rocky Mountain spotted fever, tularemia, West Nile virus

- *Inhalation*—psittacosis, blastomycosis (fungus)

- *Direct contact*—ringworm, scabies, orf, leptospirosis, cutaneous and visceral larval migrans (roundworms)

- *Ingestion of contaminated food*—toxoplasmosis

Infectious agents need a place to survive and an environment in which to reproduce. The reservoir, or habitat, for the infectious agents can be either animate (human, plant, animal) or inanimate (such as soil, air, dust, fomites, and water). The registered veterinary technician (RVT) should assess both the source of infection and the reservoir in helping to minimize these and other types of infection.

Parasitic Diseases

Larva Migrans Syndrome

Ascarids (*Toxocara canis*, *T. cati*) and hookworms (*Ancylostoma* spp.) are common intestinal parasites in both dogs and cats. Ascarids and hookworm infections occur more commonly in young pets but can be found in dogs and cats of all ages. In humans, they can cause larva migrans syndrome. Young children playing around pets are particularly at risk of coming in contact with contaminated soil containing infective eggs and larvae.

The signs and symptoms in humans are dependent on the tissues and the organs damaged during the larval migration. Most commonly, the eyes, brain, lungs, and liver are affected by ascarids. Hookworm larvae may penetrate the skin causing cutaneous larva migrans or be ingested and cause disorders related to the digestive tract.

Cerebrospinal Nematodiasis (Larva Migrans)

Baylisascaris procyonis (raccoon roundworm) is reported to infect 70 percent or more of raccoons in many urban areas. This parasite rarely causes a health problem to the raccoon. However, it can cause serious problems to humans, dogs, and 17 other species of mammals and birds. Transmission is generally by skin penetration or oral ingestion of the eggs. The larvae migrate into the eye, brain, or spinal cord and can cause blindness, central nervous system disease, and even death.

The veterinary staff can play an important role in preventing human transmission of these internal parasites. Most pet owners are unaware of the danger of these parasites, means of transmission, detection, prevention, and treatment. The veterinary staff should recommend regular fecal examinations, provide sanitation guidelines (including prompt collection and disposal of pet feces), and recommend proper anthelmintic treatments when necessary.

When handling feces, especially raccoon feces, use disposable gloves. Deworm all incoming raccoons. The ova of raccoon roundworms are resistant to many disinfectants. It is recommended, therefore, to autoclave cages and carriers that have been occupied by raccoons.

Bacterial Diseases

Mycobacterium tuberculosis, M. bovis

Historically, one of the more common causes of transmission of tuberculosis from animal to human was by the consumption of raw infected dairy products. Most industrialized nations have greatly reduced if not eliminated this particular threat by the use of pasteurization and TB eradication programs. California was tuberculosis free in 2005. However, in July of 2008, 4,800 dairy cattle were ordered to be slaughtered and another 16,000 put in quarantine when a routine inspection found a slaughterhouse cow to have TB lesions on its lymph nodes. DNA testing revealed that two of the cows that eventually tested positive shared a strain of the bacteria that originated in Mexico. California's State Department of Food and Agriculture issued a warning against contact with cattle of Mexican origins, and the USDA placed additional restrictions on the transportation of California cattle across state lines without testing by a veterinarian.

Mycobacterium bovis is pathogenic not only to cattle and humans but to deer and occasionally to carnivores. Exotic and wild animals still remain an occasional source of infection of *M. bovis*. Veterinary technicians, veterinarians, and animal handlers in the zoo must be cognizant that seals, rhinoceros, elk, and elephants can be sources for *M. bovis* exposure, too.

M. tuberculosis is more commonly associated with tuberculosis in humans. It has also been found in birds, elephants, and other mammals. In 1996, five elephants visiting California from Illinois were shown to have tuberculosis from *M. tuberculosis*. Of 22 animal handlers from the farm where the animals were raised and boarded, 12 were found positive (using the tuberculin purified protein derivative test) for *M. tuberculosis*. These and subsequent tests showed a human to animal transmission of *M. tuberculosis*. Regular skin testing is recommended for animal handlers, keepers, veterinary technicians, and veterinarians working with these and other exotic animals.[4]

Anthrax (Woolsorter's Disease)

Anthrax is a highly infectious and fatal disease caused by a large spore-forming bacterium called *Bacillus anthracis*. It is an ancient disease described in the Old Testament. The endospores produced by *B. anthracis* have been known to survive more than 60 years in arid conditions. Anthrax infection occurs in three common forms: cutaneous (skin), gastrointestinal, and inhalation. In the 19th century, European farmers sorting through their wool came in contact with the endospores causing what became known as woolsorter's disease.

Today most human cases are cutaneous. The bacterial spores enter a wound or abrasion on the skin and develop ulcers and swelling around the wound site. This type of anthrax is easily treated with antibiotics. The inhalation anthrax symptoms resemble that of the common cold initially. After several days without treatment the symptoms progress, causing severe respiratory problems and shock. In the fall of 2001, 45 percent of inhalation anthrax bioterrorist attack cases were fatal. Intestinal anthrax results from the consumption of contaminated or undercooked meat. Symptoms include vomiting of blood, severe diarrhea, and swelling about the neck. Mortality rates as high as 60 percent are common.

In cattle most outbreaks occur in areas where animals have previously died of anthrax. Cattle ingest the spores. The predominant sign is a rapid progression from a normal appearance to death in a matter of hours. Bloody discharges and rapid body decomposition are commonly observed. Vaccinations have been shown to be effective in preventing further occurrence within a herd. However, a full immunity requires 10 to 14 days postvaccination. The cattle vaccine is a

---

[4]Centers for Disease Control and Prevention. (1994). Guidelines for preventing the transmission of *Mycobacterium tuberculosis* in health-care facilities. *Morbidity and Mortality Weekly Report, 43*, 63–64.

live anthrax vaccine that stimulates immunity but does not cause disease. It is referred to as the Sterne spore vaccine. This is not the same vaccine approved for use in humans. The human vaccine requires a number of initial injections followed by yearly boosters. It is a cell-free filtrate vaccine, which means it contains no dead or live bacteria in the preparation.

In the United States, anthrax is a reportable disease. The National Center for Infectious Diseases and both local and state health departments should be notified immediately if anthrax is suspected. For all potentially exposed humans a physician should be immediately contacted for preventive antibiotics.

Brucellosis (Undulant Fever)

Transmission of *Brucella* spp. occurs by contact with body tissues, blood, urine, vaginal discharges, or the fetuses of infected animals. There are three primary *Brucella* species of concern to food animal veterinarians. *Brucella suis* predominantly affects swine and reindeer, but it can also affect cattle and bison; *Brucella melitensis* affects goats but is currently not present in the United States; and *Brucella abortus* is the most common cause of brucellosis affecting cattle. **Brucellosis** is a highly contagious disease that was once spread by the consumption of raw milk from cattle infected with the bacteria. The disease is also known as contagious abortion or Bang's disease in livestock. In humans, it is known as undulant fever because of the intermittent fever accompanying infection. It is one of the most serious diseases of livestock, because it spreads so rapidly and it is transmissible to humans. Vaccination of cattle (calves) and slaughter of infected animals are the chief methods of control.

In cattle and in bison, the disease usually localizes in the reproductive organs or the udder. Bacteria are commonly shed with the aborted fetus and afterbirth and other discharges from the reproductive tract. Brucellosis is also transmitted by consumption of raw milk or coming into contact with a contaminated environment.

The veterinary staff should wear protective gear, including disposable gloves when handling suspected or infected animals and when vaccinating using live or modified live vaccines. Heifers are usually vaccinated when they are 4 to 6 months of age. A tattoo is applied in the ear at the time of vaccination. States are designated brucellosis free when no cattle or bison are found to be infected for 12 consecutive months.

Bubonic Plague

While assisting in the treatment of wild rodents, rabbits, and mammals, the veterinary staff needs to be cognizant of parasitic fleas that may carry *Yersinia pestis* (the infectious agent causing the **bubonic plague**). Human mortality can be as high as 60 percent if untreated. Veterinary staff has been killed by this disease. Currently bubonic plague is in endemic proportions in parts of the southwestern United States.

The veterinary staff should avoid contact with fleas from wild rodents and ground squirrels in particular. Deflea all incoming wild animals.

Leptospirosis

Leptospirosis is an abortion disease of cattle, pigs, dogs, skunks, raccoons, opossums, rats, mice, and many other domestic mammals. It is spread in the urine of infected animals and can be absorbed through small cuts or abrasions on the skin or ingested. In humans it may cause fever, chills, nausea, malaise, abortion, and myalgia. In severe cases it can cause meningitis.

It is important to avoid urine contamination of feed, water, or soil and to practice good personal hygiene, especially when handling urine from wild animals or suspect animals. There are several vaccination products; however, there are more than 200 serovars of leptospirosis and vaccines for only 8 of them.

Chlamydiosis (Psittacosis, Ornithosis)

Chlamydia psittaci causes spontaneous abortion in cattle. In birds, **chlamydiosis** is a reportable, zoonotic disease. It is found in more than 100 species of wild and domestic birds. In psittacine birds (birds with a curved beak, like parrots and parakeets) it causes a disease known as parrot fever or psittacosis. In other birds and in humans it is known as **ornithosis**. Humans exposed to air- or dustborne microorganisms from infected birds or their feces exhibit clinical signs such as conjunctivitis, fever, headache, upper respiratory infection, and pneumonitis.

A dust mask, plastic face shield, and gloves should be worn when working with wild birds or suspect domestic birds. Veterinary staff members, bird banders, and rehabilitators need to practice good sanitation procedures when working with pigeons, caged birds, and wild birds, or when examining dead birds.

Salmonellosis

There are many species of salmonellae that cause **salmonellosis** and other diseases in birds, reptiles, and mammals including humans. Signs in animals include diarrhea, vomiting, mild fever, and septicemia. Its transmission to humans occurs primarily through contaminated drinking water, milk, meat, eggs, and poultry. It can also transfer to humans through the handling of reptiles and birds and ingesting the bacteria because of poor sanitation.

An estimated 2 million Americans contract this disorder annually. Symptoms in humans include gastroenteritis, headaches, high fever, dehydration, septicemia, and sometimes death. Salmonellosis is a reportable disease in many states.

Tetanus (Lockjaw)

Tetanus is an acute disease caused by toxins produced by the bacterium *Clostridium tetani*. *C. tetani* is typically found in heavily manured soils. Horses and humans are the most susceptible to this disease whereas dogs, cats, and birds are relatively resistant. When untreated the mortality rate reaches levels of 80 percent or more. Those who survive do not normally develop protective immunity to future exposure to *C. tetani*. Active immunization with tetanus toxoid can provide protection to humans for this disorder. In humans, spasms of the muscles of the head (lockjaw) making it difficult to eat occur. It is a very painful disease. Horses, sheep, pigs, and goats have clinical signs which include stiff and extended tail, prolapsed third eyelid, extension of the head and neck, increased heart rate and rapid breathing, muscle spasms, and respiratory failure (often the cause of death). Infection usually results following a puncture wound, or a surgical procedure such as docking or castration. Young goats and lambs should be treated with tetanus antitoxin providing passive immunity or protection at the time of the aforementioned surgical procedures. Horses should receive the *C. tetani* toxoid annually for maximum protection.

Viral Diseases

While in several foreign and developing nations there is still a growing threat of zoonotic viral diseases such as Dengue Virus, Japanese Encephalitis Virus, Lassa Fever and SARS Corona virus, in the United States the veterinary technician can have the greatest effect in preventing the zoonotic disorder rabies.

Rabies

A rhabdovirus is the infectious agent that causes rabies. Rabies is one of the oldest communicable diseases of humans. It is spread by bite wounds, by infected saliva entering cuts or skin abrasions, and by inhaling aerosol in bat caves containing their infected saliva. In North America, the skunk, fox, bat, and raccoon are the primary carriers of this viral disease. However, any mammal can be infected, including humans. Cats, dogs, coyotes, and cattle are also frequently infection species.

Wild carnivores and bats present the most serious risk of exposure of rabies to humans. It is generally recommended that all veterinarians, animal control officers, and rabies diagnostic workers be immunized as a pre-exposure prophylaxis. The veterinary and animal control staff should develop procedures to avoid being bitten by all animals.

The normal transmission of the rabies virus is through the bite of an infected animal. The virus is generally present in the saliva. The virus may also be introduced into fresh wounds or through intact mucous membranes. The incubation period varies between and among species. In dogs, the incubation period is approximately 20 to 80 days postexposure. Once clinical symptoms are visible, death is quite rapid (10 days or less).

Clinical signs of rabies include anorexia, irritability, behavioral changes, and paralysis. Treatment of rabid animals is usually not attempted due to its high fatal nature and its zoonotic capabilities.

Protection of veterinarians or assistants who may handle strays, wildlife, or animals exhibiting rabid characteristics or who may work in labs handling tissue from rabid animals includes the following:

- Obtain pre-exposure vaccination for rabies (series of three vaccines for the veterinary staff).

- Revaccinate any previously vaccinated animal that is bitten or scratched by a wild, carnivorous mammal or bat and observe for clinical signs for 45 days.

- Euthanize the animal if it develops any signs of rabies during the quarantine period and submit the brain for rabies testing.

Protozoal Diseases

Giardiasis

Giardiasis is a chronic intestinal protozoal infection that occurs in most domestic and wild mammals, in birds, and in humans. *Giardia lamblia* is found in beavers, muskrats, and waterfowl. It is spread to humans and domestic animals by the fecal-oral route. A zinc sulfate flotation (fecal float) is commonly used to identify *Giardia* in dogs. In humans an ELISA test can be used to detect *Giardia* antigen in the feces. Symptoms in humans include chronic diarrhea, abdominal cramps, bloating, and weight loss. Prevention is by proper sanitation measures, especially avoiding hand-to-mouth contact while handling any wildlife species.

Toxoplasmosis

Toxoplasma gondii is found in the house cat, in birds, in wild animals, and in humans. Members of the cat family, however, are the only known definitive hosts for this protozoan. They are the only animal that can pass infective oocysts in their feces. Transmission occurs through ingestion of uncooked meat containing tissue cysts or by transplacental infection in pregnant women. Pregnant women can pass the parasite to their fetus. In the United States, more than 3,000 babies are born every year with ocular lesions produced from *T. gondii*. Humans with a suppressed immune system (e.g., HIV) develop meningoencephalitis. In sheep, goats, and pigs, *T. gondii* can cause abortion and stillbirth.

To prevent toxoplasmosis in humans:

- All meat should be thoroughly cooked.
- Uncooked meat should not be fed to pet cats.
- Litter boxes should be cleaned daily.
- Hands should be washed thoroughly after changing the litter box.
- Pregnant women should avoid handling cats and cleaning their litter boxes.

Tickborne Diseases

Lyme Disease

Deer ticks are responsible for transmitting *Borrelia burgdorferi* to humans, causing the disorder Lyme disease. In humans, early signs include an expanding bull's-eye ring skin lesion, malaise, myalgia, fatigue, fever, headache, and joint pain. Small mammals and dogs act as reservoir hosts. Antibiotic therapy is effective during the early stages. Tick avoidance is the best preventive measure for Lyme disease. Killed and whole cell bacterins have been used in dogs to prevent Lyme disease; however, their efficacy is being questioned. There is currently no vaccine approved for use in humans.

CASE STUDY	**Ringworm Dermatophytosis**

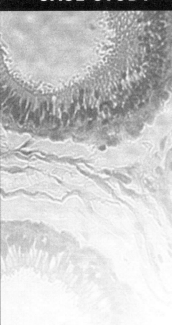

A 3-month-old kitten was presented with scaling, crusting, and alopecic lesions in the haired regions of her abdomen and neck. The lesions appeared to be erythemic at the edges. The kitten seemed to be a little itchy, but it was not intense. The little girl who lives with this kitten had a circular lesion on her arm, which could be related to the lesions on the kitten.

The following tests were performed for diagnosis. A Wood's lamp was used to check if the area fluoresced a bright yellow-green. A fungal culture was done by using a few plucked hairs at the periphery of the lesion and placing them onto dermatophyte test media (DTM). The culture is considered positive when the amber media turns red and the sample has fluffy, light-colored dermatophyte colonies. A stained slide preparation of a wet mount or Scotch tape preparation of the colony surface is done to identify the dermatophytes cultured. Dermatophytosis is a fungal infection. The most common one that affects dogs and cats is called *Microsporum canis*, which occurs in young or immunosuppressed animals. The other two are *M. gypseum* and *Trichophyton mentagrophytes*. In sheep and goats the infectious **ringworm** is called *T. verrucosum*.

Topical medications such as miconazole (Conofite) or ketoconazole are suggested for treatment. Medicated shampoos containing chlorhexidine, ketoconazole, or miconazole are also recommended. Animals with a thick, heavy coat will have to be shaved. A systemic version of griseofulvin given orally can be used to decrease the recovery time. The infected animal has to be isolated from other animals and children. This disease requires diligent cleaning, bleaching, and washing of clothes, bedding, towels, or anything else that comes in contact with the infected animal. Ringworm (Figures 10-1 and 10-2) is zoonotic, by direct contact as well as fomites, such as hairbrushes, blankets, and clothing.

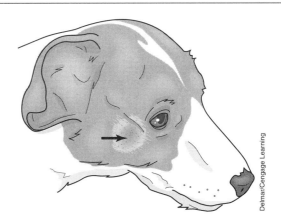

Delmar/Cengage Learning

FIGURE 10–1

Ringworm near the eye of an infected animal.

(continued)

CASE STUDY   Ringworm Dermatophytosis *(continued)*

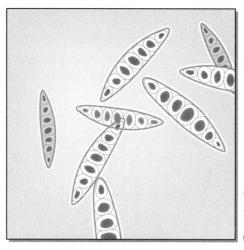

Microscopic view – Ringworm spores

Delmar/Cengage Learning

FIGURE 10–2

Microscopic view of ringworm spores.

Topical miconazole was used to treat the kitten, and the kitten was bathed weekly with a miconazole shampoo until the lesions cleared up. The kitten's environment and bedding were bleached weekly as well. The lesions cleared up after approximately 4 weeks of treatment.

Summary

Zoonotic disorders and other work-related injuries are a major concern to the veterinary practice. OSHA and other government and professional organizations have created guidelines and laws to help maximize the safety of all veterinary employees. There are more than 150 zoonotic disorders, diseases that can pass between human and animal. In addition, working with animals that bite, scratch, kick, or crush you requires specific training and equipment to minimize injury. The veterinary staff is obligated to safeguard not only the animals in its care but also the staff employed at the facilities and the public who is dependent on the educated veterinary employee for its safety.

REVIEW

FILL IN THE BLANKS

1. Under the _____ , veterinarians are obligated to provide safe and sterile working environments.

2. Psittacosis in humans is called _____ .

3. The bubonic plague is spread to humans by fleas primarily found on _____ _____ and _____ .

4. Brucellosis in cattle is _____ in humans.

5. The infectious agent *Baylisascaris procyonis* is carried by the _____ .

6. _____ is the only definitive host for *Toxoplasma gondii.*

7. _____ are the most susceptible of domestic animals to tetanus.

8. Dermatophytosis is a fungal infection commonly known as _____ .

9. Deer ticks are responsible for causing _____ disease in humans.

MATCHING

_____ 1. *Bacillus anthracis*   a. ringworm

_____ 2. *Brucella abortus*   b. ornithosis

_____ 3. *Chlamydia psittaci*   c. undulant fever

_____ 4. *Yersinia pestis*   d. woolsorter's disease

_____ 5. *Giardia lamblia*   e. Lyme disease

_____ 6. *Borrelia burgdorferi*   f. ringworm

_____ 7. *Microsporum canis*   g. intestinal protozoan

DISCUSSION QUESTIONS

1. Name five diseases that are considered zoonoses.

_____

_____

_____

(continued)

REVIEW   *(continued)*

2. How are zoonoses transmitted from animals to humans?

_____

_____

_____

3. Discuss methods practiced in a veterinary clinic to prevent zoonoses to both clients and staff.

_____

_____

_____

4. What are the most common animals to carry the rabies virus in the United States?

_____

_____

_____

PART TWO

Clinical Parasitology

Notice: The material in this study unit includes suggested treatments and procedures for controlling parasite infections. This information is presented to help veterinary technicians become familiar with how veterinarians manage such infections. Veterinary technicians should never make a diagnosis or treat any animal except under the direct orders or supervision of the veterinarian.

CHAPTER 11

The Variety of Parasites

OBJECTIVES

Upon completion of this chapter, the reader should be able to:

- Define *parasites* and *parasitology.*

- Explain the different types of animal parasites.

- List the six major parasitic groups affecting domestic animals.

- Identify the ectoparasites of both large and small animals; describe their life cycles and range of hosts; and summarize their diagnosis, treatment, and prevention.

- Identify the endoparasites of both large and small animals; describe their life cycles and range of hosts; and summarize their diagnosis, treatment, and prevention.

KEY TERMS

ectoparasite

endoparasite

flukes

host

infestation

insects

parasite

protozoan

roundworms

Infestation versus Infection

A **parasite** is an organism that lives on or within another living organism. The organism that gives a parasite a home is called its **host**. Whether parasites cause an infestation or an infection depends on what part of the host they occupy. Parasites are metabolically dependent on the host. Parasites that live on their host are called **ectoparasites** (external parasites), and their presence is called an **infestation**. They may live on the surface or in pores of the skin, or they may attach themselves to the animal's hair. Some external parasites spend only a short time on their hosts; others stay a lifetime. Lice are an example of a parasite that infests.

Parasites that live in their hosts are called **endoparasites** (internal parasites); their presence is called an infection. Each type of endoparasite occupies a particular place within the body. Hookworms are an example of an endoparasite that infects or lives in the gut.

Infections and infestations can be barely noticeable or quite pronounced. A single parasite produces only a mild disease, and the host does not appear sick. In most cases, it takes the presence of many parasites for the host to show signs of an illness.

Types of Parasites

Most of the important parasites that affect domestic animals come from six groups:

- Insects (lice, fleas, ants, bees, wasps, yellow jackets, beetles, cockroaches)
- Arachnids (mites, ticks, spiders, scorpions)
- Nematodes (roundworms)
- Cestodes (tapeworms)
- Trematodes (flukes)
- Protozoans (single-cell organisms such as coccidia, *Giardia*, amoebae, malarial organisms, and piroplasms)

The life cycle, or developmental stages, of a parasite's life differs depending on the group to which it belongs. Some of these stages are free-living stages—that is, periods of a parasite's life when it lives apart from a host. Understanding the parasite's life cycle is crucial to controlling parasite infestation and infection.

Insects

Insects are both internal and external parasites. The characteristics that identify an adult insect are three pairs of legs and three body parts (head, thorax, and abdomen). Insects can also carry parasites or other disease organisms from one host to another. (An animal that transfers an infective agent from one host to another is called a vector.) An insect's life cycle has four stages: egg, larva, pupa, and adult.

Arachnids

Spiders are familiar arachnids—animals whose adults have eight legs and two body parts (head and abdomen). The arachnid life cycle has four stages: egg, larva, nymph, and adult. The larval arachnid is unusual because it has only six legs. Most parasitic arachnids are ectoparasites.

Nematodes

Nematodes, also called **roundworms**, are relatively small, wormlike organisms covered by a tough skin called a cuticle. The nematode life cycle has three stages: egg, larval, and adult. The larval stage can be subdivided into four stages:

- First-stage larva (ready to hatch) sheds it external cuticle to grow to the next stage. The shedding of this outer layer is called molting.

- Second-stage larva eventually molts into the next stage.

- Third-stage larva is also known as the infective stage because it is infective for the definitive host.

- Fourth-stage larva occurs when the third stage molts within the definitive host. It eventually develops into the sexually mature adult stage.

Parasitic nematodes are endoparasites. However, the first three larval stages can develop either in the external environment or within the intermediate host.

Cestodes

Tapeworms, also known as cestodes, are flat, ribbonlike worms with no digestive tract. The tapeworm has a head, called a scolex, and a body made of many segments called proglottids. The tapeworm life cycle has three stages: egg, larva, and adult. All tapeworms are endoparasites, and the adult tapeworm always lives in its host's digestive tract. The larval tapeworm (also known as metacestode tapeworm) may also cause pathology in the intermediate host. The intermediate host harbors the larval, juvenile, immature, or asexual stages of the parasite. The intermediate host may be a flea, a grain mite, or a rabbit. Some tapeworms may have more than one intermediate host. Cattle can serve as an intermediate host for a tapeworm of humans, *Taenia saginata*. There are similar tapeworms of humans in other livestock as well. The definitive host harbors the adult, sexual, or mature stages of the parasite.

Trematodes

Trematodes, sometimes called flukes, are flat, leaf-shaped worms that have a mouth and gut but no anus. The free-living *Planaria* is an example of a trematode. The parasitic trematode life cycle is complex: egg, several larval stages (some of which are endoparasites of snails), and adult. Most adult parasitic trematodes are endoparasites. Exceptions include the monogenetic trematodes, which are ectoparasites of fish, amphibians, and reptiles. *Platynosomum concinnum*, a **fluke** infecting cats, has a land snail and a lizard as intermediate hosts.

Protozoans

Protozoans, single-cell animals, are a large group made up of many different types of organisms. Each type has a different life cycle, ranging from one to many stages. Parasitic protozoans are endoparasites as adults.

| CASE STUDY | Lyme Disease: *Ixodes dammini* and *Ixodes pacificus* Ticks |

The Jack Russell terrier was not acting like himself. This usually energetic and frisky dog only wanted to sleep. The owner also told the veterinarian that the dog did not want to play ball anymore; after a few throws, he started to limp on his front right limb. The veterinarian asked the owner if she and her husband had gone camping recently. The answer was yes. The next question the veterinarian asked was if their Jack Russell terrier had come in contact with any ticks. The husband confirmed that he found some ticks on the dog when they took him camping.

The veterinarian decided to run some diagnostic tests for Lyme disease, since this little dog exhibited appropriate clinical signs. A joint tap showed 60 percent neutrophils. The rest of the cells were primarily monocytes with a few lymphocytes. An ELISA serology test indicated the presence of Lyme disease. Dogs that have been vaccinated with the Lyme bacterin may be positive on the ELISA test. The antibodies will remain high for about 18 months whether or not the dog may be showing clinical signs. A Western blot test can help distinguish between vaccine-induced antibodies and the Lyme disease antibodies. A spirochete called *Borrelia burgdorferi* is the agent associated with Lyme disease. The vector for *Borrelia* is the tick *Ixodes pacificus* on the West Coast of the United States and *Ixodes scapularis* elsewhere. A polymerase chain reaction PCR or spirochete isolation (a blood culture) can be helpful, but spirochete isolation is difficult.

The clinical signs are lameness due to inflamed joints, sometimes fever, lymphadenopathy, anorexia, and lethargy. The infection is often self-limiting, and signs may subside within a few days. Certain veterinarians find treatment controversial, because the infection is self-limiting and many dogs never show clinical signs of infection. Ampicillin, amoxicillin, and tetracycline are the antibiotics of choice. Lyme disease may be prevented by removing ticks as quickly as possible. Frontline is a product used for keeping ticks off pets. A Lyme disease bacterin is available for dogs, but its efficacy has not been validated by research. The vaccination can result in antibodies that interfere with the detection of Lyme disease when using the ELISA test.

This Jack Russell was successfully treated with antibiotics, and the owners were instructed to purchase a good tick product such as Frontline to apply to their dog, especially prior to any camping trips.

Summary

Parasites live on or in another organism and are metabolically dependent on the host. Ectoparasites live on their host, while endoparasites live within. An infection is caused by large numbers of endoparasites, whereas an infestation is the result of large numbers of ectoparasites. Most parasites affecting domestic animals fall under one of six groups: insects, arachnids, nematodes, cestodes, trematodes, or protozoans. Understanding the various life cycles of parasitic organisms helps determine diagnosis and treatment.

REVIEW

FILL IN THE BLANKS

1. The animal in which a parasite lives is known as a/an _Host_.

2. A/an _ectoparasite_ is a parasite that lives on the skin of an animal.

3. A parasite that lives inside an animal is known as a/an _Endoparasite_.

4. The vector for *Borrelia* is _Ixodes pacificus_ on the West Coast of the United States and _Ixodes scapularis_ elsewhere.

For questions 5 to 10, write the name of the organism (protozoan, trematode, tapeworm, nematode, arachnid, insect) that matches the description.

5. Adult parasite with six legs and three body parts _Insects_

6. Flat, ribbonlike worm with no digestive tract _Cestodes_

7. Single-cell animal _Protozoans_

8. Adult parasite with eight legs and two body parts _Arachnids_

9. Leaf-shaped worm with a mouth _Trematode_

10. Small, cylindrical worm covered with a cuticle _Nematodes_

DISCUSSION QUESTIONS

1. Why is it more difficult to design drugs that will kill parasites but won't harm the animals they reside in or on when compared with bacteria or viruses?

   Since parasites live in or on an animal it can be difficult to get rid of parasites. If the medication does a good job killing the parasite it may cause some harm to the host as well.

MATCHING

_b_ 1. arachnid   a. fluke

_d_ 2. protozoa   b. tick

_C_ 3. nematode   c. roundworm

_e_ 4. tapeworm   d. *Giardia*

_a_ 5. trematode   e. cestodes

_f_ 6. insect   f. yellow jacket

CHAPTER 12

Ectoparasites of Large Animals

OBJECTIVES

Upon completion of this chapter, the reader should be able to:

- List the common ectoparasites of large animals.
- Describe the life cycle of the fly, flea, louse, mite, and tick.
- Describe the common diseases spread by these common ectoparasites.
- Describe briefly methods of prevention, treatment, and diagnosis.

KEY TERMS

anemia

anthelmintic

arachnid

babesiosis

bot

grub

insects

mange

mite

myiasis

nit

tabanids

warble

Common Ectoparasites

Most common ectoparasites are either **insects** or **arachnids**. Although parasitologists generally refer to any parasitic insect or arachnid as an ectoparasite, some insects and arachnids live inside their host at least part of the time. The mange mite (arachnid), for instance, lives just under the top layer of the skin; the larval horse bot, a fly (insect), lives in a horse's stomach. These insect and arachnid parasites could be called endoparasites; this chapter, however, groups them with the ectoparasites because their life cycles closely resemble those of other ectoparasites, and their treatment and control measures are the same as those for other ectoparasites.

Insects

Insects have three pairs of jointed legs and usually two pairs of wings. Some, however, are wingless. Important insect parasites come from three groups: flies, fleas, and lice.

Flies

Flies are insects with two wings. They have two large eyes, one on each side of the head. Many types of flies are external parasites of animals; most, but not all (bots, *Cuterebra*) are parasitic only as adults. The expression "wouldn't hurt a fly" suggests that the fly is a harmless annoyance, but flies damage their hosts in several ways. The annoyance itself is far from harmless, since the energy a host uses in chasing or running away from flies often causes weight loss. Flies that bite (like mosquitoes) can cause blood loss and direct tissue damage. In addition, they transmit bacterial, viral, and parasitic diseases.

Mosquitoes, which are among the most important parasitic flies, have a typical insect life cycle (Figure 12-1). They lay their eggs on water or on a surface that gets wet when it rains. The larvae and the pupae live in the water and are free-living organisms, as are adult males. Only adult females are parasites. Mosquitoes are vectors for many diseases, including heartworm disease.

Fleas

Fleas are small, narrow insects about 4 mm long. Their back legs are designed for jumping on and off their hosts. Fleas are parasitic only during their adult stage. The adult flea's mouthparts are made for sucking blood. The free-living stages of the flea (egg, larva, and pupa) greatly resemble the free-living stages of the fly. The adult female flea lays her eggs on the host. The eggs then fall off the host. The larva, which hatches from the egg, eats all sorts of organic material but prefers the feces of adult fleas that fall off the host. Like the fly pupa, the flea pupa is enclosed in a hard case and does not eat. The life cycle of a flea is illustrated in Figure 12-2.

Lice

Lice (singular, louse) spend their entire life cycle on their host. There are two types of lice. Chewing lice (Mallophaga), also known as biting lice, have mouthparts made for chewing their food. They feed on skin or hair. They have rounded heads and are usually yellow in color. Biting lice infect dogs, cats, cattle, sheep,

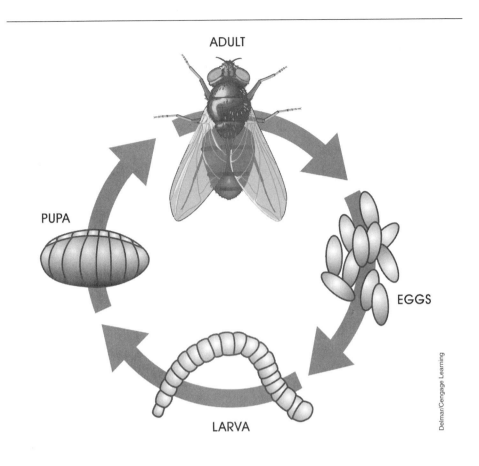

FIGURE 12–1

Life cycle of a fly.

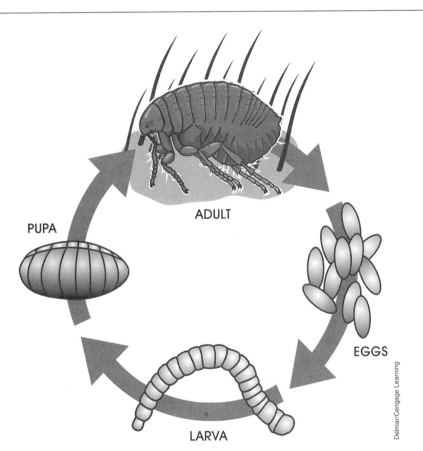

FIGURE 12–2

Life cycle of a flea.

goats, and horses. Sucking lice (Anoplura) have mouthparts designed for sucking blood. Their heads are narrower, and they are red to gray in color. Both types of lice are small (1–2 mm) and flat, with claws on the ends of their legs. These claws grip the hair of their host.

All stages of lice live on the host. The life cycle of lice (Figure 12-3) differs from that of other insects in that lice undergo no larval or pupal stages. Instead, they undergo several nymphal stages. The nymph looks just like the adult but is smaller. The female louse glues her eggs, called **nits**, to the hairs or feather shaft of the host. When the nits hatch, the nymphs emerge and develop into adults in 2 or 3 weeks. The adults live for about 1 month. Lice usually move to another host only when an infested host comes in close contact with an uninfested host. Nymphs and adults do not live longer than 7 days if they are removed from the host.

FIGURE 12-3

Life cycle of a louse.

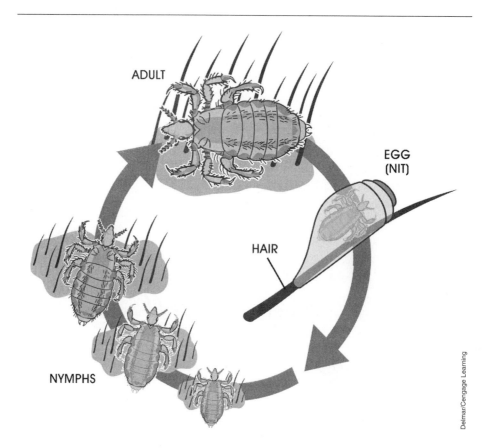

ADULT

EGG (NIT)

HAIR

NYMPHS

Delmar/Cengage Learning

Arachnids

Arachnids are arthropods with four pairs of legs and simple eyes. A typical parasitic arachnid lives on or in its host's skin. Ticks and mites are the most prevalent parasitic arachnids.

Ticks

Ticks are the largest parasitic arachnids. There are two important families of ticks in the veterinary field: hard (ixodid) and soft (argasid) ticks. An example of the hard tick is the brown dog tick, and an example of the soft tick is the fowl tick. Larval, nymphal, and adult ticks are ectoparasites, and all are bloodsuckers. They hold on to their hosts much tighter and longer than other parasites by inserting their mouthparts into the skin and staying there, sucking blood until they fill up. This process may take an adult tick anywhere from 1 to 3 weeks. When female ticks are feeding, they may increase in size from 3 to 10 mm or more. There are four major stages of development in ticks: egg, larva, nymph, and adult.

All ticks lay their eggs on the ground, but the number of hosts they feed on varies. A one-host tick spends all three of its feeding stages on one host. A two-host tick stays on one host for two of its life stages and then feeds on a second host during its third feeding life stage. A three-host tick feeds on three different hosts—one for each life stage. Larva, nymph, and adult may all feed on the same type of host, or the host may change. For example, the adult American dog tick (*Dermacentor variabilis*) feeds on dogs, but the larva and nymph feed on mice and other rodents. One-host ticks are the easiest to control because they are the easiest to find. Figure 12-4 illustrates the various stages in the life cycle of a three-host tick.

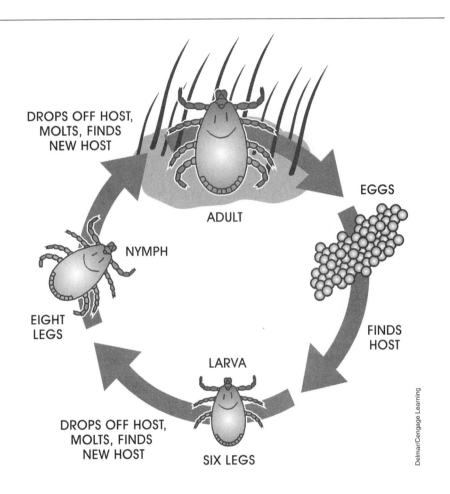

FIGURE 12–4

Life cycle of a three-host tick.

Mites

Mites, the other common parasitic arachnid, are small, external parasites. The adult is about 0.25 to 1.3 mm long—so small that it cannot be seen with the naked eye. Mites can cause a skin disease known as mange. The legs of the mange mite are very short, and spines stick out from the side. Sarcoptiform mites have pedicels or stalks on the tips of their legs. The pedicel can be long or short or jointed. This description is used in identifying the type of mite. The female mites burrow into the skin to lay their eggs. The eggs hatch, and the larvae come to the top of the skin and wander around. All of this activity causes intense itching in the skin of the host.

The mite life cycle (Figure 12-5) is similar to that of the tick, except that mites may undergo more than one nymphal stage. The developmental stages of the mite are an egg stage followed by a six-legged larval stage, an eight-legged nymphal stage, and finally the eight-legged adult stage.

FIGURE 12-5

Life cycle of a mite.

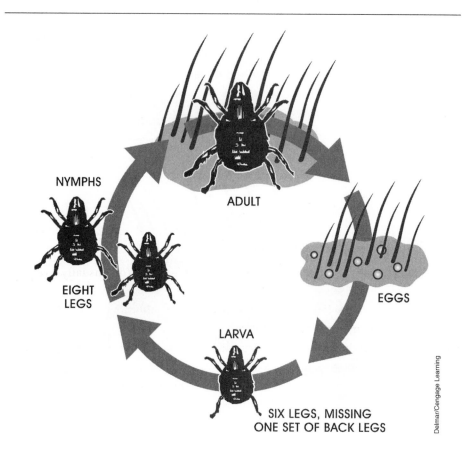

Ectoparasites of Cattle

All of the ectoparasites discussed in this section prey on cattle. Some also attack other animals as noted.

Flies

Tabanids (Tabanus *species*)

Tabanids (Figure 12-6) are large flies commonly called horseflies, deer flies, or green-head flies. Only the female fly feeds on blood, and it prefers the blood of large mammals such as horses, cattle, deer, and humans. Tabanids ingest too little blood to harm their hosts, but their bite is very painful. Since the host instantly tries to dislodge it, the fly must attack several times to get sufficient blood to produce her eggs. A horse or cow fighting off tabanids stops grazing. Many tabanids in a grazing area can reduce a herd's weight gain and lower milk production.

Tabanids are also disease vectors. Because the host drives the female tabanid away before she is full, she will likely feed on several different hosts in rapid

Delmar/Cengage Learning

FIGURE 12-6

Tabanids have clear wings with characteristic veins that may be used to tell them apart from other related flies.

succession. If one of the hosts has a bloodborne disease, the fly may carry the disease to its next host. Tabanids are vectors for such bloodborne diseases as anthrax, tularemia, bovine anaplasmosis, and equine infectious anemia.

Because tabanids need water to complete their life cycle, they usually live near rivers, streams, and swamps. The female fly lays her eggs on plants near the water. When the eggs hatch, the larvae fall into the water, where they develop while feeding on insects, snails, and other animals. Adult flies die during the winter, but the larvae survive by hibernating. In the spring and summer, when the larvae are completely developed, they crawl out of the water and onto dry land. They then burrow a few inches underground and develop to the pupal stage, which lasts for 1 to 3 weeks. Within the pupal cocoon, they transform to the adult stage. When conditions are right, the adult tabanids emerge.

Tabanid control is very difficult. The flies do not stay on the host long enough for pesticides to have much effect. Pasturing horses and cows far from swampy areas and streams may not be practical, and tabanids are strong fliers. Tabanids will not feed indoors, so some people keep their animals indoors during the times of day when tabanids are most active. Fly traps can be used when other control measures are not practical.

Stable Fly (Stomoxys calcitrans)

The stable fly (Figure 12-7) is a blood-feeding fly that looks like a housefly. Both male and female stable flies suck blood; they usually bite the lower part of a horse's or cow's legs. Stable flies feed indoors, as their name suggests, but they also feed on pastured animals. Like the tabanids, the stable fly is unlikely to take a complete blood meal before the host chases it away. The fly therefore attacks the same host or several hosts many times before it is full. These attacks, especially when there are many flies in the area, can so annoy horses and cattle that they stop grazing. Experts estimate that in grazing areas with many stable flies, fly annoyance can cause meat and milk production to decline by 20 percent. The stable fly can also carry such diseases as bovine anaplasmosis and equine infectious anemia.

FIGURE 12–7

The stable fly looks much like a common housefly. It is gray and approximately 6 to 8 mm long. Its mouthparts form a stout needle that the fly thrusts through the skin of its host.

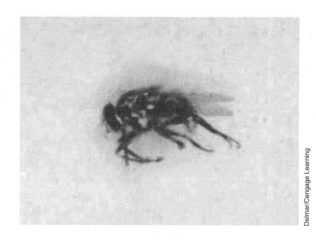

Delmar/Cengage Learning

The female stable fly lays her eggs in moist plant material like manure and damp hay, which she prefers. The larvae and pupae live where they hatch. Control measures include keeping hay dry and removing and drying manure. It also helps to put screens on stable and barn windows.

Horn Fly (Haematobia irritans)

The horn fly, a gray bloodsucking fly about half the size of a housefly, is an ectoparasite of pastured cattle. Both male and female horn flies suck blood, taking up to 20 small blood meals a day. They tend to feed on the backs and sides of cattle and spend their entire life on or near the cow. When there are many horn flies on a cow, their constant biting causes a condition known as fly worry, so named because the cow spends its time shooing off the flies instead of grazing. A cow with fly worry will not gain weight as fast as it should.

The female horn fly lays her eggs only in fresh manure. The larvae feed on the manure; the pupae live either in the manure or in the soil near it. Several measures control the horn fly. Because the flies are unwilling to leave the cow, insecticide applied to the cow's skin kills them. Other insecticides can be fed to the cattle to kill the horn fly larvae in the manure. Thus, there will be no new adult horn flies to bother the cattle.

Face Fly (Musca autumnalis)

The face fly (Figure 12-8) looks much like a housefly. The male face fly feeds on flowers, and the female feeds mainly on cattle, although it also feeds on horses. This fly does not bite; instead, it laps up fluids, especially those around the host's eyes, nose, and mouth. As its name suggests, the face fly hovers around the cow's face. A cow will not eat if too many face flies are feeding around its eyes. The face fly's habit of flying from cow to cow makes it an excellent vector of bovine pinkeye, which can lead to blindness.

Face flies are a parasite of pastured cattle; they enter buildings only in the autumn when they are looking for a place to spend the winter. Like the horn fly, the female face fly lays her eggs in fresh manure. Thus, insecticides fed to cattle, which are passed through manure, may help control face flies. Because face flies spend most of their time off the host, insecticides applied to the cow's skin have little effect.

Delmar/Cengage Learning

FIGURE 12–8

Face flies feed on the secretions in and
around the face of their host.

Screwworm Fly (Cochliomyia hominivorax)

Adult screwworm flies are blue-green and twice the size of a housefly. The adult
screwworm fly is not parasitic, but causes **myiasis**, a disease caused by larval
flies. The female fly lays her eggs at the edge of a wound on a warm-blooded
animal. Cattle, pigs, sheep, and goats are this fly's favorite hosts. The larvae
hatch and move into the wound to feed on the tissue around it. This enlarges the
wound and makes it give off a bad smell, which attracts other flies, which then
lay their eggs near the wound.

The first screwworm fly may be drawn to a small wound caused by a tick or tabanid
bite. Once larvae enter this wound they grows rapidly. In sufficient numbers, screwworm fly larvae can do great damage. Cattle have lost ears, tails, and even hooves to
screwworm fly larvae. After feeding in the wound for about a week, the larvae drop
to the ground and become pupae. Pupae become adults in about a week.

Scientists eradicated the screwworm fly from the United States in the 1950s.
Capitalizing on the fact that the female screwworm fly mates only once, scientists released millions of sterile male flies in areas where the screwworm fly
was a problem. This made sterile males far outnumber normal males, so female
flies usually mated with the sterile ones and produced eggs that did not hatch.
Several years of this sterile breeding eliminated the screwworm fly from the
United States, but it still lives in South and Central America and occasionally
travels into the United States with cattle imported from these areas. Wounds left
by husbandry practices such as branding, dehorning, and castration expose animals to the screwworm fly. Therefore, these procedures should be done during
the winter or dry season, when the number of screwworm flies is lowest.

Cattle Grub (Hypoderma lineatum *and* Hypoderma bovis)

Cattle **grubs** (Figure 12-9) look like large bees but belong to a group of flies
known as botflies. **Bots** are maggots of flies that infest animals. Only the maggots are parasitic; adult cattle grubs do not feed. Although cattle grubs live only

FIGURE 12–9

The females of the species *Hypoderma bovis* lay their eggs on the hair shafts of a cow. Larvae developing from these eggs penetrate the skin and migrate in the connective tissue, where they grow and develop.

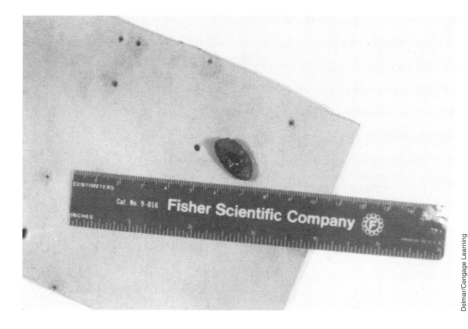

Delmar/Cengage Learning

a few days, that is enough time for the female to attach her eggs to the hair of cattle. She prefers to lay her eggs on the hair of the legs, and she prefers young cows to older ones. Although the egg-laying process is painless, the cattle instinctively fear this fly and try to run away from an adult cattle grub buzzing around their legs, a behavior called gadding. Gadding interrupts grazing and may result in lower weight gains and decreased milk production among cattle that graze in areas with many cattle grub adults. These cows may also hurt themselves while blindly running around in panic.

The adult fly interferes with grazing, but only the larval cattle grub is truly parasitic. When the eggs hatch, the larvae crawl down the hair shaft and burrow into the skin. Unlike screwworm fly larvae, cattle grub larvae can penetrate intact skin. The larvae then migrate through the body of the cow, feeding on host tissue as they go. *Hypoderma lineatum* larvae spend the winter in the wall of the cow's esophagus; *H. bovis* larvae live in the fat around the spinal cord. The larvae that spend the winter inside the cow are about one-half inch long. In late winter, the larvae migrate toward the back of the cow. When they reach the area under the skin of the back, they make a small breathing hole in the skin. The skin over the larva swells, producing a lump called a **warble**. The larvae grow in the warble for 1 or 2 months and change in color from white to dark brown. The full-grown larva, about 1 inch long, emerges through the breathing hole and drops to the ground. Once on the ground, the larva pupates; about a month later, the adult fly emerges.

The damage the larvae in the warble do can cause the cattle rancher problems. The inflammation that forms around the larva destroys some meat, thus decreasing the value of the carcass. The holes left by the emerging cattle grubs heal but leave scars that decrease the hide's value. Insecticides kill larval cattle grubs, but if the larvae die in the esophagus or spinal cord, the dead larvae cause inflammation. Inflammation around the esophagus could make eating difficult; inflammation around the spinal cord could paralyze the cow. The time to treat cattle with insecticide is just after the adult flies have finished laying eggs. Another good time is in the spring, after the larvae leave their winter homes but before warbles form.

Ticks

A number of different ticks parasitize livestock. Ticks are an important livestock parasite because they transmit several protozoal, bacterial, and viral diseases of cattle. Heavy tick infestations may also cause **anemia**, a disease in which there is insufficient blood to carry enough oxygen to support the tissues of the body. Some of the most significant cattle ticks follow.

Spinose Ear Tick (Otobius megnini)

The spinose ear tick is a one-host soft tick, which spends its larval and nymphal stages in a cow's ear (Figure 12-10). Adult ear ticks live on the ground and do not feed. Like all other ticks, the ear tick lays its eggs on the ground. The larvae hatch and climb to the top of grass or weeds. When the cow comes to graze on the grass, the larvae detach themselves from the plant and attach themselves to the cow. They then make their way into the cow's ears, where they feed on blood. The larvae stay in the ears and molt to the nymphal stage. The nymphs also feed on blood. When ready to molt, the nymph leaves the ear and drops to the ground. In response to the irritation of feeding larvae and nymphs, the cow shakes and rubs its head, trying to rid itself of the ticks. An ear with too many ticks may become inflamed. This inflammation may damage the eardrum, resulting in deafness. Feeding ticks may also leave wounds prone to bacterial infection. If the cow has only a few ear ticks, they can be removed by hand. An infestation of more than a few ear ticks may require application of an approved pesticide.

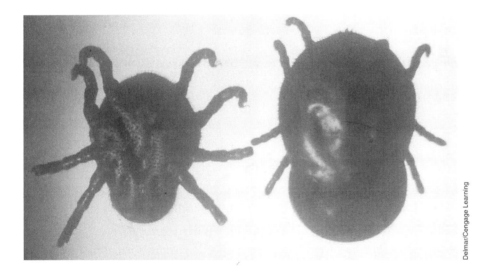

Delmar/Cengage Learning

FIGURE 12–10

The spinose ear tick is a soft tick most often found in the ears of many domestic animals, including dogs, horses, cows, goats, pigs, and cats.

Cattle Tick (Boophilus annulatus)

The cattle tick is a one-host hard tick, which has been eliminated from much of the United States, but sometimes turns up on cattle in Texas. It's common on cattle in Mexico and travels to Texas on deer and stray cattle. This tick transmits *Babesia bigemina*, which causes bovine **babesiosis**, a protozoal disease also called tick fever or Texas cattle fever. A single application of a veterinarian-approved pesticide kills all stages of one-host ticks in cattle and slows the reproduction of the entire tick population. This widespread pesticide treatment is the strategy that eliminated the cattle tick from the United States. If diagnosed, this disease-spreading tick must be reported to both federal and state authorities.

B. annulatus is still found in Mexico and can easily come across the border with livestock, feral animals, pets, and humans.

Cattle that cross legally from Mexico must be treated to eliminate ticks. However, stray or smuggled livestock from Mexico, and the native and exotic wildlife that cross back and forth, can carry fever ticks. If these animals cross the river carrying "hot" fever ticks that complete their life cycle on the U.S. side, the potential exists to introduce a highly virulent disease into a cattle population unprepared to handle it.

In 2008, Mexico reported acaricide-resistant ticks. The Texas Animal Health Commission (TAHC) has devoted much of its resources trying to keep the cattle fever tick contained within the permanent quarantine zone along the Rio Grande, with the ultimate goal of pushing cattle fever ticks back across the river into Mexico. At one time, cattle fever ticks spread north to Kansas and east to Southern California causing as high as a 90 percent death rate in some areas. Through the joint efforts of the USDA's cattle tick eradication program, the veterinary professional can assist in the elimination of this threat to our livestock industry.

Lone Star Tick (Amblyomma americanum)

The lone star tick is a three-host hard tick that feeds on small mammals and birds as a larva and a nymph. The adults prefer to feed on large mammals such as cattle and horses. This tick is named for a single white dot in the center of the adult female's back. Adult ticks prefer to feed on the cow's neck and on the area where the cow's legs meet its body. It has long mouthparts, which can cause deep, irritating wounds. If enough ticks are feeding on a cow, the cow can develop anemia.

Gulf Coast Tick (Amblyomma maculatum)

The Gulf Coast tick is a three-host hard tick. As its common name suggests, it lives in the United States along the Gulf of Mexico. The larva and nymph prefer to feed on such ground-dwelling birds as quail. The adult ticks feed on cattle, horses, and other large mammals. They prefer to feed in and around the base of the ears. Like the lone star tick, the Gulf Coast tick has long mouthparts that cause a nasty wound.

Rocky Mountain Wood Tick (Dermacentor andersoni)

The Rocky Mountain wood tick is a three-host hard tick that lives in the western United States. The larva and nymph feed on wild rodents, while the adult prefers large animals such as cattle and horses. The adults prefer to feed on the host's belly. This tick is a vector of bovine anaplasmosis, a bacterial infection characterized by fever, jaundice, and emaciation. As it feeds, the Rocky Mountain wood tick injects a toxin (poison) that may cause tick paralysis, a spreading paralysis that may eventually impair the muscles involved in breathing. Fortunately, tick paralysis rapidly subsides once the tick is removed. Not every feeding tick causes paralysis, but no one knows why some Rocky Mountain wood ticks cause paralysis and others do not.

Winter Tick (Dermacentor albipictus)

The winter tick is a one-host hard tick that feeds on large mammals like cattle, horses, moose, and deer. The larval winter tick begins to seek a host in

September or October. The adult ticks develop on the host by early winter and drop off to lay their eggs by February or March. Heavy winter tick infestations may severely weaken their hosts and may be difficult to detect under the thick winter coats of horses and cattle.

Deer or Blacklegged Tick (Ixodes scapularis *and* Ixodes pacificus)

The deer tick is a three-host hard tick that feeds on birds and small mammals as larvae and nymphs (Figure 12-11). Adults feed on large mammals such as cattle, horses, dogs, and deer. The adult ticks have long mouthparts, which can cause considerable damage while the tick feeds. The adult blacklegged ticks usually feed on the head or neck of their host, where there is good blood supply. These ticks can be vectors of bovine anaplasmosis and Lyme disease. They may also cause tick paralysis.

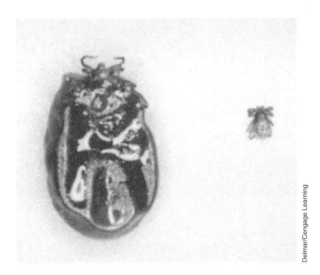

Delmar/Cengage Learning

FIGURE 12–11

The deer tick is a hard tick associated with the transmission of Lyme disease. *Left,* the adult deer tick. *Right,* the deer tick in its larval stage.

Veterinarians treat light infestations of ticks on cattle by removing the ticks one at a time. Heavier infestations need approved pesticides applied to the cattle. This treatment is sufficient to control one-host ticks, but three-host ticks need a more thorough approach because not all of them are on the cattle. To kill three-host ticks on their cattle, ranchers apply three-host tick pesticides every week during the tick season (late spring to early fall). It is important that the veterinary staff educate the cattle rancher on the required withdrawal times for all pesticides used. To kill ticks that are off the cattle, ranchers often burn pastures during the dry season, till pastures, or rotate pastures (move cattle to a new pasture until the old pasture's ticks die).

Mites

Mites that cause **mange** in cattle, sheep, and goats are *Sarcoptes, Demodex, Chorioptes,* and *Psoroptes.* The life cycles of the *Sarcoptes* and *Demodex* are covered in Chapter 13.

The *Sarcoptes* mite causes sarcoptic mange (scabies), a potentially serious disease of cattle characterized by intense itching and a large loss of hair in the affected areas. Veterinarians treat it with a single injection of ivermectin, a broad-spectrum **anthelmintic** (a substance destructive to worms). Scabies is highly contagious and can be spread by direct contact and by fomites. Because it is so highly contagious and is a zoonotic disorder, it is a reportable disease. Diagnosis is made by skin scrapings or skin biopsy, or both.

Demodex mites (Figure 12-12) cause demodectic mange, a disease characterized by the formation of small nodules (growths) in the skin. Though veterinarians rarely treat demodectic mange, ivermectin kills the *Demodex* mites. When there is an increase in numbers of these mites, the clinical disease is called demodicosis. *Demodex* mites are host specific and most commonly infect dogs with low immune systems.

FIGURE 12–12

Demodex mites.

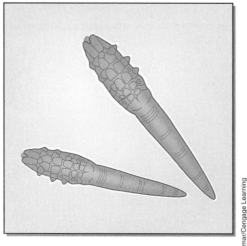

Demodex mites

Psoroptes is a nonburrowing mite that can also harm cattle. All the stages of this mite live on the host's skin. *Psoroptes* mites have chewing mouthparts, which severely damage the host's skin and cause psoroptic mange, a disease of cattle characterized by vesicles (small blisters) on the skin that cause intense itching. The itching makes cattle so restless that they gain weight very slowly. Ivermectin also kills psoroptic mites. Even though this disorder is not zoonotic, it is still a reportable disease.

Chorioptes mange (leg mange) mites can also infest cattle. In cattle they are found in the tail region as well as the hind legs. Similar to the *Psoroptes* mange mite, the *Chorioptes* mite does not burrow and spends its entire life cycle on the host. Chorioptic mange is like psoroptic mange, but usually much milder. Ivermectin kills chorioptic mites. Chorioptic mange is also a reportable disease but does not affect humans. It is the most common type of mange in cattle in the United States. These mites can be identified by their short, unjointed pedicels with suckers on the ends of some of their legs.

Ectoparasites of Sheep

Flies

Sheep Ked (Melophagus ovinus)

The sheep ked (Figure 12-13) is a wingless fly that spends its entire life cycle on the sheep. The female fly lays a fully developed larva in the sheep's wool.

Delmar/Cengage Learning

FIGURE 12–13

The sheep ked has heavy claws on stout legs, which it uses to hold on to the sheep.

The larva pupates almost immediately, gluing the pupal case to the wool in the process. The pupa develops into an adult fly in about 3 weeks. Contact between sheep spreads the sheep ked.

Adult flies of both sexes suck blood, which causes sheep a good deal of irritation. Sheep often stop grazing to bite and rub at the areas where sheep keds feed. Thus, a sheep infested with sheep keds may not gain weight and will damage its wool, which is also devalued by the pupae glued to it. Heavily infested sheep may become anemic.

Veterinarians who notice sheep scratching themselves check the wool for the easily visible sheep keds and pupae. Spring shearing removes many sheep keds. Approved insecticides applied to the sheep's wool and skin also kill adult sheep keds.

Sheep Nasal Botfly (Oestrus ovis)

Like other botflies, the sheep nasal botfly's larva is parasitic. The adult nasal botflies do not feed. The female nasal botfly squirts a stream of fluid into a sheep's or goat's nasal opening. This fluid contains up to 25 small (1 mm) larvae (Figure 12-14), which migrate deeper into the nasal passages, feeding on mucus as they go. The larva moves around the nasal cavity until it enters a sinus cavity, where it matures in about 1 month. The mature larva then leaves the sinus, migrates out of the nose, and drops to the ground, where it pupates. The adult fly emerges from the pupal case in 1 to 2 months. Larval nasal bots, which enter a sheep's nose in late autumn, stay the entire winter and can remain in the sheep for up to 9 months. Anyone familiar with sheep can easily spot a flock under attack from nasal botflies. Recognizing the adult flies, the sheep bunch together and try to hide their noses by pushing them against other sheep or onto the ground. While the sheep hides its nose, the fly cannot lay its larva in the nostril, but a sheep hiding its nose cannot eat.

Most infections from botflies are light and cause sheep to sneeze and rub their noses on fences or other objects. Sheep with heavy infections are unthrifty (do

FIGURE 12–14

Larvae of the nasal botfly can damage the nasal passage and sinuses, making it difficult for the sheep to breathe.

not put on weight), may be uncoordinated, and may have trouble breathing. The larval nasal bots irritate the lining of the nose, causing a runny nose and sneezing. The sinus where the larva is growing may fill up with pus. Sheep nasal botflies are difficult to control, though ivermectin kills the larvae.

Blowfly (Phormia, Lucilia, Calliphora, and Phaenicia species)

Adult blowflies are metallic black or green. They feed on moist, decaying matter (like dead animals), open festering wounds, and soiled wool. The female lays her eggs in these sites. The blowfly larvae hatch and start feeding, in some cases invading the skin. Ranchers call the larvae in a sheep's decaying wool fleeceworms and wool maggots. Even if they feed only on wool soaked with urine or feces and on the skin's moisture, this causes sheep a major problem: The wool rots and falls out, and the rotten odor that feeding larvae produce attracts other blowflies. Blowfly infestations not promptly treated may kill the sheep. Insecticide applied to the sheep kills the larvae. The ranching practices of docking (removing the tail) and crutching (shaving the wool away from the area below the tail) aim to remove the sites most likely to become soiled. These measures and any others that keep the sheep clean and dry help to prevent blowfly myiasis.

Mites

Demodectic, sarcoptic, and psoroptic mange mites all occur on sheep. Refer to the section on cattle mites previously in this chapter for descriptions of these mites.

Demodectic mange of sheep is a mild disease and very rare. Sarcoptic mange of sheep normally starts on the face or other areas that have little wool. Infested sheep suffer intense itching and tend to lose weight because of the time and energy they spend rubbing against objects. Sarcoptic mange in sheep is also rare.

Psoroptic mange, also known as sheep scab, is the most important mange of sheep. It is also a reportable disease. Similar to the psoroptic mange that cattle suffer, this disease causes itching so intense that sheep rub the wool from large areas of their bodies. They are so restless that they may lose weight. The worst cases of sheep scab usually occur during the winter. Psoroptic mange of sheep is rare in the United States.

Veterinarians diagnose the type of mange by identifying the mites from skin scrapings. Dipping the sheep—submerging the animal in a bath containing a recommended pesticide—controls mange. Ivermectin controls sheep mange, but it is not licensed everywhere for this purpose. Sarcoptic mange and demodectic mange of goats are similar to the diseases in cattle. Psoroptic mange in goats is usually limited to the ears, face, and neck. Pesticides, recommended by a veterinarian, can be applied to goats to control mange.

Ectoparasites of Swine

Flies

No fly is a particular problem for swine. Mosquitoes and horseflies feed on swine and cause annoyance. The larvae of the rodent botfly (*Cuterebra*) sometimes turn up in the skin of hogs and pet swine (Figure 12-15). The best treatment is surgical removal.

Delmar/Cengage Learning

FIGURE 12–15

The rodent botfly is large, about 20 mm or more in length, with a beelike body and tiny mouthparts.

Lice

The hog louse (*Haematopinus suis*), a large (6 mm) sucking louse, is the only louse that feeds on swine. Heavy hog louse infestations cause irritation that may lead to a poor weight gain. This louse is easy to see with the naked eye. Veterinarians diagnose hog louse infestation by finding the lice or their eggs. They treat infested pigs by spraying them with an insecticide or giving them ivermectin. Because the eggs take 2 to 3 weeks to hatch, the infestation may require several treatments.

Fleas

The cat flea (*Ctenocephalides felis*) sometimes feeds on swine, especially pet pigs that share a home with dogs or cats. The infested pig should be treated with an insecticide, and its holding area should be first cleaned and then treated with an insecticide. Refer to the section on dog and cat fleas in Chapter 13 for more information about this flea.

Ticks

The spinose ear tick, the Gulf Coast tick, the lone star tick, the blacklegged tick, and the American dog tick all feed on pigs. Refer to the sections on cattle ticks (previously, in this chapter) and dog ticks in Chapter 13 for details on the life cycles of these ticks.

A pig infested with ticks can be treated with a pesticide to kill the attached ticks. If the infestation is light, the individual ticks can be removed as described in Chapter 13 for dog and cat ticks.

Mites

The sarcoptic mange mite (*Sarcoptes scabiei*), the most important ectoparasite of commercially raised swine, is a fairly common ectoparasite of pet pigs. The section on mites of the dog and cat in Chapter 13 describes this mite's life cycle. This mite causes pigs intense itching in the infested area. The infested skin thickens and forms a scablike crust. The sarcoptic mange mite usually infests pigs on the face or around the neck. Two doses of ivermectin, given 2 weeks apart, treat sarcoptic mange mites in pigs. An infested herd should be treated together to prevent untreated pigs from reinfesting treated pigs. All new pigs should be examined and treated for sarcoptic mange before they are added to the herd.

Ectoparasites of Horses

Flies

Many of the flies that attack cattle, such as the horsefly and the stable fly, also attack horses. (Refer to the section on cattle flies previously in this chapter to review these flies.) Mosquitoes are usually a minor annoyance to horses but can carry equine encephalitis, a severe viral brain disease. The most pathogenic viruses for horses are Eastern equine encephalomyelitis virus, Western equine encephalomyelitis virus, Venezuelan equine encephalomyelitis virus, and West Nile virus.

The nonfeeding adult horse botflies (*Gasterophilus* sp.) attach their eggs to the hairs of a horse's forelegs (Figure 12-16). The horse nose bot places her eggs around the horse's lips, while the horse throat bot places hers under the horse's jaw. The eggs hatch in about a week. The eggs of the common horse bot hatch when the horse licks its legs; the newly hatched larvae (Figure 12-17) burrow

FIGURE 12–16

The adult horse botfly is brown and hairy. The adult female lives only a short time, during which she deposits her eggs on the hairs of a horse.

Delmar/Cengage Learning

Delmar/Cengage Learning

FIGURE 12–17

The larvae of the horse bot are parasitic maggots that attach themselves to the stomach lining of horses.

into the tongue. The eggs of the nose and throat bots hatch without the horse's licking them. The larvae from these eggs migrate to the mouth and burrow into the tongue. As the larvae grow, they move from the tongue to the spaces between the horse's upper molars. The larvae stay in the mouth for 3 to 4 weeks and then move to the back of the mouth and are swallowed. The swallowed larvae attach themselves to the stomach wall (nose bot larvae to the small intestine wall) and grow. They remain attached for 10 to 12 months, growing 1 to 2 cm long. When the larvae mature, they detach themselves from the stomach and pass out with the feces. They pupate on the ground; adults emerge in 1 to 2 months. Adult horse botflies live only a few days.

Because the throat and nose bots lay their eggs on the horse's head, they annoy the horse more than the common bot does. The horse's grazing is interrupted by the adult botfly annoyance. If this annoyance persists, it may affect the horse's weight. The larvae normally cause very little damage but in rare cases can cause stomach ulcers. Although larvae in the stomach cannot be diagnosed, botfly eggs on the horse's hairs mean the horse has bots.

Certain drugs (ivermectin being the most effective) kill the larval bots in the stomach. The best time to treat the horse is in the winter, when adults have died and all the eggs should have hatched. In the summer or autumn, eggs on the horse's legs can be killed with warm water containing an insecticide. The warm water hatches the eggs, and the insecticide kills the emerging larvae.

Lice

Two types of horse lice are the *Bovicola equi* and the *Haematopinus asini*. *B. equi* is a chewing louse; *H. asini* is a sucking louse. Both types irritate the horse. A horse heavily infested with either of these lice will rub or bite at the infested area and damage its hair. Many times a horse is infested with both lice at the same time. Heavy infestations usually occur during the winter when the horse has a longer coat. Diagnosing these lice involves combing the horse and looking for them.

The infested area is treated with an application of an insecticide. Horses infested with the horse-sucking louse can be treated with ivermectin.

Fleas

The cat flea (*Ctenocephalides felis*) sometimes feeds on young foals. The infested horse should be treated with insecticide. The stable should be cleaned first and then treated with insecticide. Refer to the section on dog and cat fleas in Chapter 13 for more information about this flea.

Ticks

The blacklegged tick, the Rocky Mountain wood tick, the winter tick, the Gulf Coast tick, and the lone star tick all infest horses. Refer to the section on cattle ticks earlier in this chapter to review these ticks.

Mites

Sarcoptes, Psoroptes, and *Chorioptes* mange mites feed on horses and cause diseases similar to those that dogs and cattle suffer. *Demodex* mange mites also feed on horses, but like those on cattle, they cause little harm.

CASE STUDY	Cattle Grubs: Hypodermosis of Cattle

A Wyoming cattle rancher has noticed hypodermal rashes on the backs of his cattle. Six of the steers are having dysphagia (difficulty eating) and are drooling. The cattle rancher notices a few cysts (warbles) that have holes in them.

DIAGNOSIS

During the third stage, the larvae of cattle grubs can be differentiated between the two species, *H. lineatum* and *H. bovis*. The *H. lineatum* has spines on the tenth segment and a flat spiracular plate. The *H. lineatum* is also smaller. The *H. bovis* are larger without spines on the tenth segment and a funnel-type spiracular plate. Heel flies in the pastures or around the legs of the cattle can be a good indicator.

TREATMENT

The first important treatment is control of the heel fly population. Systemic insecticides such as ivermectin are used but are contraindicated in lactating dairy cows. Coumaphos, trichlorfon, phosmet, and ivermectin can be used for pour-on treatments. Sprays and dips such as coumaphos and phosmet can be used for control of cattle grubs. Ivermectin injected subcutaneously or given as an oral paste has proven effective against cattle grub larvae. The larvae wander through the subcutaneous connective tissues, including areas of the spinal canal. It is important not to use systemic insecticides during the time the larvae are in the region of the spinal canal. If they die while in this region, they could cause severe reactions, including death. The best treatment time with the least possible adverse effects is as soon as possible after the end of the heel fly season. Cattle should not be treated later than 8 to 12 weeks before the anticipated first appearance of grubs in the backs.

Summary

The most common ectoparasites of large animals are insects or arachnids. Flies are two-winged insects that are capable of transmitting bacterial, viral, and parasitic diseases to livestock. Several parasitic flies also infest livestock, causing millions of dollars in lost production every year. Fleas are jumping, bloodsucking insects. Ticks are the largest parasitic arachnids and can be one-host, two-host, or three-host species. The mite causes the skin disease known as mange. Eradication efforts by both private and governmental sources are needed to reduce disease-causing levels of parasites that commonly exist within and near our borders.

REVIEW

FILL IN THE BLANKS

1. To reduce the stages of the tick population that is off the host, a rancher can
   _Burn_____, _Till_____, or
   _rotate_____ the pasture.

2. When the female cattle grub fly tries to lay her eggs on a cow, the cow may try to avoid the fly, a behavior known as
   _gadding_____.

3. If a horse is treated only once a year for horse bots, it should be treated during the
   _Winter_____.

4. _Dipping_____ an animal is a process whereby the animal is submerged in a bath containing a recommended pesticide.

5. _Docking_____ is the shortening of the tail in sheep and helps prevent
   Blowfly _myliasis_____.

6. _Cattle Tick (Boophilus annulatus)_ is the ectoparasite associated with causing bovine babesiosis.

MATCHING

Match the parasite on the left with the problem it causes on the right.

_B_ 1. horn fly   a. mange

_D_ 2. tabanid   b. fly worry

_e_ 3. face fly   c. fleeceworms

_a_ 4. *Sarcoptes*   d. vector of bovine anaplasmosis

_C_ 5. blowflies   e. vector of bovine pinkeye

CHAPTER 13

Ectoparasites of Small Animals

OBJECTIVES

Upon completion of this chapter, the reader should be able to:

- List the common ectoparasites of small animals.
- Review the life cycle of the fly, flea, louse, mite, and tick from Chapter 12.
- Describe the common diseases spread by these common ectoparasites.
- Describe (briefly) methods of prevention, treatment, and diagnosis.

KEY TERMS

acaricides

Anoplura

chigger

demodicosis

ectotherms

erythema

Mallophaga

pediculosis

pruritus

pustules

scabies

Ectoparasites in Companion Animals

Persistent ectoparasites do not restrict themselves to large animals. Their effect on the host, however, can be surprisingly different. Some seemingly innocent ectoparasites can be quite dangerous to small animals. Others that sound dangerous actually are not. As we encroach upon and reduce our available wildlife areas, there will be an increase in cross-infestation from wildlife to companion animals. It is important to educate the companion animal owner/caretaker about the importance of separating wildlife areas from our pet population.

Ectoparasites of Dogs and Cats

Flies

Most flies cause few problems for dogs and cats. Tabanids and other bloodsucking flies may feed on dogs and cats. The larvae of some flies may cause myiasis in dogs and cats. One such fly is *Cuterebra*. *Cuterebra* larvae (also known as wolf or wolf worm) infest the skin of rabbits, rats, mice, squirrels, chipmunks, dogs, and cats. In cats, which are not a natural host for the botfly, the larvae usually find their way from the nose lining to the area under the skin. There they grow normally. In some cases, though, the larvae get lost in the cat and may end up in the brain or in other tissues. In the brain, the larvae may cause enough damage to kill the cat. Rodent botfly infestation is almost impossible to diagnose unless the larva ends up under the skin. Bots under the skin of any animal leave a breathing hole. Treatment consists of surgically removing the bot from its pocket under the skin. Care must be taken not to crush the bot as it can cause anaphylaxis to the animal itself. Some drugs kill the larvae, but the dead bots cause an inflammatory reaction that may prove more harmful than the live parasite.

Cuterebra infections are rare in indoor cats and even rarer in dogs. The larvae of flesh flies sometimes infest wounds in the skin of dogs and cats and in cats and dogs seeking out rodent prey in their burrows. The stable fly (*Stomoxys calcitrans*), also known as the biting housefly, feeds on domestic animals. In particular, they feed on the tips of the ears of dogs with pointed or raised ears such as Dobermans and German shepherds. Both the male and female fly are avid blood feeders.

Myiasis (Fly Strike)

Fly larvae (larval dipterans) may develop in the subcutaneous tissues of the skin of domestic and wild animals. Flies are attracted to moist wounds, skin lesions, soiled hair coats, dead animal tissue, and feces. The hatched eggs (maggots) ingest dead cells and secretions and sometimes travel through healthy layers of skin, producing exudates of their own. Thousands of maggots can occupy large cavities in the skin. The animal may die of shock, intoxication, histolysis, or infection if the infestation is not arrested. This condition is known as fly strike or facultative myiasis.

Lice

Both sucking or **Anoplura** lice (*Linognathus setosus*, Figure 13-1A) and biting or **Mallophaga** (*Trichodectes canis*, Figure 13-1B) lice feed on dogs. Eyes in

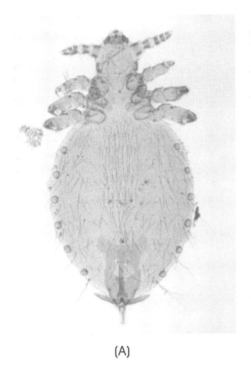

(A)   (B)

Delmar/Cengage Learning

FIGURE 13–1

A. The sucking louse has a head that is more slender than its thorax, with mouthparts designed to suck its host's tissue, body fluids, and blood. B. The biting louse has no eyes, and its body is broad and flat. The head of a biting louse is wider than its thorax and as wide as its abdomen. The mouthparts of a biting louse are designed to bite and chew. They feed on debris on the surface of the skin rather than suck like the other group of lice.

Mallophaga are either absent or reduced, and the mandibles are developed for chewing. Only a biting louse (*Felicola subrostratus*) feeds on cats. These lice travel from host to host by direct host contact or fomites (grooming clippers, brushes, blankets, bedding). Heavy sucking lice infestations can cause anemia. The dog-sucking louse usually feeds on the neck or shoulder region. Chewing lice cause skin irritation and itching on both cats and dogs. Pets with a heavy infestation of biting lice may scratch so much that they remove their hair and cut the skin. Sucking lice can ingest enough blood to cause severe anemia packed cell volume (PCV can drop by as much as 20 percent in young animals). Infestation by lice of either type (biting or sucking) is referred to as **pediculosis**. Dogs or cats (and sometimes humans) can be the definitive host to *Dipylidium caninum* (the cucumber or "double-pored" canine tapeworm), and the flea or louse the intermediate host. The dog's or cat's perianal region becomes contaminated with eggs of the tapeworm as feces are passed. The flea or louse ingests the eggs, and the dog or cat ingests the flea or louse. Although it is much more common with fleas, the louse is still an important intermediate host to consider.

A dog or cat that scratches constantly should be examined for adult lice or nits attached to the hairs. Powders or shampoos containing an insecticide can treat lice infestations. The veterinarian can best prescribe, after diagnosis, the proper treatment. The dog or cat must be retreated after 2 weeks because pesticides do not kill the nits. In 2 weeks, the nits hatch; retreatment kills the nymphs before they develop into egg-laying adults. Shaving a cat before treating it reduces the number of eggs, which are found on the hairs.

Fleas

Ctenocephalides felis (Figure 13-2) is by far the most common flea on dogs and cats. *C. canis* (Figure 13-3) also feeds on dogs and cats but is rare. Although fleas lay eggs on the dog or cat, the eggs soon fall off. When the eggs hatch

FIGURE 13–2

This flea is similar to the dog flea in appearance. Although it is usually unnecessary to differentiate between the dog flea and the cat flea, it may be done by examining the size and shape of the head and the size of the spines on the mouth comb.

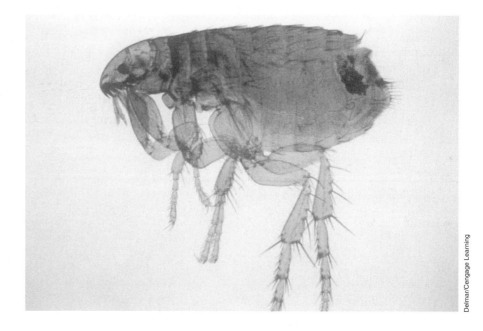

FIGURE 13–3

This dog flea is small (4 mm long), with powerful legs for jumping.

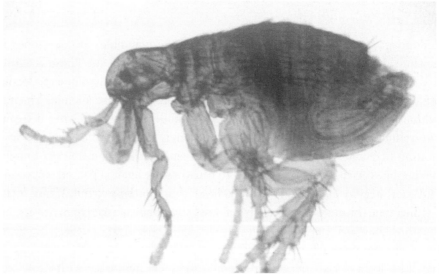

in 1 or 2 weeks, the larvae feed on organic material. The larval flea eventually spins a silklike cocoon and becomes a pupa. The time from the hatching of the egg to the hatching of the adult from the cocoon may be as short as 3 weeks in hot weather or as long as 2 years in very cold weather.

The adult flea jumps onto a dog or cat, where it lives up to 2 months. Because many eggs drop where the animal spends a lot of time and the larval flea stays close to where it hatches, most adult fleas do not have long to wait before a host comes along.

Once on the host, the adult flea keeps moving, stopping only for a few minutes at a time to take a blood meal. This constant moving around and feeding irritates the host and causes it to scratch. As with lice infestations, the host's constant scratching can remove hair and cut the skin. Heavy flea infestations can cause anemia, which can be fatal to kittens and puppies. Some pets develop an allergy to the flea's saliva. The bite of even a single flea can cause severe itching in an allergic dog or cat. The flea saliva contains proteolytic enzymes and histamine-like substances that cause **pruritus** and hypersensitivity to the bite. Flea allergy dermatitis (FAD) is the most common dermatological disease of domestic dogs in the United States.

On pets, fleas are more difficult to control than lice. Like lice, adult fleas on the host succumb to shampoos and powders that contain insecticides; flea collars contain a slow-release insecticide that kills fleas on the pet for several months; and other insecticides placed on the pet's back kill adult fleas for 1 to 3 months. But killing adult fleas on the host solves only half the problem. The insecticide applied to the pet does not kill eggs, larvae, and pupae not on the host. Drugs are available that prevent flea eggs on pets from hatching. These drugs should be prescribed by the veterinarian.

It is important to clean thoroughly those areas where eggs may have dropped off the pet, especially the pet's bedding. All rugs and areas under the furniture, as well as cracks and crevices in the floor, should be vacuumed, and insecticide should be applied to these areas. Ultimate flea control may require several applications of insecticides to both the pet and its surroundings.

Ticks

Brown Dog Tick (Rhipicephalus sanguineus)

The common dog tick (Figure 13-4) feeds on dogs as a larva, a nymph, and an adult. The parasite's host, however, is a different dog in each stage. The larval tick feeds and then drops off the dog. The larva molts on the ground and becomes a nymph. It then climbs to the top of a blade of grass or bush and waits for another dog to come along. After the nymph finds a host and feeds, it repeats the process. The adult tick finds a third dog. Adult ticks mate on the dog. After feeding, the female tick drops off to lay her eggs on the ground. Unfed larval, nymphal, and adult ticks can survive for months while waiting for a host.

FIGURE 13–4

The brown dog tick is a hard tick with a hard dorsal shield and a rather plain appearance.

Delmar/Cengage Learning

Ticks can harm their hosts beyond the blood loss of feeding; they commonly irritate the skin around their bite, causing the dog or cat to scratch much as it would if it had fleas or lice. Ticks can also carry serious diseases. Because a tick parasitizes several hosts during its life cycle, it can readily transmit viruses, bacteria, and protozoan parasites. The brown dog tick is an intermediate host for *Babesia canis*, the etiological agent that causes canine piroplasmosis (babesiosis), a protozoan disease, and canine ehrlichiosis, a bacterial disease. It also causes tick paralysis in some dogs.

Removing the ticks with tweezers controls a light infestation of adult ticks. Each tick is grasped close to its head, and its mouthparts are pulled from the skin. The tick is removed without squeezing its head to avoid injecting parasitic agents into the host. Larval and nymphal ticks are small and difficult to find, so tweezers do not control them effectively. To control infestations involving larval and nymphal ticks or heavy infestations of adult ticks, a pesticide or tick collar should be used. Because female ticks normally lay eggs outdoors, pesticides are not normally used indoors. However, kennels do use pesticides indoors to control the brown dog tick. Since all three of its parasitic stages feed on dogs, this tick can thrive in kennels. A kennel with an infestation should be treated with a pesticide along with its dogs, with treatment repeated regularly until the infestation is under control.

Blacklegged or Deer Tick (Ixodes scapularis)

The blacklegged tick is another important vector of disease in the United States. (Review its life cycle in the section on cattle ticks in Chapter 12.) Both the nymph and adult of this tick feed on dogs.

The damage this tick causes is the same as that caused by the brown dog tick. It can transmit Lyme disease, a bacterial disease, from mice to dogs. Lyme disease can also be spread to people. The past few years have seen dramatic increases in the number of human Lyme disease victims in the northeastern United States.

Blacklegged tick infestations are usually light in dogs or cats, so individual ticks can be removed with tweezers. Pesticides (**acaricides**) in collars or applied to the skin of the pet prevent or control tick infestations.

American Dog Tick (Dermacentor variabilis)

The American dog tick (Figure 13-5) is a three-host hard tick. The larvae and the nymphs prefer to feed on rodents and rabbits. The adults feed on dogs, cats, and such wild carnivores as foxes and coyotes.

American dog tick infestations are usually light and cause damage similar to that of the brown dog tick. The American dog tick carries canine piroplasmosis (babesiosis) and Rocky Mountain spotted fever, a bacterial disease. Infestations of the American dog tick are treated in the same way as those of the blacklegged tick.

Mites

All mites commonly affecting dogs and cats spend their lives (egg, larva, nymph, and adult) on one host. Mites can transfer from one host to another during close contact.

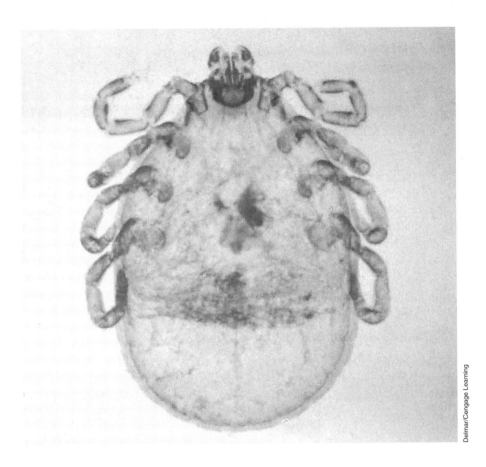

Delmar/Cengage Learning

FIGURE 13–5

The American dog tick is most numerous in the spring and summer months, since it needs moisture in the environment to survive.

Demodectic Mange Mite (Demodex canis)

Demodex canis (Figure 13-6), a common mite of dogs, is cigar-shaped and lives in hair follicles. Suckling transmits these mites from the mother dog to her pups. They normally cause no disease but in large numbers can cause mange. **Demodicosis** is the clinical disease caused by many of these mites infecting the fauna of the skin.

Demodectic mange can be a mild, localized disease or a serious, generalized disease with lesions covering large areas of the body. The localized disease may cause slight hair loss, usually on the muzzle, face, and forelimbs. These mild lesions should still be treated. Cats have *Demodex* mites but rarely get demodectic mange. When they do, it is usually the localized type on the face.

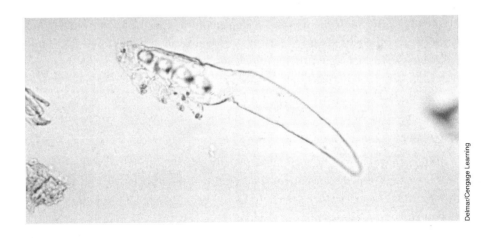

Delmar/Cengage Learning

FIGURE 13–6

The demodectic mange mite lives in the hair follicles and sebaceous and sweat glands in the skin of the dog.

Generalized demodectic mange occurs in dogs whose immune system is suppressed. This is common when confounded with another disease, a genetic defect, or drugs given to the dog for a different condition. Immunosuppressed dogs with this more serious mange have hair loss and thickening of the skin in the area of the lesions. If these lesions become infected with bacteria, **pustules** (pimple-like growths containing pus) may form. These lesions cause little or no itching.

Veterinarians diagnose demodectic mange by scraping the skin in the area of the lesions and finding larval, nymphal, and adult mites. *Demodex* mites are present in normal dogs, but the presence of many different stages indicates that the mite population is increasing rapidly and therefore causing the lesions.

Treatment of demodectic mange (demodicosis) is a difficult and long process. The mites are so deep in the skin that pesticides may not reach and kill them immediately. It can take months of repeated treatment to cure the disease. Amitraz (Mitaban) dips are usually done for 4 to 6 weeks at 2-week intervals. Antibiotics are given for prevention of secondary infections.

Sarcoptic Mange Mite (Sarcoptes scabiei)

Sarcoptes scabiei (Figure 13-7) is a small, round mite that can also cause mange (**scabies**) in dogs and other animals. Humans can also become infected with this mite. Thus, scabies is a zoonotic concern. In humans, it causes papules or vesicles to appear on the trunk, arms, abdomen, face, and genitalia. The lesions are extremely pruritic. This type of mange is rare in cats. Sarcoptic mites prefer the skin of the face, ears, and forelimbs, but in heavy infestations they can be found anywhere on the body.

FIGURE 13-7

Sarcoptic mites are round to oval in shape and have short legs with spines that stick out from the sides. The pedicel can be long or short, it is unjointed, and a sucker is on the end of some of the legs.

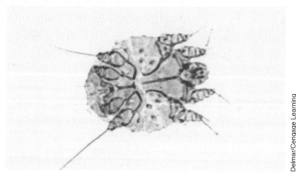

Delmar/Cengage Learning

The life cycle from egg to adult takes about 17 days. The adult burrows into the skin, forming a tunnel. The female lays eggs there, and the larvae and nymphs stay there until the adult stage. The female dies after laying her eggs. The larvae then form side tunnels from their tunnel of birth. Some larvae leave the tunnel and wander on the skin. Each larva makes a small pocket in the skin in which it feeds and molts. This burrow becomes the start of a new tunnel when the mite becomes an adult. While the larvae are on the skin, they can be transferred to a new host.

The first lesions of sarcoptic mange are usually seen on the edges of the ears. The lesions are usually crusted over, and the hair is lost from the area. The burrowing of the adult mites causes an intense itching, unlike in demodectic mange. Sarcoptic mange is very contagious; infested dogs must be kept away from other animals.

Veterinarians diagnose *S. scabiei* mites by finding them from a skin scraping of the host's lesions. Some treatment is still similar to that of demodectic mange. It was generally recommended (for the treatment of scabies) to use lindane, an effective pesticide, weekly for 4 weeks to kill newly hatching larvae. The National Pediculosis Association and the FDA now recommend using it only as a last resort. A growing body of evidence shows that lindane can cause serious side effects, including death in humans. It should not be used on or near pregnant or nursing women. Ivermectin also treats sarcoptic mange, but it is not labeled for this use. More common treatments used today include clipping the hair, removing the crusts and dirt, and providing a soaking with a good antiseborrheic shampoo and an acaricidal dip. Several dips given 5 days apart are most effective in treating S. mange.

Notoedric Mange Mite (Notoedres cati)

Notoedres cati (Figure 13–8) causes mange in cats. A small mite, it looks and acts like *Sarcoptes scabiei*. Notoedric mange, marked by dry, encrusted lesions and thickened skin, usually shows up on the ears and the back of the neck.

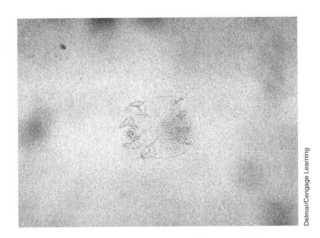

Delmar/Cengage Learning

FIGURE 13–8

The notoedric mange mite causes mange-like skin irritations usually around the ears and neck but may extend to the face and feet of the cat.

Like sarcoptic mange, notoedric mange causes intense itching and is very contagious. Since many pesticides are toxic to cats, the veterinarian should prescribe the most appropriate treatment. There are several options:

- Dipping in lime sulfur baths (but it leaves an objectionable smell on the cat)
- Amitraz (Mitaban) dips
- Ivermectin injected weekly for one month
- Selamectin (Revolution), a topical medication, although it is not labeled for this use

Ear Mite (Otodectes cynotis)

The ear mite (Figure 13–9) lives deep in the ear, usually both ears, of dogs and cats. Its feeding causes skin irritation and an increase in ear wax, and a crusty material (which looks something like coffee grounds) builds up in the ear. The animal may scratch at its ears and shake its head as it tries to get rid of the irritation. Scratching the ear can lead to swelling (hematoma) and infections. This mite can cause convulsions in cats.

FIGURE 13–9

The ear mite, which infects both dogs and cats, burrows into the skin of the ear canal, causing intense irritation, redness, and swelling.

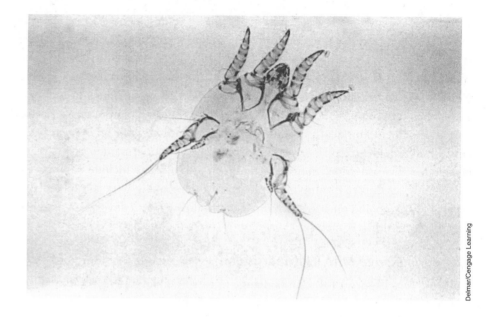

Veterinarians diagnose ear mites by use of an otoscope and by finding the mites in material removed from the ear and examining the debris under a microscope. The mites have short pedicels with suckers on the end of the front legs. Their body can resemble a bell pepper.

Treatment involves flushing out debris (along with the mites) and putting in ear medication specifically formulated for that purpose. The medications usually contain anti-inflammatories to help reduce the inflammation in the ears, antibiotics to minimize any secondary bacterial infections, and an antiparasitic to help kill the mites. It sometimes helps to put mineral oil in the ear to loosen the wax and crusty material before using the medication. Ivermectin also treats ear mites but is not labeled for this use. Other common treatments include the following:

- Selamectin (Revolution) applied topically, with three treatments 2 weeks apart

- Frontline spray (fipronil) with drops placed into the ear canal

- Acarexx (0.01% ivermectin) applied twice to the ear canal

The mosquito is an important dog and cat ectoparasite because it carries heartworm disease. (Chapter 15 covers heartworm disease.)

Ectoparasites of Rabbits and Rodents

Flies

Rodent Botfly (Cuterebra *species*)

The adult rodent botfly is a large (1 inch long), hairy fly that looks like a bumblebee. The female fly lays her eggs near a rodent's burrow. When these eggs hatch, the small larvae (1–2 mm) wait for a host. When a small mammal comes by, the larva attaches itself to the host's hair and moves toward the nose and mouth. The larva enters either the nose or mouth of the host and burrows through the lining. It then moves to a position under the skin and cuts a small breathing hole in the skin. The larva stays in this space for 2 to 3 weeks, growing to more than

1 inch. The mature larva enlarges the hole in the skin and leaves the host. The larva pupates on the ground, where the adult eventually will emerge. Although adult flies live only a few days, the pupa can survive the winter.

The rodent botfly can infest mice, rats, chipmunks, squirrels, rabbits, and many other mammals. Rodent botflies commonly infest pet rabbits kept outside. Even cats and (more rarely) dogs have been infected. The bots do little harm to their host while under the skin, but their emergence from the skin leaves an open wound prone to bacterial infection and to maggots (especially in rabbits).

Flesh Flies (Sarcophaga *species and* Wohlfahrtia vigil)

The adults of these flies are large (up to ½ inch long) and feed on decaying meat (Figure 13-10). The female flesh fly produces live larvae, which she usually lays on dead animals. The larvae feed on the rotting meat for about a week and then move to the ground, where they pupate. In some cases, the flesh fly lays her larvae in a wound or sore on any kind of animal. The larvae then feed on the tissue surrounding the wound and may end up under the skin.

Flesh fly infestations are treated by removing as many of the larvae as possible and putting an approved insecticide on the wound. Immediately treating all wounds and applying an insecticide approved for wound treatment prevents infestation.

Delmar/Cengage Learning

FIGURE 13–10

This *Sarcophaga* has a gray checkerboard pattern on its abdomen. It feeds and flies around during the day and can cover long distances because of its strong flying ability.

Lice

Rabbit Louse (Haemodipsus ventricosus)

The rabbit louse is a member of the insect order Anoplura (the sucking lice). The adult rabbit louse usually feeds on the rabbit's back, sides, and groin. The bloodsucking activity of this louse causes an intense itching. The rabbit rubs and scratches, often removing all hair from the areas where the lice are feeding. The rabbit louse prefers young rabbits as its host. A heavy infestation on a young rabbit may lead to poor growth. Although *Haemodipsus ventricosus* is not considered a zoonotic louse, it does carry *Francisella tularensis*, the etiological agent of tularemia, a bacterial disease of both rabbits and humans.

Insecticides applied to the rabbit kill the adult and nymphal lice; an infested rabbit must be isolated during treatment because the lice travel through direct host contact. Eggs, which are attached to the rabbit's hair, hatch in 7 days, and thus some insecticidal treatments require repeat applications weekly for 2 or 3 weeks. The rabbit's cage must be cleaned to remove eggs that may be attached to the hair the rabbit scratched off.

Guinea Pig Lice (Gliricola porcelli *and* Gyropus ovalis)

The guinea pig lice belong to the insect order Mallophaga (the chewing or biting lice). These are normal-sized lice; adults measure about 1 to 1.5 mm long. *Gliricola porcelli* is known as the "slender guinea pig louse"; its head is longer than it is wide, and its abdomen is only slightly wider than its head. *Gyropus ovalis* is called the "oval guinea pig louse" because its abdomen is much wider in the middle than at either end. These lice feed on skin debris and generally cause their host no problems. Guinea pigs with heavy infestations may scratch infested spots bald.

Veterinarians diagnose guinea pig lice infestations by finding either the lice or their eggs in the host's fur. Guinea pig lice are species specific and thus do not cross-infect other animal species, including humans. Treatment is the same as for the rabbit louse.

Rat and Mouse Lice (Polyplax *species and* Hoplopleura *species*)

Different species of *Polyplax* (anopluran or sucking lice) and *Hoplopleura* feed on mice and rats, but these lice are grouped together because their life cycles are similar and their treatment is the same. Infested rats and mice appear restless, have ruffled fur, and may constantly scratch. A heavy infestation of these bloodsucking (insect order Anoplura) lice may cause anemia and even death. The rat louse carries endemic typhus (an acute infectious disease caused by *Rickettsia typhi*) and other rat diseases. The mice louse carries several bacterial diseases of mice.

Veterinarians diagnose infestation by finding the lice or their eggs in the fur. Rat and mouse lice are treated the same as rabbit lice: The infested rodents are isolated and treated with an insecticide weekly for 2 or 3 weeks. The cages are cleaned at the same time as the animals are treated.

Ticks

Although many ticks feed on wild rodents and rabbits during their larval and nymphal stages, ticks rarely turn up on pet rabbits and rodents because these animals are not allowed to run free through the areas where ticks lurk. Rabbits kept outside should be regularly examined for ticks. The ticks should be removed by a veterinary technician or veterinarian (their skin is fragile and can tear easily).

Mites

Fur Mites of Mice (Myobia musculi *and* Radfordia affinis)

Fur mites are common on wild and pet mice. These mites, which live on the skin and in the fur, feed by sucking fluids from the skin. Light infestations cause the mouse no problems. Heavy infestations may cause hair loss, itching, and damage to the skin from the mouse's scratching. Eggs, attached to the bottom of hairs, take about 8 days to hatch. The fur mite travels from mouse to mouse through direct contact.

Veterinarians diagnose fur mite infestations by finding the mite or its eggs in the mouse's fur. Plucked fur or fur removed by the use of cellophane tape is placed in a drop of mineral oil on a glass slide and examined for mites or mite eggs (see "How to Do a Cellophane Tape Preparation for Lice or Mites" in Appendix I). The infested mouse must be isolated to prevent transmission to other mice and then a pesticide is applied to the mouse. Tick powders containing the pesticide permethrin can be placed on the mouse or in its bedding. Pesticide treatment must be continued for at least 4 weeks as the eggs resist treatment and they must hatch. Alternatively, the mouse can be dosed with ivermectin. This drug also fails to kill the eggs, so ivermectin should be given twice, about 2 weeks apart.

Rat Fur Mite (Radfordia ensifera)

Like the fur mites of the mouse, the rat fur mite spends its entire life on its host. Transmission occurs when two rats are in direct contact. Rats with heavy infestations may scratch so much that they damage their skin. The measures that control mouse fur mites also work for rat fur mites.

Guinea Pig Fur Mite (Chirodiscoides caviae)

Like the rat and mouse fur mites, the guinea pig fur mite spends its entire life cycle on its host. Transmission is by direct contact between an infested guinea pig and the new host. Infestations by this mite usually cause the guinea pig no problems. They may go undetected until the owner notices these small mites walking on the fur. On rare occasions, heavy infestations may cause hair loss and itching. Veterinarians diagnose mites by finding them in the fur and by microscopic examination. A pesticide is applied to the guinea pig, with the application repeated every week for 3 weeks. Ivermectin may also be used for guinea pig fur mites.

Fur Mite of Rabbits (Cheyletiella parasitovorax)

This mite, like the fur mites of rodents, spends its entire life cycle on its host and travels by direct contact between hosts. This mite's life cycle takes about 2 weeks to complete. The rabbit fur mite (Figure 13-11) usually feeds on the skin of the rabbit's back and shoulder region. Occasionally it turns up on the skin of the belly. Light infestations have no clinical signs; heavy infestations may cause hair loss and scaliness.

Delmar/Cengage Learning

FIGURE 13–11

The fur mite of rabbits has a small, round body, less than 1 mm in size.

Veterinarians diagnose this infestation by finding the mite in the rabbit's fur and also by microscopic examination. Infested rabbits are treated with a pesticide, with treatment repeated in a week to allow the eggs time to hatch. Because some of the stages of this mite can survive off the rabbit for a few days, the rabbit's cage should be cleaned at the same time as the rabbit is treated.

Mouse Mange Mite (Myocoptes musculinus and Myocoptes romboutsi)

These mites are common on wild and pet mice. Many mice have both fur and mange mites at the same time. The mouse mange mites feed on the mouse's skin tissue, and all stages live on the mouse. It takes about 2 weeks for mouse mange mites to complete their life cycle. Diagnosis and control of these mites is the same as that for the mouse fur mites.

Mange Mite of Guinea Pigs (Trixacarus caviae)

This mange mite is similar in both life cycle and appearance to the sarcoptic mange mite. It lives in the skin of guinea pigs and may cause hair loss, crusty skin, and itching. It travels by direct contact between hosts. Veterinarians diagnose infestation by finding the mite or its eggs in a skin scraping of the infested area. Treatment is usually ivermectin. This mange mite may infest humans who handle an infested guinea pig. Although this infestation of humans will cause an itchy rash, it will go away without treatment, because the guinea pig mange mite cannot complete its life cycle on humans. Eventually all of the mites on the human will die off.

Demodectic Mange Mites of Gerbils (Demodex meroni) and Hamsters (Demodex criceti and Demodex aurati)

The demodectic mange mites are uncommon on gerbils and hamsters. Even when they are present, they rarely cause disease. All stages of the demodectic mange mite live on their host. Transmission is by direct contact between hosts, with the primary route being from mother to suckling young. There is usually no cross-species contamination. Like the demodectic mange mite of dogs, both *Demodex meroni* of the gerbil and *D. aurati* of the hamster live in the hair follicles. *D. criceti* is unusual in that it lives in small pits in the skin instead of in hair follicles. In the gerbil, *D. meroni* can cause hair loss and scaly skin on the tail and legs. The demodectic mange mites of the hamster usually cause disease only in very old hamsters under some form of stress. These hamsters suffer hair loss and scaly skin. Usually these signs appear on the infested hamster's back.

Veterinarians diagnose infestation by finding the mite in scrapings of the skin's diseased areas. The mites live deep in the hair follicles and sebaceous glands. Mitaban (Amitraz) dips (in a diluted dosage) are sometimes effectively used in treating gerbils.

Ear Mange Mites of Rodents (Notoedres muris and Notoedres notoedres)

The ear mange mite of the rat (*Notoedres muris*) and the ear mange mite of the hamster (*N. notoedres*) are small mites that look like the ear mite of cats. They live in the skin of a rat's or hamster's ears, nose, and tail. All stages of these mites live in the skin. It takes about 2 weeks for the mite in the hamster to complete its life cycle; the mite in the rat takes an additional week. These mites travel by direct contact between hosts. Infestations with the ear mange mite of rats cause the rat's skin to become red and thickened. There may be wartlike growths on the rat's ear or nose. The disease in the hamster is the same as in the rat.

Veterinarians diagnose infestation by scraping the affected areas of the skin. This scraping contains all stages of the mite from eggs to adults. The infested rat or hamster is treated with injections of ivermectin and must be kept from other animals during treatment to prevent transmission of the mite.

Ear Mite of Rabbits (Psoroptes cuniculi)

The ear mite is the most common ectoparasite of rabbits. This mite lives mainly in the rabbit's ear, but in heavy infestations it may travel to the skin of the head and neck. All stages of the mite live in the rabbit's ear. Transmission is by direct contact between rabbits. This mite irritates the skin of the ear, causing tissue fluids to leak through the skin. The mite feeds on these fluids, causing a dry, crusty material (coagulated serum) to build up in the rabbit's ear. The ear becomes very itchy; the rabbit may shake its head and scratch its ears. This scratching may cause cuts prone to bacterial infection. Most commonly associated as a secondary infection with these mites is *Pasteurella multocida* (pasteurellosis, also known as snuffles).

Veterinarians diagnose rabbit ear mites by finding them in some of the crusty material taken from the rabbit's ears. The standard treatment for infestations of the rabbit ear mite is to clean the crusty material out of the rabbit's ears and put mineral oil in the ear daily for about 3 to 4 weeks. The mineral oil smothers the mites. Heavier infestations may require medicated ear drops. Ivermectin has also been shown effective for ear mites. Any scratches are treated with an antibiotic ointment. Since the crusty material from the ear may contain mites or mite eggs, the rabbit's cage must be cleaned to eliminate any of this material that the rabbit may have scratched out of its ear.

Tropical Rat Mite (Ornithonyssus bacoti)

The tropical rat mite lives on rats, mice, hamsters, and many other mammals—even some birds. The larvae and adults of this mite suck blood. When not sucking blood, these mites leave the host and hide in the bedding or elsewhere in the cage, where they lay their eggs. The life cycle takes about 2 weeks to complete. Heavy tropical rat mite infestations can result in anemia and even death from blood loss. In the wild, this mite often carries murine (pertaining to or affecting mice or rats) typhus (*Rickettsia typhi*) and plague (*Yersinia pestis*). These mites attack humans who handle an infested rodent. Some humans are allergic to the tropical rat mite's painful bite.

Veterinarians diagnose rat mites by finding them on the rodent or in its cage. These mites are large (about 1 mm long) and white. Engorged with blood, they become red to black in color.

Tropical rat mite extermination concentrates on the animal's environment since the mite spends much of its time off the host. To kill any feeding mites on the infested animal, the animal is treated as one would an animal infested with mouse fur mites. The treated animal is moved to a clean cage at the start of treatment. The old cage and the area around it are cleaned thoroughly and treated with a pesticide.

Ectoparasites of Ferrets, Foxes, and Mink

Flies

The fox maggot fly (*Wohlfahrtia vigil*) is a flesh fly that lays larvae on the intact skin of young foxes and mink. The larvae can penetrate the skin and feed inside the host. As few as five larvae in a body cavity will kill a young fox. Mink and fox ranchers have found that they can prevent fox maggot attacks by putting an insecticide in the animal's bedding. The young fox or mink rolls around in its bedding, distributing the insecticide throughout the fur. The insecticide then kills any larvae the flesh fly lays on the skin. Placing screening around the animals' cages helps keep the adult fox maggot fly from the young animals in the first place.

The rodent botfly (*Cuterebra* species) infests ferrets and often feeds on mink and foxes. The treatment and control measures are the same as those for infestations of the rabbit.

Fleas

Ferrets can be infested with the same fleas (*Ctenocephalides* species) as dogs and cats. (Refer to the earlier section on ectoparasites of dogs and cats for the life cycle of these fleas.) A ferret infested with enough fleas has scaly skin and shows signs of being itchy. The infested ferret may scratch off the hair from some areas of its body. Ferret flea infestations are treated like those of cats. Both the animal and the house should be treated. A mild flea shampoo or powders can be used for dusting the ferret, but do not use a collar containing dichlorvos (an insecticide) and anthelmintic, which can be toxic to ferrets. The skin of the ferret is sensitive, and thus the ferret can become toxic easily.

Ticks

The same ticks that infest dogs and cats can also infest ferrets, foxes, and mink. Ferrets allowed to roam outside are more likely than an animal kept indoors to get ticks. Careful, regular examination of the ferret's fur will reveal any attached tick. The ticks can be removed manually.

Mites

Ferrets are susceptible to the ear mite (*Otodectes cynotis*) and the sarcoptic mite (*Sarcoptes scabiei*), which also feed on dogs and cats. Ferrets pick up these mites by contact with infested animals. Some ferrets with ear mites may display the same clinical signs as cats do; others show no outward signs. A ferret infested with ear mites, however, has a buildup of brownish black wax in its ears. Ferrets infested with sarcoptic mites develop itchy, scabrous skin lesions that cause hair loss. These lesions appear on the body (sometimes in just a few places, sometimes covering the body) or only on the feet. Severe mite infestations on the feet can make the ferret's claws fall out.

The infested animal should be isolated during treatment to keep the mites from spreading to other animals. The same pesticides used in cats work with ferrets, but care should be taken because some are not proven to be safe for ferrets. The lesions caused by the sarcoptic mite can be treated with corticosteroid ointment to stop the itch.

Ectoparasites of Reptiles

Flies

Adult flesh flies of turtles (*Cistudinomyia cistudinis*) look like houseflies. The adult female fly lays live larvae into wounds in the turtle's skin; a favorite site is the hole left by a tick. The larvae feed on the turtle's skin, eventually growing to 15 mm in length. The larvae feed on the turtle for about 2 months before they drop off and pupate on the ground. The damage these larvae do to the turtle's skin can be considerable and can kill the turtle. This fly is common in the southeastern United States. Live larvae can be found on box turtles, painted turtles, and gopher turtles in this region. To treat flesh fly infestations, the larvae must be removed from the skin wound. The wound is washed with an antiseptic and antibiotic is applied. Myiasis of turtles can be prevented by keeping them indoors or in a fly-proof cage.

Cutaneous myiasis is frequently caused by the botfly in turtles. The fly creates a cutaneous wound and lays its eggs, which then hatch and mature in the wound site. Veterinary technicians or veterinarians should remove the bots with forceps, flush the wound site, and apply antibiotic ointment. Prevention is similar to that described previously for flesh flies.

Ticks

Iguana Tick (Amblyomma dissimile)

This hard tick attacks only **ectotherms** (cold-blooded animals). The iguana tick often turns up on iguanas caught in the wild and also on rattlesnakes, gopher snakes, and eastern fence lizards caught in the southeastern United States. This tick is seen rarely in domestic iguanas.

This tick irritates the skin and may leave an open wound. It is removed by grasping it with blunt forceps as close to the head as possible and pulling straight out. Treating the wound with antibiotic ointment prevents infection.

Gopher Tortoise Tick (Amblyomma tuberculatum)

This is one of the largest hard ticks. It lives on mammals and birds as a larva and on the gopher tortoise as a nymph and adult. Gopher tortoises captured in the wild often carry the gopher tortoise tick. A fully fed adult female gopher tortoise tick can grow to 25 mm long. Like the iguana tick, the gopher tortoise tick irritates the skin and may leave an open wound. Treatment is the same as for the iguana tick: The tick is grasped with blunt forceps as close to the head as possible and pulled straight out, with the wound treated to prevent infection.

Other reptile ticks belong to the *Aponomma* species and *Hyalomma aegyptium*. Hard ticks of the genus *Aponomma* often turn up on snakes and lizards imported from Africa and southern Asia. *H. aegyptium* is a hard tick that favors Greek tortoises imported from southern Europe. All of these ticks can irritate the skin. Treatment is the same as that for the iguana tick.

Mites

Snake mites (*Ophionyssus natricis*) are found on many captive snakes. This mite sucks blood as a nymph and adult, when it lives under the scales and around the eyes of its host. The female adult mite lays her eggs on the ground, where the larvae hatch and live. The complete life cycle takes 2 to 3 weeks. These mites irritate the snake; heavy infestations may cause anemia or even death. This mite also carries *Aeromonas hydrophila*, which causes a bacterial disease.

Veterinarians diagnose snake mites by finding them on the snake or in their water dish; the first tipoff is usually the mite's white feces. An infested snake is treated with ivermectin. Its cage must be thoroughly cleaned to eradicate eggs and larvae, and all beddings and shavings must be disposed of.

Ectoparasites of Chickens, Turkeys, and Other Birds

Flies

Although flies are not generally a problem with chickens, many bloodsucking flies (like black flies and mosquitoes) feed on birds and can transmit disease. Flesh flies lay their eggs around wounds on birds, and the larvae can cause myiasis. If flies present a problem, chickens and other birds should be kept in screened cages.

Lice

Chicken Body Louse (Menacanthus stramineus)

This chewing louse is common on chickens, which transmit it by direct contact with each other. It lives and feeds on the skin of the breast, thighs, and anus region. The female glues its eggs in masses around the base of the feathers. The chicken body louse sometimes chews through the skin and causes bleeding. Infested chickens do not gain weight as fast because of the mite's irritation. Heavy infestations may kill the chicken. Veterinarians diagnose infestation by finding the louse or its eggs on the chicken. Infested birds are isolated and treated with an insecticide (dusting powder, sprays, and ivermectin).

Shaft Louse (Menopon gallinae)

The shaft louse is a common ectoparasite of chickens and other fowl. It lives on the feathers of the breast and thighs. This louse usually causes the bird no problems, but heavily infested birds may become restless. Shaft louse infestations are treated the same as chicken body louse infestations.

Fleas

The stick-tight flea (*Echidnophaga gallinacea*) turns up on chickens and turkeys in the southern United States, usually in older, poorly managed poultry houses. This flea can also feed on dogs and cats that live in or near a ranch farmhouse. Stick-tight fleas are small—less than 2 mm long. The female attaches itself permanently to the chicken, usually around the bird's eyes, comb, and wattles. The eggs of the female stick-tight flea fall to the ground, where they hatch. The larvae and pupae live in the chicken house litter. Adults emerge from the pupal case, wait for a host to come by, and then jump onto it. The embedded flea can cause skin irritation and localized swelling. If enough fleas infest a chicken's eye area, this tissue reaction can cause blindness. Heavily infested chickens can also suffer anemia and decreased egg production.

Veterinarians diagnose stick-tight flea infestation by finding the female fleas embedded on the host's head. To control stick-tight flea infestations, infested birds are treated with an insecticide (dusting powder). The chicken house floor is cleaned and insecticide is applied to kill the larvae. Repeat treatment may be necessary until all eggs have hatched or been removed.

Ticks

The fowl ticks (*Argas* species) are soft ticks that feed on chickens and other birds. The larval fowl ticks feed for about 1 week on the bird. The larvae then drop off and molt into nymphs. Both nymphs and adult fowl ticks feed on the bird at night and then leave it to hide in nearby cracks and crevices during the day. The adults may feed up to eight times (this may be over several days as they are preparing to reproduce). The female lays a batch of eggs in the hiding place after each blood meal. The fowl tick can live 2 to 4 years without a blood meal.

The feeding ticks annoy the birds, causing egg production to decrease. The fowl tick also causes anemia; heavy infestations may even kill the bird. The adult fowl tick can cause tick paralysis in some of its host chickens. Fowl ticks carry avian spirochetosis (*Borrelia anserina*), a bacterial disease of birds. Veterinarians diagnose fowl ticks by finding the feeding ticks on the bird. Since the nymphs and adults feed at night, the veterinarian may have to examine under the bird's wings, the nest, or any cracks or crevices near the bedding in order to find the ticks.

Chickens are treated with pesticides to kill any larval fowl ticks on them. The chicken house is cleaned and treated with a pesticide to kill the adults and nymphs. Particular attention should be paid to the cracks and crevices of the chicken house. Since the eggs take from 1 to 3 weeks to hatch, treatment may have to be repeated several times.

Mites

Northern Fowl Mite (Ornithonyssus sylviarum)

This mite, the most common ectoparasite of U.S. poultry, feeds on domestic fowl and wild birds. The adults are about 1 mm long and dark red when engorged with blood. The entire life cycle takes place on the host over the course of about 1 week. Female mites glue their eggs to the feathers; these eggs may hatch in 1 or 2 days. The larvae do not feed. Adult mites may not be on the host if the bird is heavily infested. An adult northern fowl mite can survive off the host for about 1 month. They feed intermittently off the host.

The northern fowl mite likes to feed around the vent of the bird. The feathers in this area become matted and discolored. Heavily infested chickens may show a decreased egg production, anemia, and scabs in the area where the mites are feeding. Infested chickens are treated with a pesticide; if the infestation is heavy, the chicken house should be treated as well. Wild birds, which may be infested with this mite, should be kept away from the chicken house. Farmers often enclose the chicken house with screen for this purpose.

Chicken Mite (Dermanyssus gallinae)

This bloodsucking mite attacks poultry and many different wild birds. Only the nymphs and the adult chicken mites feed on birds. This mite lives off the host during the day, returning to the bird at night to feed. Like the northern fowl mite larva, the chicken mite larva does not feed. The chicken mites live in cracks and crevices of the chicken house during the day and they lay their eggs here as well. This mite may also feed on rats and other rodents if birds are not present. The chicken mite will also bite humans. Chickens infested with this mite may have decreased egg production and anemia. Heavily infested chickens, especially young birds, may die. Like the fowl tick, the chicken mite carries *Borrelia anserina*. Veterinarians diagnose chicken mite infestation by finding the mites on the birds at night. To control the chicken mite, the chicken house is cleaned and treated with a pesticide.

Scaly-Leg Mite (Knemidocoptes mutans)

The scaly-leg mite lives on poultry and domestic and wild birds. All stages of this mite live on the bird; it travels from bird to bird by direct contact between hosts. This is a small mite that looks like *Sarcoptes scabiei*. The scaly-leg mite lives in the skin of the bird's legs and feet. It lays its eggs in tunnels in the skin; all stages live in these tunnels. The burrowing mites cause irritation, which leads to swollen legs with crusts on their surface. If untreated, this irritation may deform the leg and cause lameness.

Veterinarians diagnose the presence of the scaly-leg mite by loosening the crust on the leg with warm vegetable oil, scraping the crust, and finding the mites. Infested birds are treated by applying warmer vegetable oil and removing as much of the crust as possible, then treating the infested area with a pesticide. Birds should be isolated and treated with ivermectin. This treatment is repeated 7 days later.

Turkey Chigger (Neoschoengastia americana)

A **chigger** is a larval mite that feeds on animals. As this ectoparasite's name suggests, only its larvae feed on turkeys; all other stages are free-living organisms. The chigger feeds on the turkey for 4 to 6 days and then drops off. Turkey chiggers prefer to feed on the breast and thighs and under the wings. After they drop off, scabs may develop where they fed. The larval stage is the only developmental stage of the chigger that can cause parasitic disorders in domestic animals, wild animals, and people. Treatment for chiggers usually concentrates on the soil where the adults and nymphs live. A pesticide is applied to the turkey yard.

CASE STUDY	Canine Demodicosis (*Demodex*)

A 5-month-old puppy from the pound has bald spots by its ear, eye, and right shoulder and it is very thin. The owner, who just adopted the puppy, is concerned about the bald spots and wonders if they are contagious. There is also some crusting and redness by the bald spots.

DIAGNOSIS

A skin scraping is done to check for eggs, nymphs, and adult *Demodex* mites. Scraping should be done until the capillary blood oozing is seen because these mites live deep in the hair follicles and sebaceous glands. The sample is placed on a slide in mineral oil and examined under the microscope for eggs, nymphs, and adult mites.

The clinical signs of demodicosis are alopecia, crusting, and **erythema** (redness of the skin due to capillary dilatation) of the skin. In severe cases, there may be secondary pyoderma. There are two types of demodicosis. In localized *Demodex,* the clinical signs are in just a few places, such as the head, forelimbs, ears, or trunk of the dog. In generalized *Demodex,* lesions may be all over the dog's body. These dogs usually develop this condition because they are immunosuppressed or immunodeficient. The immunosuppression allows the mites to increase in numbers. Demodex mites are part of the normal fauna of the skin. *Demodex* is not contagious. The mother can transfer the mites to the puppies within the first 3 days of life. Stillborn puppies will not have mites. In this case study, the owner of the puppy cannot get *Demodex.*

Large animals will have nodular lesions, which should be incised with a scalpel. The caseous material from the nodular lesion is smeared on a slide that has mineral oil. A cover slip is used to cover the sample on the slide and examined microscopically for *Demodex* eggs, nymphs, and mites.

TREATMENT

Shampoo, gels, or ointments with benzoyl peroxide, such as Pyoben, provide good follicular flushing action (flush out the mites and debris), which can help prevent secondary bacterial infection. Antibiotics can be used in cases with secondary pyoderma. A miticidal dip, amitraz (Mitaban), is done every 2 weeks for 6 weeks. This can vary depending on how severe the case is. Treatments are continued, until two consecutive skin scrapings are negative. The skin scrapings are done 1 month apart.

Summary

Although many of the same parasites that infest large animals also infest smaller companion animals, their effect on them is quite different at times. Pet owners need to be concerned about zoonotic disorders that many times pets spread to owners and their families. There are many new diagnostic procedures and treatments available for parasitic infestations of small animals. Powders, shampoos, as well as topical, injectable, and oral compounds are used to control many ectoparasite infestations. Separating wildlife from companion animals will also greatly reduce reinfestation.

REVIEW

MULTIPLE CHOICE

1. Maggots found under the skin, near an open wound, on a rabbit are probably the larvae of a

    a. flesh fly.

    b. horn fly.

    c. botfly.

    d. rodent botfly. *(circled)*

2. The best way to treat guinea pig lice is to

    a. apply insecticide to the guinea pig weekly for 2 or 3 weeks.

    b. pick off visible lice and apply insecticide.

    c. apply insecticide to the guinea pig and isolate it for 2 or 3 weeks.

    d. apply insecticide to the guinea pig, isolate it, and clean the cage. *(circled)*

3. The deer tick can be the vector of

    a. Rocky Mountain spotted fever.

    b. Lyme disease. *(circled)*

    c. cycle fever.

    d. sarcoptic mange.

4. A dog with mange has alopecia only on its face. The mite infesting the dog is most likely

    a. *Demodex canis.* *(circled)*

    b. *Otodectes cynotis.*

    c. *Sarcoptes scabiei.*

    d. *Notoedres cati.*

FILL IN THE BLANKS

1. _____Demodex canis_____ mange in dogs is not contagious in an animal with a healthy immune system.

2. _____Northern fowl mite_____ is the most common ectoparasite of U.S. poultry.

3. A favorite laying site for the adult flesh fly of turtles is a hole left by a/an _____Tick_____.

4. _____ear mites of rabbits_____ is the most common ectoparasite of rabbits.

5. *Cuterebra* larvae, also known as _____Rodent Botfly_____, are a newly hatched form of a large, hairy fly that looks like a bumblebee.

6. _____Pustules_____ are pimple-like growths containing pus.

7. _____Chigger_____ is a larval mite that feeds on animals.

8. _____Erythemia_____ is an abnormal redness of the skin caused by dilatation of blood vessels.

DISCUSSION QUESTIONS

1. To control fleas on a dog that spends most of its time in the house, is it sufficient to kill the adult fleas on the dog? Why or why not?

   No, the killing the fleas on the animal only solves part of the problim. There may be eggs around. You must clean the area and their bedding.

2. How might a veterinarian diagnose the presence of the scaly-leg mite in avian species?

   loosening the crust on the leg with warm vegetable oil, scraping the crust and finding mites.

CHAPTER 14

Endoparasites of Large Animals

OBJECTIVES

Upon completion of this chapter, the reader should be able to:

- Describe the life cycle of the nematode, cestode, trematode, and *Giardia*.

- Describe common diseases caused by endoparasites in large animals.

- Describe common diagnostic tools and procedures for endoparasites in large animals.

- Describe common treatments and preventive measures for endoparasites in large animals.

KEY TERMS

arrested larvae

ascarid

cestode

coccidia

Cryptosporidium

cyst

fluke

Giardia

lungworm

nematode

Ostertagia

proglottid

strongyles

tapeworm

threadworm

trematode

trophozoite

whipworm

Endoparasites of Laboratory and Farm Animals

An endoparasite lives inside its host. Although many endoparasites live in the gut of their host, they can turn up in the lungs, the blood, or almost any other area of the body. The level of internal parasitism can vary from year to year depending on various weather and feeding patterns. Realizing that diagnosing and deworming procedures can be costly to livestock producers, the veterinary staff needs to be cognizant of the economics and economic benefits with each recommendation made. Advice should be based on sound scientific data made about a producer's method of operation and after monitoring the herd/flock's specifically timed, fecal egg per gram (epg) (quantitative) counts. This advice will help improve livestock production and reduce the cost related to the deworming program.

Fecal egg counts estimate pasture contamination and the need for deworming and/or pasture rotation. The McMaster slide technique is one of several tests used to determine epg counts. It is an easy technique available to most veterinary clinics.

Nematodes

Nematodes, or roundworms, are generally small, wormlike organisms with a tough skin (cuticle). They can be found in almost any tissue of the body. Some common areas are the skin, kidneys, lungs, urinary bladder, nervous system, blood, and, the most common area, the intestines. The nematode life cycle (Figure 14–1) has three stages: egg, four-larval, and adult. Depending on the species, adults may measure from 2 mm up to 30 cm long. Nematode larvae grow to the next stage by molting. Parasitic nematodes are endoparasites as adults.

Tapeworms

Tapeworms, or **cestodes,** are flat, ribbonlike worms with no digestive tract. The tapeworm has a head, or scolex, and a body composed of many segments called **proglottids.** The tapeworm life cycle (Figure 14-2) has three stages: egg, larva, and adult. Depending on the species, an adult tapeworm may measure from 5 mm to more than 7 meters long. All tapeworms are endoparasites with two hosts in their life cycle: the intermediate host, where the larval tapeworm lives, and the definitive host, where the adult tapeworm lives. The adult tapeworm always lives in its host's intestines. Tapeworm proglottids have muscles that help them to move about.

Trematodes

Trematodes, or **flukes,** are flat, leaf-shaped worms with a partial digestive tract. They have a mouth and an intestine but no anus. They have a complex life cycle (Figure 14-3): egg, several larval stages (some of which are endoparasites of snails and crayfish), and adult. The adult trematode can be from 2 mm to 10 cm long. Most are endoparasites. A few adult trematodes are ectoparasites of fish. *Gyrodactylus,* which lives on the skin of fish, is an example of an ectoparasite trematode. *G. salaris* resides on the skin of Atlantic salmon,

FIGURE 14–1

Life cycle of a nematode.

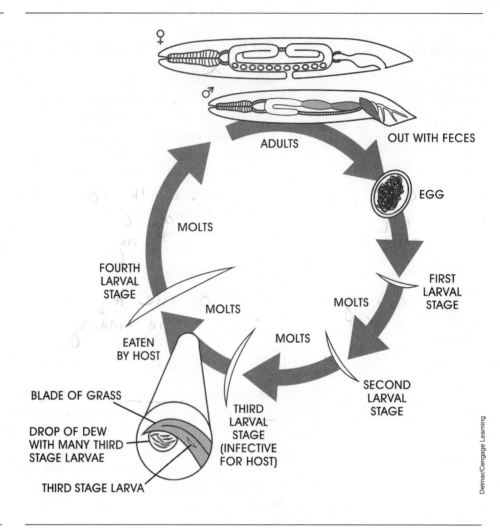

FIGURE 14–2

Life cycle of a tapeworm.

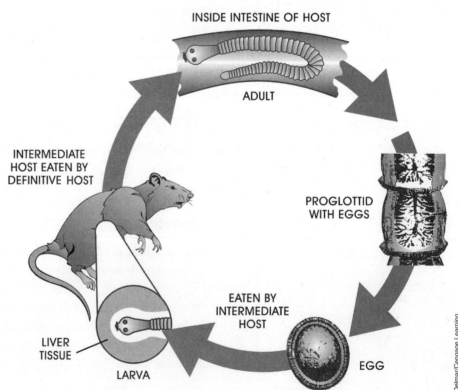

FIGURE 14–3

Life cycle of a trematode.

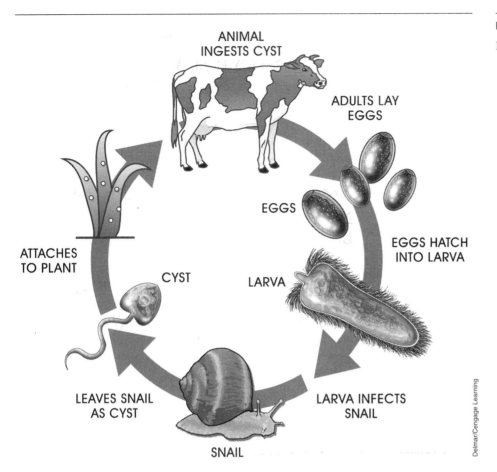

ANIMAL
INGESTS CYST

ADULTS LAY
EGGS

EGGS

EGGS HATCH
INTO LARVA

ATTACHES
TO PLANT

CYST

LARVA

LEAVES SNAIL
AS CYST

LARVA INFECTS
SNAIL

SNAIL

Delmar/Cengage Learning

rainbow trout, lake trout, and grayling. It receives its nourishment from epithelial cells and mucus on the skin surface. It destroys the skin surface, and eventually the host dies.

Protozoans

Protozoans, single-cell animals, comprise a large group of organisms made up of many different types. Most protozoans are microscopically small. Each type of protozoan has a different life cycle, ranging from one to many stages. Figure 14-4 shows the life cycle of a *Giardia*. Because protozoans are only one cell, they have no adult or egg stages. The ameboid, vegetative, or asexual stage of a protozoan is sometimes called a **trophozoite** and the infectious stage is sometimes called a **cyst**. Parasitic protozoans are endoparasites. A fecal float using zinc sulfate is sometimes used to detect protozoan cysts (Figure 14-5). Lugol's iodine can be used to aid in the visualization of the internal structures of trophozoites and cysts (Figure 14-6).

Endoparasites of Ruminants (Cattle, Sheep, and Goats)

Because the major endoparasites of cattle, sheep, and goats include a number of parasites that infect all three types of animals, these hosts are grouped together.

FIGURE 14–4

Life cycle of *Giardia* (protozoan).

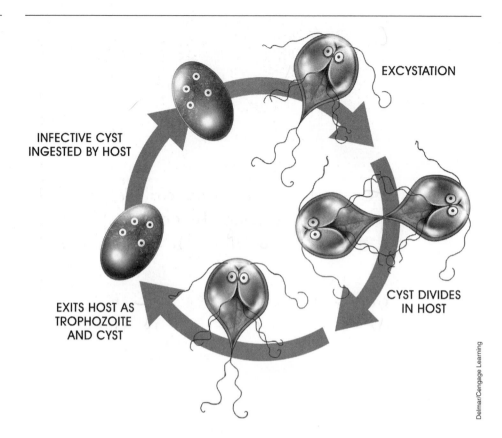

INFECTIVE CYST
INGESTED BY HOST

EXCYSTATION

CYST DIVIDES
IN HOST

EXITS HOST AS
TROPHOZOITE
AND CYST

Delmar/Cengage Learning

FIGURE 14–5

A fecal float using zinc sulfate is
sometimes used to detect protozoan
cysts.

Delmar/Cengage Learning

Nematodes

Abomasal Worms (Haemonchus *sp.,* Ostertagia *sp., and* Trichostrongylus *sp.*)

Ostertagia, Haemonchus, and *Trichostrongylus* are important endoparasites of the abomasum (stomach) of cattle, sheep, and goats. The life cycles of these three nematodes are similar. The first- and second-stage larvae are free-living organisms. The host eats the third-stage larvae, starting the infection. Adults of all three worms live on the surface of the mucosa (the lining of the stomach).

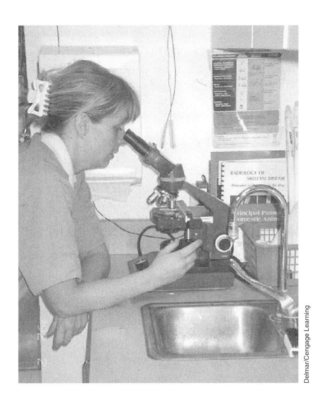

Delmar/Cengage Learning

FIGURE 14–6

Lugol's iodine can be used to aid in the visualization of the internal structures of trophozoites and cysts.

Both the larvae and adults of *Haemonchus* suck blood. A thousand *Haemonchus* adults can suck 50 ml of blood per day, causing severe anemia. A heavy *Haemonchus* infection (20,000–30,000 worms) can kill a sheep very quickly. Sheep and goats of all ages are susceptible to *Haemonchus* infection; cattle over the age of 2 usually resist it.

Trichostrongylus larvae greatly damage the stomach's mucosa. The young adults may cause bleeding and small ulcers when they break out of the stomach lining. Heavy infections cause diarrhea and weight loss. Even light infections can decrease growth and diminish appetite.

Ostertagia is probably the most important roundworm parasite of livestock. *Ostertagia* larvae invade the stomach's gastric glands. There the worms molt and develop and then return to the stomach. The larvae and young adults destroy some gastric gland functions, including acid production. Without acid, the stomach cannot digest food. Severe diarrhea and a large loss of weight are signs of *Ostertagia* infection. Heavy infections may be fatal. To survive harsh conditions that exist outside their host, the fourth-stage *Ostertagia* larvae may slow their development. These are known as **arrested larvae**. In the northern United States and Canada, *Ostertagia* larvae arrest just before winter and resume development in the early spring. In the southern United States, the larvae arrest in the late spring and resume development in the autumn. These patterns of arrest ensure that adults do not lay eggs when free-living larvae would be unable to survive: in the cold winters of the North or the hot, dry summers of the South. These developmental patterns result in two types of *Ostertagia* disease: type I and the more severe type II.

Type I disease is caused by unarrested larvae, a few at a time, destroying the gastric glands. It normally occurs during the first grazing season.

Type II disease is caused by a group of arrested larvae destroying the gastric glands upon resuming their development. Type II normally occurs in older animals during the northern spring and the southern autumn.

Ostertagia, Haemonchus, and *Trichostrongylus* infections often occur simultaneously in an animal. Since their life cycles are similar, control measures for all three parasites are the same. To prevent heavy infections, veterinarians recommend prophylactic doses of anthelmintics, timed to prevent heavy buildup of larvae on the pasture. In the northern United States and Canada, for example, veterinarians prevent severe type II disease by prescribing anthelmintics from late spring to midsummer. This ensures that adult worms die before they lay eggs; thus, no arrested larvae hatch the following spring. Another control method is to move the livestock to a different pasture every year. Arrested larvae in the vacated pasture do not survive long enough to infect a host.

Intestinal Strongyles (Cooperia *sp. and* Nematodirus *sp.*)

The free-living portion of this nematode's life cycle resembles that of the abomasal nematodes. They lay eggs in the gut; the eggs pass out with the feces. The first- and second-stage larvae are free-living organisms. The host eats the third-stage larvae, starting the infection. Adult nematodes live in the host's small intestine. These parasites do not usually cause their hosts many problems, but heavy infections, especially in a host with other parasites, may cause diarrhea and weight loss. Veterinarians diagnose **strongyles** by finding the eggs in the feces. Anthelmintics treat the infection. The control measures and prophylactic treatments for abomasal nematodes also work for intestinal nematodes.

Lungworms (Dictyocaulus *sp.,* Protostrongylus rufescens, *and* Muellerius capillaris)

All three types of **lungworm** live in sheep and goats, but only *Dictyocaulus* is found in cattle. Adult lungworms live in the host's lungs. *Dictyocaulus* is a large nematode, up to 8 cm long. *Protostrongylus* is a moderate-sized worm, up to 3 cm long. Both of these lungworms live in the bronchi (lung's air passages). *Muellerius* is also known as hair lungworm of sheep and goats. *Muellerius,* at about 2 cm, is also a moderate-sized worm; it lives in the lung tissue. All these lungworms lay eggs that hatch in the host's lungs. The first-stage larvae migrate up or are coughed up the trachea (windpipe). When they reach the back of the mouth (the top of the trachea), the host swallows them. The first-stage larvae then pass through the gut and exit the animal with the feces.

Dictyocaulus larvae grow and molt on the ground until they become third-stage larvae. Third-stage larvae climb up onto a blade of grass, to be swallowed by the host when it eats the grass. Slugs or snails, while feeding on grass, eat the first-stage *Protostrongylus* and *Muellerius* larvae. The larvae grow and molt inside the slug or snail (the intermediate host) to the third stage. The sheep or goat inadvertently eats the slug or snail as it feeds. Digestion frees the third-stage lungworm larvae from their intermediate host. They burrow into the intestinal wall and get into a blood vessel. The blood eventually carries them to the lungs, where they break out of the blood vessel and develop to adult lungworms.

In cattle, light lungworm infections usually cause few problems. The cow may cough a little and breathe faster than usual. Heavy infections usually happen only to calves. Heavily infected calves may have trouble breathing because the large worms may block the air passages of the lungs. The calf will breathe rapidly and cough frequently, and it may have so much trouble breathing that it does not have time to feed. Lungworm infections in sheep and goats are usually light and cause few problems. Goats sometimes suffer heavy *Muellerius* infections. A goat with a heavy lungworm infection may have a persistent cough and trouble breathing.

Veterinarians diagnose lungworm infection by finding the first-stage larvae in the feces. In a heavily infected calf, the worms may block the air passages before they lay eggs. In these cases, diagnosis depends on the signs (coughing and rapid breathing) alone. The veterinarian can give the infected animals anthelmintics to kill the adult worms. Control of snails and slugs on the pasture may prevent *Protostrongylus* and *Muellerius* infection. Calves can be protected from heavy lungworm infections by placing them in a pasture that has not been grazed in the current year. Such a pasture should have few, if any, larvae.

Tapeworms

Tapeworms have two hosts in their life cycle: the intermediate host, where the larval tapeworm lives, and the definitive host, where the adult tapeworm lives. Ruminants are the definitive hosts for some tapeworms and the intermediate hosts for others.

Ruminant Tapeworm (Moniezia *sp.*)

The adult ruminant tapeworm lives in the small intestine of cattle, sheep, and goats. It is a large worm—up to 600 cm long and 2.5 cm wide. Like most of the other common tapeworms, its last proglottid, when full of developed eggs, breaks off the body and passes out with the feces. Often the proglottid breaks up, releasing its eggs, as the host passes the feces. The feces thus have both intact proglottids and eggs. Free-living mites, which are the ruminant tapeworm's intermediate host, eat the eggs on the ground. These mites live on the ground and on blades of grass, where a ruminant inadvertently eats them as it grazes. If the ruminant eats a mite with a larval tapeworm in it, the larva develops to the adult stage in the ruminant's small intestine. The ruminant tapeworm turns up in young animals during their first grazing season.

This tapeworm usually causes its host no serious problems. Heavily infected young animals may be unthrifty and suffer diarrhea. Veterinarians diagnose tapeworms by finding the eggs or proglottids in the ruminant's feces. Anthelmintics treat tapeworms. Farmers and ranchers with ruminant tapeworm problems often plow the pasture in the fall, killing many of its resident mites. The grass that grows back the following spring will have few infected mites to be ingested by the new calves, lambs, or kids.

Taeniid Tapeworms (Taenia *sp.*)

Ruminants are the intermediate hosts for a number of taeniid tapeworms, whose definitive host is usually a dog. The typical larval taeniid, about the size of a large pea, lives in a ruminant's internal organ or muscle. The beef tapeworm of humans is a taeniid tapeworm whose larvae reside in cattle muscles. As the name suggests, adult taeniid tapeworms can live in humans as well as dogs.

The taeniid tapeworm reaches its definitive host as follows: A ruminant inadvertently eats the taeniid's eggs or proglottids; then a dog (or a human in the case of the beef tapeworm) eats raw meat infected with the larvae. Tapeworm adults develop in the definitive host's small intestine. Proglottids and eggs pass out of the definitive host in the feces.

Although the larvae of these tapeworms cause ruminants no problems, meat inspectors condemn larva-infested meat, making it worthless. Worse, a taeniid tapeworm of sheep has a larva that travels to the sheep's brain, where it can grow to the size of a golf ball. This larva destroys brain tissue as it grows, eventually affecting the sheep's movement. The circling and lack of balance that result are called true gid.

Since there is no easy way to diagnose larval tapeworms before slaughter and no treatment for the infection, the farmer or rancher must try to prevent the problem. Since the dog is the definitive host for most of these taeniid tapeworms, dogs not needed to herd sheep should be kept off the pasture; this keeps the tapeworm proglottids and eggs they may carry off the pasture as well. Dogs needed for sheepherding respond well to prophylactic treatment for the adult tapeworm, as do humans. Treating infected cowhands before they go to the pasture has virtually eliminated the beef tapeworm from the United States. Another prophylactic measure is to keep dogs from eating raw meat; cooking kills the larvae.

Hydatid Tapeworm (Echinococcus granulosus)

The hydatid tapeworm's definitive host is the dog. Ruminants, horses, and humans that inadvertently eat the hydatid tapeworm's eggs all serve as its intermediate host. Sheep raised on pasture with sheepherding dogs are prone to larval hydatid infection. This tapeworm is named for its larval stage: the hydatid cyst, a large (up to 10 cm), fluid-filled ball typically in the liver or lungs of the intermediate host. The small larvae grow on the inside wall of this cyst. Ruminants with only a few cysts rarely show signs of infection, but their livers are condemned at slaughter.

Hydatid cysts are difficult to detect before slaughter. Ultrasound finds cysts in living animals but is too expensive to be practical except for small herds at high risk for the parasite. The hydatid cyst, like the taeniid tapeworm, has no practical treatment in ruminants, but hydatid cysts can be prevented with the same measures that prevent taeniid tapeworm infections.

Trematodes

Liver Fluke of Ruminants (Fasciola hepatica)

This large trematode can grow up to 3 cm long. Liver fluke adults, as their name suggests, reside in the liver, specifically the bile ducts, of a ruminant, horse, or human. Trematode eggs (Figure 14–7) travel with the bile to the intestine, eventually leaving the host in the feces. The eggs hatch only in water, where the first larval stage infects a snail. The larvae go through several stages in the host snail, multiplying at each stage. Many larvae then leave the snail, find a plant growing

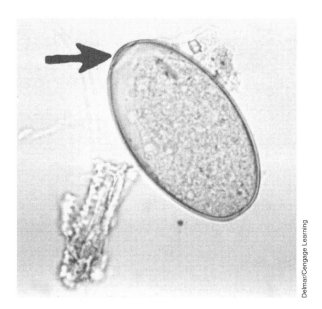

FIGURE 14–7

The trematode egg is spindle shaped with a lid on one end (arrow). It has a smooth, thin shell.

in the water, and climb above the water level on its stem. There they change into a cyst. A ruminant feeding on plants at the pond's edge eats the cysts along with the plants. The cysts hatch in the ruminant's small intestine and burrow through the intestinal wall. They travel to the liver, penetrate it, and move through the liver until they find a bile duct, where they mature to adulthood.

A sheep can suffer fatal liver damage if it eats many cysts in a short time. Fortunately, this rarely happens. Most sheep and other ruminants suffer only minor damage as a few larvae at a time move through the liver. Over time, however, many larvae move through the liver, leaving scar tissue that takes up much of the liver and reduces its function. The larval fluke can also transmit black disease, a bacterial liver infection. Meat inspectors condemn fluke-infected livers at slaughter.

Veterinarians diagnose the liver fluke by finding its eggs in the feces. Anthelmintics kill liver flukes. Wet areas where the intermediate host snail lives can be fenced off to prevent ruminants from eating cyst-carrying plants.

Deer Liver Fluke (Fascioloides magna)

The deer liver fluke has the same life cycle as the liver fluke of ruminants. The deer liver fluke makes its home in deer, cows, and sheep and behaves differently in each. The large (up to 10 cm long) adult flukes live in cysts attached to a deer's bile ducts. In the cow's liver, these cysts do not attach themselves to the bile ducts. In the sheep, the larva wanders through the liver, growing but never forming a cyst or reaching adulthood. Although meat inspectors condemn fluke-infected cow livers at slaughter, these flukes cause deer and cattle no problems. A sheep, however, can die from even one deer liver fluke larva. This happens because the larva grows large and continually moves through the sheep's liver, destroying much of the liver tissue. It is difficult to raise sheep in areas where deer are infected with this fluke.

Veterinarians cannot diagnose deer liver fluke infections in cattle or sheep; the eggs do not come out with the cow's bile, and no eggs form in the sheep. Some anthelmintics kill the adult fluke in deer and cattle. This parasite is controlled as you would the liver fluke of ruminants: Livestock can be kept from feeding on plants that may carry the cysts, and efforts can be made to keep deer from the livestock pastures.

Protozoans

Coccidia (Eimeria *sp.*)

The **coccidia** of ruminants live in the cells of their host's intestine. There are several different species of coccidia, which can live in each of the different ruminants. Some of these cause disease; others do not. All coccidia are species specific—that is, an *Eimeria* species that infects cattle does not infect any other host, and the same is true for the coccidia of other animals.

Coccidia have a complicated life cycle. They go through many stages and invade many different cells of the host's intestine. During most of these stages, the coccidia multiply, so a few coccidia can start an infection that grows to millions of parasites. When the coccidia finish multiplying in the host, they form oocysts that pass out with the feces. The oocyst is the infective stage. When a new host accidentally eats it, the oocyst hatches in the new host's small intestine and then invades the cells of the intestinal wall.

Coccidiosis usually turns up in younger animals, especially those under stress. Disease outbreaks commonly strike cattle in feedlots and sheep just after shipping. Severe winter weather may be all it takes to turn a calf with a light coccidia infection into a calf suffering from bloody diarrhea. Infected sheep and goats may develop diarrhea, but it is usually bloodless. At greatest risk are young goats; heavy coccidia infections can kill them.

Veterinarians diagnose coccidial infection by finding the oocysts in the feces. Because the diarrhea may occur before oocysts are being shed, however, a veterinarian may diagnose coccidiosis based on only the age of the host and the diarrhea. Several drugs treat coccidiosis. Veterinarians give them to young ruminants about to be stressed (shipped or put into a feedlot, for example) to prevent disease.

Cryptosporidium

Cryptosporidium (Figure 14-8) is another coccidian parasite that infects many different mammals, including ruminants, horses, and humans. It has a complicated life cycle during which it lives in cells of its host's intestine.

Infected animals may suffer a very watery diarrhea. The oocysts that leave the host in the feces are small. Examination by a direct fecal smear using a modified acid-fast stain is helpful to see the oocysts and parasite ova. A fecal float is also used to detect sporulated oocysts. Currently no good treatment exists for *Cryptosporidium*. Supportive therapy such as rehydration and correction of acidosis are helpful in recovery efforts.

Tritrichomonas foetus

This protozoan parasite lives in a cow's vagina and uterus and on a bull's penis. Like all other protozoans, it multiplies in its host. Even one organism can lead to a heavy infection. Transmitted during sexual intercourse, this protozoan does not harm the bull but can cause the cow to suffer early abortion and infertility.

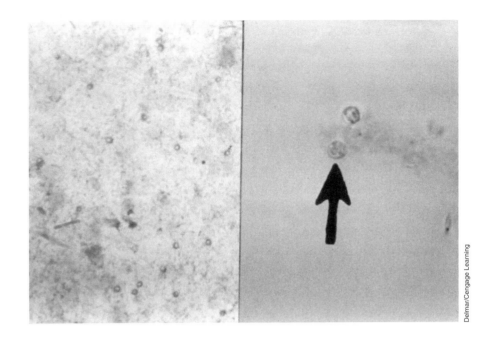

Delmar/Cengage Learning

FIGURE 14–8

Cryptosporidium is a protozoan parasite of the intestine of large animals. It causes severe diarrhea.

Veterinarians diagnose *Tritrichomonas foetus* by finding it on the bull's penis, in the cow's vagina, or in an aborted fetus's tissues. The drugs that kill this protozoan may not always work in bulls. This is one of the few parasites for which there is a vaccine. The animal should receive the vaccine before it is first bred. Cows artificially inseminated with semen from an uninfected bull do not contract this parasite, since it is a sexually transmitted disease. If artificial insemination is not an option, the rancher should use a young bull to service the herd. Young bulls are less likely to be infected.

Giardia

Giardia, described in more detail in Chapter 15 in the section on the protozoans of dogs and cats, can live in a ruminant's small intestine and can cause diarrhea in young animals.

Sarcocystis

Sarcocystis is a protozoan that lives in the muscles of ruminants. It uses the ruminant as an intermediate host. The dog is the definitive host for *Sarcocystis*. (See the section in Chapter 15 on the protozoans of dogs and cats.)

Endoparasites of Swine and Pet Pigs

Nematodes are the major endoparasites of swine. The **ascarid** (roundworm), the **whipworm**, and the nodular worm of pigs are the most common endoparasites of pet pigs. Commercially raised swine have few parasites because they are raised indoors on hard floors, and their feces and urine are removed daily.

Nematodes

Pig Ascarid (Ascaris suum)

Ascaris suum (Figure 14-9) has a life cycle almost identical to the *Parascaris equorum* of horses. In heavy infections, the larvae migrating in the liver may cause

FIGURE 14–9

The egg of *Ascaris suum* is brown and contains a central mass (more easily seen at higher magnification, as on the bottom).

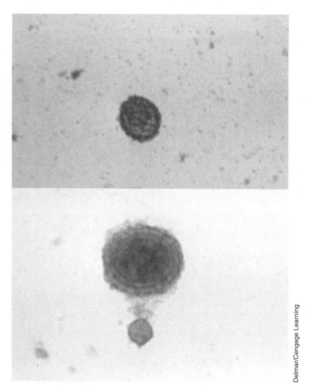

Delmar/Cengage Learning

white spots to appear on its surface. Adult worms are as large as the horse ascarid and can compete with the pig for food. A heavily infected pig does not put on weight as easily as a healthy pig.

Though *A. suum* causes its host little harm, the farmer trying to sell the pig for food wants to maximize its body weight and prevent spots on its liver. Prevention is important from the farmer's point of view. If the pigs are being raised on concrete, it is easy to remove feces before the eggs become infectious. Several anthelmintics kill the pig ascarid, which may be a problem for young pet pigs raised with older infected pigs. Older pet pigs are usually the only pig in the household; thus, they are unlikely to become infected since they do not encounter the worm's eggs.

Trichinella spiralis

Trichinella spiralis (Figure 14-10) is another economically important nematode parasite of swine. This parasite causes the disease trichinosis in humans and other mammals. Adult *Trichinella* live in the mucosa of the small intestine. Female worms give birth to first-stage larvae, which enter blood vessels, travel to the body's muscles, and penetrate a skeletal muscle cell. There the larvae coil up and may live for many years. The larvae enter a new host's digestive tract when the host eats this muscle. The host digests the muscle tissue, thereby freeing the larvae. The larvae molt four times and then become adults.

Pigs are infected when they eat uncooked meat containing first-stage larvae. Infection spreads through the herd in a number of ways. Infected meat can come from the uncooked garbage some swine are fed. Pigs raised in crowded conditions may bite off and eat each other's tails. They may eat other pigs that have died in the pen. Finally, pigs may eat live or dead infected rats.

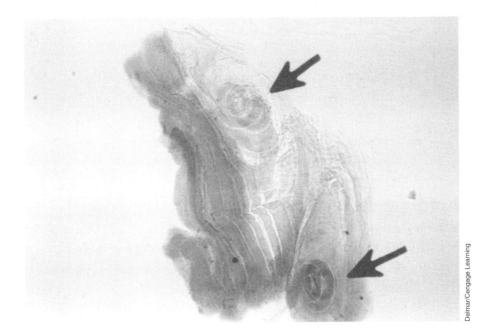

FIGURE 14–10

A crushed piece of muscle tissue in which two larvae of *Trichinella spiralis* of pigs are encysted (arrows).

Trichinella infections are normally light, causing the pig little harm. But farmers cannot afford to allow *Trichinella* into the herd, since pork containing *Trichinella* larvae is not fit for human consumption. Anthelmintics kill *Trichinella*, but infected pigs are hard to identify since the infection causes the pig no problems. Rat control helps prevent this infection; so does cooking garbage to kill any larvae it contains before pigs eat it. *T. spiralis* is not a problem in pet pigs.

Nodular Worm of Pigs (Oesophagostomum *sp.*)

These small nematodes (about 12 mm long) live in the pig's large intestine. The pig eats food off the ground and inadvertently consumes the infective larvae. The larvae burrow into the wall of the large intestine and cause nodules to form in the wall. The larvae live in these nodules for about a week. Then they emerge from the nodules and develop to adults in the intestine. Eggs pass out with the feces and hatch on the ground. The larva goes through several free-living stages before it becomes infective.

Light to moderate infections of the nodular worm cause the pig few problems. Heavy infections can cause diarrhea and anorexia (the pig does not want to eat). Young pigs with heavy infections may show severe signs; some may die. This worm can infect any livestock or pet pigs raised outside. Veterinarians diagnose nodular worms by seeing the clinical signs and finding the eggs in the feces. The infection responds to anthelmintics. The same anthelmintics prevent infection in pigs sharing an area with infected pigs.

Pig Whipworm (Trichuris suis)

The whipworm is a common parasite of livestock and pet pigs raised outside. The adult whipworms are moderate-sized nematodes (5 cm) that live in the large intestine. The adult whipworm is so named because its thick back (one-third of its body) and narrow front (the remaining two-thirds) make it resemble a whip: The back is the whip's "handle," the front its "lash." The adult whipworm burrows into the intestinal wall until only its "handle" remains in the intestine.

The whipworm feeds on blood from the intestinal wall. The whipworm's eggs, which pass out in the feces, are easy to recognize: They are brown and lemon shaped, and have a plug on either end. The eggs mature on the ground until the infective larvae develop inside the egg. When the pig inadvertently eats the egg off the ground, the egg hatches in the intestine, and the larva develops into an adult in 6 to 7 weeks.

Whipworm infections can cause diarrhea and blood in the feces. The infection responds to anthelmintics. Because the eggs can survive on the ground for several years, the pigs may need regular retreatment.

Lungworm of Pigs (Metastrongylus sp.)

The adult pig lungworm is a moderate-sized nematode (3–5 cm long) that lives in the air passages of the pig's lung. There, the lungworm lays eggs, which the pig coughs up and swallows. The eggs pass out of the pig with the feces. The larvae in the eggs on the ground develop only when eaten by the lungworm's intermediate host, an earthworm. In the earthworm, the larvae develop to the infective stage. The infective-stage larvae can live in the earthworm for many years. When the pig eats the infected earthworm, the larvae enter the intestinal wall and get into the blood. The blood carries them to the lungs, where they break out of the blood vessels and go to the air passages.

Pig lungworm disease is usually mild. The pig may cough and become unthrifty. The lungworm, since it needs an earthworm intermediate host, turns up only in pigs raised outside on dirt. Veterinarians diagnose lungworm infection by finding the eggs, each of which has a larva, in the feces. Anthelmintics kill lungworms. Lungworm infections can be prevented by keeping the pig away from earthworms.

Stomach Worm of Pigs (Hyostrongylus rubidus)

The adult stomach worm is a small nematode (7–10 mm long) that lives in a pig's stomach, where it lays eggs that pass out with the feces. The eggs hatch on the ground. The larvae undergo two free-living stages and then become infective. The pig inadvertently eats the infective larva, which enters the wall of the pig's stomach and develops to adulthood. The adult worm comes out of the stomach wall and lives in the stomach, where it sucks blood from the stomach wall.

The feeding adults and embedded larvae may damage the stomach lining and cause ulcers. Heavily infected pigs may be unthrifty and may have diarrhea. The infection responds to anthelmintics. Stomach worm infections are prevented by raising pigs off dirt and removing the feces regularly.

Kidney Worm of Pigs (Stephanurus dentatus)

The adult of the kidney worm is a moderate-sized nematode (3–5 cm) that lives in the kidney and walls of the ureters, the vessels that drain the urine from the kidney. Kidney worm eggs pass out of the pig in the urine. The free-living stages are the same as those of the stomach worm. The infective larva may enter the pig by accidental ingestion or by burrowing through the skin. No matter how they enter, the larvae travel to the pig's liver. There they move around for more than 3 months while they mature. Once mature, the larvae travel from the liver to the peritoneal cavity and then on to the kidney and walls of the ureters, where they mature to adulthood. This process may take anywhere from 6 to 19 months because of the long time the larvae spend moving through the pig.

Though the adult worms usually cause no problem, the larvae can damage the liver while moving through it. Liver damage can lead to unthriftiness and, in heavy infections, to ascites (fluid in the peritoneal cavity). If mature worms are present, veterinarians can diagnose kidney worm infection by finding the eggs in the urine. The long developmental period and the damage done by the larvae, however, may result in unthrifty pigs with no eggs in their urine. In these cases, veterinarians base their diagnosis on the history of the herd: If the kidney worm has turned up previously, it may still be in the pigs. This infection responds to anthelmintics. Free-living larvae are killed by drying out the soil in which they live.

Threadworm of Pigs (Strongyloides ransomi)

The parasitic adult threadworm is a small (3–4 mm) nematode that lives in the small intestine. Parasitic adult threadworms are all female; they do not need males to produce fertile eggs. Threadworms have an unusual life cycle for a parasitic nematode. Eggs pass out in the feces and hatch. Threadworms grow to male and female adults in moist soil. These free-living adults lay eggs that hatch in the soil and develop to infective larvae. The infective larvae burrow through the pig's skin. Some of the larvae that enter the pig become dormant in the tissues. The rest burrow into a blood vessel and travel to the lungs. There, these larvae break out into the air spaces and then migrate up the trachea. When they reach the back of the mouth, the pig swallows them. In the small intestine, the infective larvae mature into female adult worms. Dormant larvae in a pregnant pig may "wake up" and find their way into the small intestine of the fetus, maturing to adults after the piglet is born. Dormant larvae in a nursing pig may also "wake up," make their way to the mammary gland, and enter the piglet in the milk. Larvae passed in the milk go right to the small intestine and mature.

The adult worms in the intestine can cause diarrhea and anorexia, especially in young piglets. Heavy infections can kill suckling pigs. Veterinarians diagnose threadworm infection by finding the egg in the feces. The threadworm egg has a larva in it when it is passed in the feces. There are anthelmintics that kill the adult threadworms in the small intestine but none that kills dormant larvae. If a pregnant pig receives certain anthelmintics just before she gives birth, however, they kill the larvae that have awakened and started moving toward the fetus or the mammary gland. Since free-living threadworms need moist soil, pigs raised on a dry surface do not suffer from threadworm infection.

Tapeworms

There are no important tapeworms that live in the pig as adults. One tapeworm of humans, the pork tapeworm (*Taenia solium*), however, uses the pig as an intermediate host. The pork tapeworm lives in the small intestine of humans as an adult. The larval stage lives in the muscles of the pig. The pig gets the infection by eating the eggs or proglottids with eggs in them. The proglottids are found in the feces of infected humans, who get infected by eating raw pork. This infection in pigs is rare in the United States, since good sanitation facilities prevent the swine from encountering the proglottids. The infection in pigs causes them no problem.

Veterinarians cannot diagnose larval infection in a living pig, nor is there any treatment for it. Meat carrying pork tapeworm larvae is condemned at slaughter.

Protozoans

The coccidia of pigs (*Eimeria* species and *Isospora suis*) have life cycles like the coccidia of the ruminants. Also like the ruminant coccidia, not all of the coccidia of swine cause disease. *I. suis* can cause severe diarrhea in newborn pigs. Though veterinarians diagnose coccidia by finding the oocysts in the feces, sometimes a newborn's diarrhea occurs before the oocysts are shed. In those cases, finding the oocysts in the mother pig's feces or in the feces of other pigs in the same area may explain what is happening to the newborn pigs. There are drugs that treat coccidia infections in pigs. Good sanitation and sufficient living space help prevent severe disease.

Endoparasites of Horses

Nematodes

Ascarid of Horses (Parascaris equorum)

Parascaris equorum, the ascarid of horses, is very large. Males may be up to 30 cm, and females can grow to 50 cm. Adults of *P. equorum* live in the horse's small intestine, following a life cycle similar to that of dog and cat ascarids.

Eggs develop on the ground for 10 to 14 days. Eggs containing the second-stage larva can infect the horse that eats them. The roundworm eggs hatch in the small intestine. The larvae then burrow through the intestinal wall and into a blood vessel. They travel first to the liver; up to 2 weeks later, they reach the lungs. The larvae break out of the lungs and migrate up the trachea, where the horse swallows them. Back in the small intestine, the larvae develop into adults.

Adult horses usually develop immunity to ascarids and rarely have more than a few worms in their intestines. Roundworms usually cause disease only in young horses. The larvae migrating through the lungs and trachea may cause coughing. Moderate to heavy ascarid infections are a major cause of unthriftiness in foals. The adult ascarids are so large and numerous that they rob the foal of enough food, causing weight loss. In rare cases, there may be so many adult ascarids in the foal that they block the passage of food through the small intestine.

To diagnose infection, veterinarians depend on finding *P. equorum* eggs (Figure 14-11) in the feces. Many different anthelmintics kill the adult ascarids. It is better, however, to prevent heavy infections in foals. Pregnant mares are treated with anthelmintics before they give birth and are moved to a clean stall. Manure is promptly removed before the eggs become infective.

When the mare and foal are ready to be put on pasture, they should be released on a clean pasture—one on which no infected horses have grazed. *Parascaris* eggs may survive for several years, however, so finding a truly clean pasture is difficult. Therefore, infections in foals may also be prevented through prophylaxis, with anthelmintics administered when the foal is 4 weeks old and the medication repeated once every 6 weeks until the foal is 2 years old.

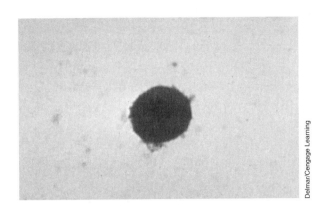

Delmar/Cengage Learning

FIGURE 14–11

The egg of *Parascaris equorum* is round and usually brown.

Strongyles

The strongyles, another group of nematodes, are also important internal parasites of horses. Strongyles come in two groups: large (2–5 cm) and small (less than 1.5 cm). The large strongyles cause more harm. Adult strongyles, small and large, live in the large intestine and cecum (a pouch at the beginning of the large intestine) of horses. They feed by biting off pieces of mucosa. A horse with a strongyle infection usually has a mixture of large and small strongyles.

The worm lays eggs in the intestine, which pass out with the feces and hatch on the ground. The first- and second-stage larvae are free-living organisms. The third-stage larvae can infect the horse. They migrate onto the blades of grass that the horse eats. At this point, the life cycles of the various strongyles begin to differ. Once swallowed, the larvae of all the small strongyles, and one of the large strongyles, molt in the large intestine and then invade the intestinal wall. The fourth-stage larvae live in the intestinal wall. The larvae eventually come back out into the intestine and become adults.

The life cycles of the three large strongyles are all slightly different. *Strongylus vulgaris* is the most important large strongyle. Once the horse swallows the third-stage *S. vulgaris* larvae, the larvae penetrate the wall of the large intestine and molt. The fourth-stage larvae enter small arteries and migrate to the cranial mesenteric artery, which supplies the large intestine with blood. The larvae live in this artery for several months before they molt to the adult stage. Through the artery, the young adults return to the wall of the large intestine. There, a nodule forms around the worm. This nodule eventually ruptures, releasing the adult into the intestine.

Neither of the other two large strongyles—*S. edentatus* and *S. equinus*—enter blood vessels. After penetrating the intestinal wall, one migrates in the abdominal cavity, while the other stays in the intestinal wall. In both cases, the young adults live in nodules. These nodules rupture, releasing the worm into the intestine.

Both larval and adult strongyles cause disease in horses, usually minor. The larvae of *S. vulgaris*, however, can cause great harm. During their migration in the cranial mesenteric artery, the larvae can damage the blood vessel. Its wall can thicken, and blood clots can form. This damage can lead to colic (severe pains in the abdominal region), bloat, and even death. By feeding on mucosa, small and large strongyle adults can cause ulcers in the large intestine. These ulcers may cause blood loss, which may lead to anemia. Both larval and adult strongyles cause unthriftiness.

Horses of any age that graze on pasture can be infected with strongyles. Foals, however, usually have more strongyles in their large intestine than do older horses. By the end of the summer, the pasture on which infected foals have been grazing is heavily contaminated with strongyle eggs. Newborn foals and their mothers should, if possible, graze on pasture the previous year's foals have not used.

Veterinarians diagnose strongyle infection by examining the feces for strongyle eggs (Figure 14-12). The eggs of all large and small strongyles look the same; they resemble the hookworm egg (Figure 14-13). The best way to control strongyles is to give prophylactic drugs. The anthelmintics dichlorvos and ivermectin kill both roundworms and horse bots. All horses should receive these or other anthelmintics every 4 to 6 weeks to control strongyle infections.

FIGURE 14–12

The egg of *Strongylus vulgaris* is similar to any other strongyle-type egg.

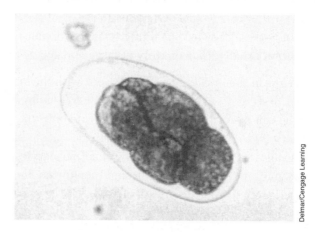

FIGURE 14–13

Compare this hookworm egg to the strongyle egg in Figure 14-12.

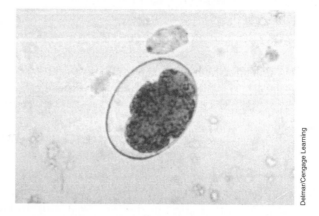

Pinworms (Oxyuris equi)

The pinworm is another nematode that lives in a horse's large intestine and cecum. Female pinworms may grow to 10 cm in length, though male pinworms rarely grow larger than 1 cm. Horses become infected by eating eggs containing third-stage larvae. The eggs hatch in the small intestine; then the larvae move to the large intestine. There, the larvae molt, and the fourth-stage larvae feed on the mucosa. In heavy infections, this may irritate the intestinal wall. Fourth-stage larvae molt and develop into adults, which feed on the contents of the intestine. Females migrate to the anus to lay eggs. They extend their bodies to the outside and lay eggs on the skin around the anus. The irritation caused by the movement of the female and the itching caused by eggs stuck around the anus are the major harm that pinworms bring. The infected horse tries to scratch by rubbing against its stall or other objects. Tail hair may break, and hair around the rump may rub off. Scratching may also damage the skin of the rump.

Veterinarians diagnose pinworm infection by observing the horse's scratching behavior and finding eggs around its anus. Sometimes the eggs turn up in the feces. The same anthelmintics that kill ascarids and strongyles kill pinworms. Prophylactic doses of drugs for ascarids or strongyles kill pinworms as well. The stable area should be cleaned regularly to remove pinworm eggs.

Threadworm of Horses (Strongyloides westeri)

The **threadworm** of horses has a life cycle much like that of the threadworm of pigs. The only difference is that the horse threadworm does not infect the fetus. The major problems threadworms cause happen to foals infected through the milk. The infected foal may develop severe diarrhea. Heavy threadworm infections can kill foals. Adult horses generally have very light infections and show no clinical signs. The same anthelmintics that kill ascarids also kill thread-worms. Feces must be removed from the foal's stall to reduce the free-living stages. Mares should be treated with a safe anthelmintic just before giving birth to reduce the number of larvae they pass to the foal through the milk.

Tapeworms

Horse tapeworms (*Anoplocephala perfoliata*, *A. magna*, and *Paranoplocephala mamillana*) are short tapeworms (5–80 cm) that live in a horse's small intestine. *A. perfoliata* lives at the junction between the small and large intestines and sometimes in the large intestine. The life cycle of these tapeworms is the same as that of the ruminant tapeworm.

P. mamillana does not cause disease in the horse. Heavy infections with *A. magna* or *A. perfoliata* can cause bleeding ulcers in the small intestine. Heavy infections with these two worms can cause unthriftiness, colic, and diarrhea. Veterinarians diagnose such infections by finding the eggs (Figure 14-14) or proglottid in the feces. There are anthelmintics that treat horse tapeworm infections. Prophylactic treatment of horses can prevent heavy infections.

Protozoans

Coccidia of Horses (Eimeria leuckarti)

The life cycles of the horse coccidia and the ruminant coccidia are the same. Heavy infections of these coccidia may cause diarrhea in young horses. Veterinarians diagnose coccidia infection by finding the oocyst in the feces. The oocyst of these coccidia does not float in a saturated salt flotation, so veterinarians or veterinary technicians must use a sugar or zinc sulfate flotation or centrifuge technique or a direct smear. Veterinary prescription for individual treatment may be warranted.

FIGURE 14–14

Eggs of the horse tapeworm. Notice the triangular shape, with the rounded corners.

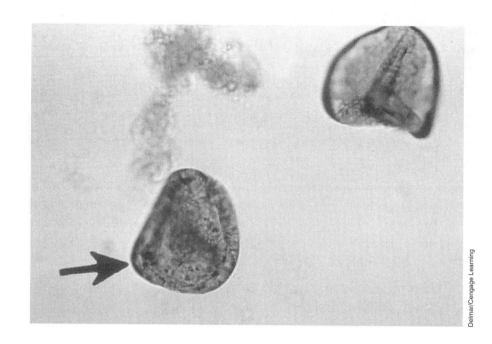

Delmar/Cengage Learning

Giardia *and* Cryptosporidium

Both of these protozoans can be found in horses. *Giardia* life cycle is described in Chapter 15. *Cryptosporidium* life cycle is described earlier in this chapter. *Giardia* causes no problems in horses. *Cryptosporidium* can cause diarrhea in young horses.

| CASE STUDY | Roundworms in Pigs (*Ascaris suum*) |

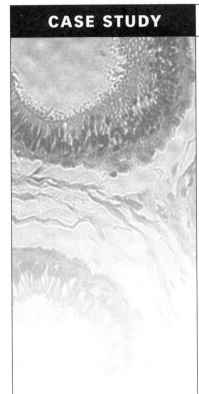

A farmer in Iowa is having problems with his herd of pigs. The pigs show signs of weight loss, difficult respiration, and unthriftiness. The farmer has two pigs with pneumonia, and one pig had died. The pig that died was icteric (yellow).

DIAGNOSIS

Ascaris suum (roundworm) are among the most destructive parasites of swine. They are common in swine, especially if housing conditions are contaminated with feces. Eggs in the feces can be diagnosed during the patent period. The clinical sign in young pigs is their difficulty in respiration during the *A. suum* prepatent period. Damage to the lungs causes a high risk for bacterial and viral pneumonia. A necropsy of a pig would show immature worms in the intestines, lung tissue, or bile ducts. The adult worms cause icterus in the pig by migrating into and occluding the bile ducts, causing liver damage.

TREATMENT

The first form of treatment is good hygiene in the pig unit. Control of endoparasites is based on anthelmintic treatments, such as ivermectin, pyrantel, dichlorvos, levamisole, or benzimidazoles. The anthelmintics can be placed in their feed.

Summary

The endoparasites of large animals are vast in both number and variety. Nematodes or roundworms, tapeworms, flukes or trematodes, and various protozoans are a few of the common groupings of endoparasites in livestock. Many of the endoparasites suck blood and can thus cause severe anemia when present in large numbers. Other parasites invade the lung, liver, heart, and stomach. The veterinary staff is well equipped to diagnose the particular parasite and then prescribe appropriate treatment. Fecal floats are just one common tool used in diagnosing the parasite.

REVIEW

MULTIPLE CHOICE

1. All but _____ are nematodes found in the abomasums of the cow.

    a. *Haemonchus*

    b. *leuckarti*

    c. *Ostertagia*

    d. *Trichostrongylus*

2. The deer liver fluke poses the greatest threat to

    a. deer.

    b. horses.

    c. sheep.

    d. humans.

3. A nematode that can be transmitted to a foal in its mother's milk is the

    a. threadworm.

    b. ascarid.

    c. pinworm.

    d. tapeworm.

MATCHING

Match the parasite on the left with the problem it causes on the right.

C   1. *Ostertagia*   a. coughing in calves

a   2. *Dictyocaulus*   b. broken tail hair on a horse

e   3. *Tritrichomonas foetus* c. destroyed gastric glands of the stomach

B   4. pinworms   d. diarrhea

d   5. *Cryptosporidium*   e. infertility in cows

FILL IN THE BLANKS

1. The ___tapeworm___ has a head, or scolex, and a body composed of many segments called proglottids.

2. ___Trematodes — flukes___ are flat, leaf-shaped worms with a partial digestive tract.

3. ___Protozoans___ are single-cell animals and thus have no adult or egg stages.

4. __Ostertagia__ is probably the most important roundworm parasite of livestock.

5. The hydatid tapeworm's definitive host is the __Dog__ .

6. __Nematodes__ are the major endoparasites of swine.

7. __Ascites__ is an accumulation of fluid in the peritoneal cavity.

8. One tapeworm of humans, *Taenia solium,* uses the __Dog__ as an intermediate host.

9. Pigs infected with *Trichinella spiralis* are difficult to identify. To minimize infection the veterinary staff should recommend __Killing rats or cooking garbage.__

CHAPTER 15

Endoparasites of Small Animals

OBJECTIVES

Upon completion of this chapter, the reader should be able to:

- Compare the life cycle of the nematode, cestode, trematode, and *Giardia* in small and large animals.

- Describe common diseases caused by endoparasites in small animals.

- Describe common diagnostic tools and procedures for endoparasites in small animals.

- Describe common treatments and preventive measures for endoparasites in small animals.

KEY TERMS

ELISA

heartworm

hookworm

host

prepatent

trematode

Endoparasites of Dogs and Cats

Important parasites of dogs and cats come from all five groups (nematodes, cestodes, trematodes, protozoa, insects) of the parasitic animals discussed in earlier chapters. Some of these parasites are **host** specific—that is, they infect only one type of host. Others are not host specific and live in both dogs and cats.

Nematodes

A number of nematodes can infect dogs and cats. Gastrointestinal nematodes are some of the most important dog and cat parasites.

Ascarids of Dogs and Cats (Toxocara canis, Toxocara cati, and Toxascaris leonina)

Ascarids of dogs and cats are large nematodes (up to 3–18 cm) that live in the small intestine of dogs and cats. *Toxocara canis* (Figure 15-1) lives in the dog, and *T. cati* (Figure 15-2) lives in the cat. *T. leonina* (Figure 15-3) can be found in both dogs and cats.

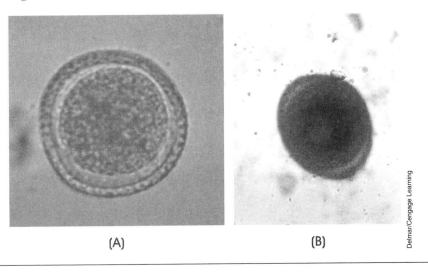

(A)   (B)

Delmar/Cengage Learning

FIGURE 15–1

A and B. The egg of *Toxocara canis*, which measures about 80 μm (micrometers). The egg in *B* focuses on the surface of the egg's shell. The egg of *Toxocara cati*, the ascarid of cats, looks exactly like that of *Toxocara canis*, except that it is a little smaller, about 70 μm. The eggs are round to slightly oval in shape. They have a rough shell and a brown central mass that fills most of the shell.

Animals become infected when they eat a *Toxocara* egg that contains a second-stage larva. An egg with a first-stage larva is not infective and does not become infective until it is outside the host for 4 weeks. It reaches the outside by means of the host's feces. An egg eaten before the 4 weeks are up does not develop into a second-stage larva.

The life cycle of *Toxocara* differs according to the age of the host. In young puppies and kittens, usually under 3 months of age, the process is as follows:

1. Infective eggs hatch in the small intestine.

2. Second-stage larvae burrow through the intestinal wall and enter a vein.

FIGURE 15–2

The anterior (front) end of *Toxocara cati*. Note the wide "wings" that run along the side of the head.

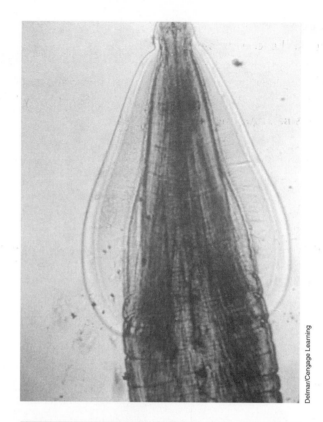

FIGURE 15–3

The egg of *Toxascaris leonina* has the same shape as that of *Toxocara*, but the surface is smooth and the central mass does not fill all of the space in the shell.

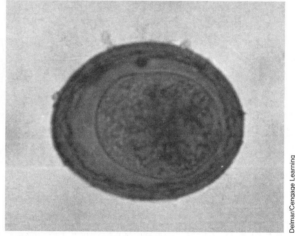

3. Blood carries the larvae through the liver and into the lungs.

4. The larvae molt in the lungs and become third-stage larvae.

5. The third-stage larvae crawl from the lungs up the trachea.

6. The larvae reach the throat, and the host swallows them.

7. The *Toxocara* return to the small intestine and stay there. They molt two more times and become adults.

It is important for the veterinary technician to understand the **prepatent** period. This is the time from the point of infection until a specific diagnostic stage can be identified. The veterinary technician is usually the individual who will explain to the client the cycles involved with these parasites. The prepatent period for *T. canis* is 21 to 35 days, and for *T. leonina* it is 72 days.

The adults mate in the small intestine. Veterinarians diagnose the infection by finding eggs in the feces. In older animals, most of the second-stage larvae do not molt in the lung. Instead, the blood carries them to any of several different organs. The second-stage larvae usually stay in that organ until they die. Some larvae do molt in the lung, however, and an older animal may have a few adult ascarids in its small intestine.

Second-stage larvae that get into the mammary glands of a nursing mother molt to third-stage larvae and then pass with the milk into the suckling pup or kitten. This type of larval passage is called transmammary transmission. Transmammary-transmitted third-stage larvae do not invade the pup's intestinal wall. Instead, they develop into adults in the intestine.

If the second-stage larvae migrate into the uterus of a pregnant dog, they can invade the lungs of the fetus, where they molt into third-stage larvae just before the pup's birth. They complete their life cycle in the newborn puppy and become adults in the small intestine. This is known as prenatal transmission.

Cats and dogs may become infected with *Toxocara* by eating its transport host, an infected rodent or bird. When a rodent or bird eats infective eggs, the second-stage larvae migrate to various organs, where they stay and do not develop. When a cat or dog eats the infected rodent or bird, the second-stage larvae develop into adults in the new host's small intestine. These larvae do not travel to other body tissues.

Light to moderate *Toxocara* infections cause little damage to the host. In heavy infections, larvae passing through the lungs may make the host susceptible to pneumonia. Coughing is one of the symptoms of this stage of the infection. Adult worms may irritate the intestine, causing diarrhea. Young puppies or kittens with heavy infections may appear potbellied, and they may not gain weight as fast as they should. Intestinal obstruction or colic that results may be life threatening to young puppies.

A fecal flotation and a direct smear are used to detect *Toxocara* eggs. *Toxocara* is zoonotic. Children are susceptible to infection because of their habit of putting objects in their mouths and thus predisposing themselves to eating dirt or soil. When they ingest the soil or dirt that contains roundworm eggs, they have a great risk of developing visceral larval migrans (VLM). The zoonotic potential to human adults is rare but can happen, resulting in VLM as well. The larvae migrate in the tissue of the abdominal organs, causing damage or even death. The larvae do not develop into adults in humans. Ocular larval migrans results when the larvae invade the eye. Faulty vision or blindness may occur. An enzyme-linked immunosorbent assay (**ELISA**) test can be used for detecting human exposure to roundworms.

Anthelmintics kill adult worms. Pups and kittens should be treated at 2 and 4 weeks of age to eliminate prenatally acquired worms and again at 2 months of age to kill any worms acquired by transmammary transmission. The mother should be treated at the same time as the pups. In addition, feces should be removed and pets prevented from eating transport hosts. Some common examples of effective anthelmintic drugs are fenbendazole (Panacur), pyrantel pamoate (Nemex), febantel plus praziquantel (Vercom), and milbemycin oxime (Interceptor).

Hookworms (Ancylostoma *species*)

The **hookworm** is another important gastrointestinal roundworm parasite of dogs and cats. *Ancylostoma caninum* is the hookworm of dogs, raccoons, wolves, foxes, and coyotes. *A. tubaeforme* is the hookworm of cats (both wild and domestic). *A. braziliense* infects both dogs and cats. *Uncinaria stenocephala* is the northern canine hookworm, which also occurs in dogs, cats, foxes, coyotes, and wolves. Hookworms, which are small nematodes (1–2 cm), get their name from their hooklike appearance. The head bends at an angle to the rest of the body.

The hookworm's life cycle begins when the egg (Figure 15-4), containing an embryo, passes onto the ground in the feces and hatches. The first- and second-stage larvae are free-living; the third-stage larvae are infective. These larvae can penetrate the skin of the host.

FIGURE 15–4

The egg of a hookworm of dogs and cats has a thin shell and a mass of cells in the center. Hookworm eggs measure 55 to 77 mm in length.

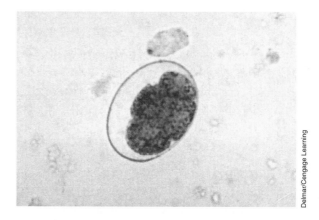

Delmar/Cengage Learning

In the skin, the third-stage larvae enter a blood vessel and travel to the lungs, where they molt and become fourth-stage larvae, which migrate up the trachea. The host swallows them, and the larvae then attach themselves to the lining of the small intestine and suck blood. The larvae molt and become adults, which also suck blood.

Hookworm infections in cats always start with penetration of the skin by third-stage larvae. *A. caninum*, however, can infect puppies by prenatal or transmammary transmission. This happens when third-stage larvae go to the dog's muscles instead of its lungs. These larvae remain dormant in the muscles. If the dog becomes pregnant, these larvae "wake up" and go to the mammary glands; in rare cases, they cross the placenta to the fetus. There is no prenatal or transmammary transmission with either *A. tubaeforme* or *A. braziliense.*

Hookworm infections in cats are not as severe as those in dogs. A heavy infection can cause severe anemia, which can be fatal in both kittens and puppies. Hookworms can also cause diarrhea that contains blood and mucus. Hookworm infection is diagnosed when eggs are found in the feces. Anthelmintics are given every 2 weeks until 2 weeks after the pup is weaned. Commonly used anthelmintics include fenbendazole (Panacur), albendazole, ivermectin and praziquantel, and pyrantel pamoate. This kills the worms that entered the pup by the transmammary route. Because they are also treating for ascarids at the same time, veterinarians use an anthelmintic that kills both hookworms and ascarids.

Whipworms (Trichuris vulpis)

Whipworms are the other major gastrointestinal nematode of dogs. Adult dog whipworms look like pig whipworms, and their life cycle is the same. Cats are rarely infected with whipworms.

Heavy infections can irritate the large intestine and cause diarrhea, sometimes with blood in it. Veterinarians diagnose infection by finding whipworm eggs (Figure 15-5) in the feces. Anthelmintics kill the adult worm. To ensure that the infection does not return, the dog must be kept away from contaminated soil. Whipworm eggs can survive for up to 2 years on the ground. A dog that keeps getting reinfected with whipworms should be put on prophylaxis. Since the whipworm has a 3-month prepatent period and the anthelmintics kill only the adult worms, the dog should be treated prophylactically every 3 months.

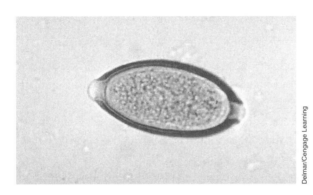

FIGURE 15–5

The surface of the shell of a whipworm egg, which is smooth.

Threadworms (Strongyloides stercoralis)

Like the threadworm of swine, this nematode is very small (2 mm), and its adult females live in the small intestine. *Strongyloides stercoralis* infects dogs, foxes, cats, humans, and nonhuman primates. The *S. stercoralis* life cycle differs from that of the swine threadworm in three ways. First, the eggs hatch in the small intestine, and first-stage larvae (Figure 15-6) leave the host in the feces. Second, some of the larvae may develop to the infective stage while they are in the host's large intestine. These larvae can burrow through the intestinal wall and find their way back to the small intestine, where they become adult females, a process called autoinfection. Finally, there are no dormant larvae in the dog's tissues. In addition, infective larvae that enter a nursing dog may find their way to the mammary glands and be passed in the milk to the puppy.

In cats and dogs over 6 months old, threadworm infections are mild, autoinfection usually does not take place, and the host's immune system kills the worms

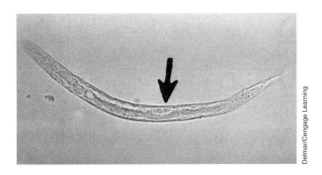

FIGURE 15–6

The first-stage larva of *Strongyloides stercoralis*. Note the tapered tail and the oval-shaped group of cells near the center of the worm (arrow).

in about 3 to 4 months. In puppies, however, autoinfection takes place. The migrating larvae may damage the lungs, causing bleeding into the air spaces. Many adult worms eventually develop and damage the intestinal wall, causing diarrhea. Veterinarians diagnose threadworm infection by finding the larvae in the feces. The prepatent period varies depending on the species of threadworm and means of transmission but generally is from 8 to 14 days. Animals become infected initially via the skin, the transmammary route, or across the placenta. A number of anthelmintics kill the adult worms. Since these anthelmintics do not kill the autoinfective larvae, however, the puppy must be retreated in 7 to 10 days.

Heartworm (Dirofilaria immitis)

One of the most dangerous nematode parasites of dogs is the **heartworm**. These long, slender worms (20–30 cm) live in the host's pulmonary artery. In heavy infections, they migrate to the right ventricle of the heart. Cats rarely get heartworm; when they do, it is a serious disease.

Female worms release embryos, known as microfilariae (Figure 15-7), into the blood. These microfilariae are taken up by a mosquito and develop into first-stage larvae. After molting twice, the larvae move to the mosquito's mouthparts. When the mosquito bites a dog or cat, the third-stage larvae enter the animal's skin. In the tissues under the skin, the larvae molt twice over several months. The young adult worms then move into veins and travel to the heart (Figure 15-8). In the pulmonary artery, they grow into full-sized adults. The prepatent period in dogs can be as long as 6 months.

FIGURE 15–7

A direct blood smear containing the microfilaria of *Dirofilaria immitis*, which is fairly large (300–320 mm long) (arrow).

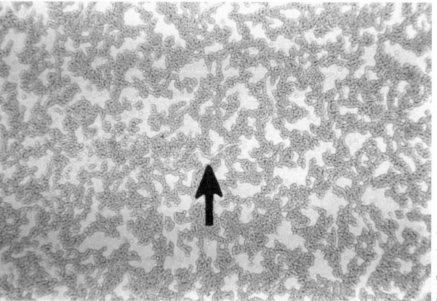

Delmar/Cengage Learning

In heavy infections, the worms in the pulmonary artery may block the flow of blood to the lungs. Because the heart must work harder to get the blood to the lungs, heart failure may occur. Signs of heartworm infection in dogs include listlessness, breathing difficulty after exercise, and coughing. Veterinarians confirm heartworm infection by finding the microfilaria in the blood, but sometimes

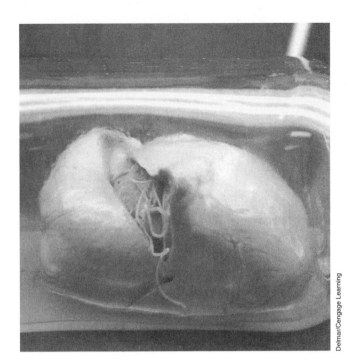

Delmar/Cengage Learning

FIGURE 15–8

Young adult worms move into veins and travel to the heart.

microfilariae are not visible in the infected dog's blood. In these cases, veterinarians complete a serological test (a test performed on blood serum) to diagnose the disease by finding heartworm antigens. A modified Knott's technique can detect microfilariae and allow for the differentiation between the *Dirofilaria immitis* and *Dipetalonema reconditum*.

Treating heartworm infection is very difficult. Anthelmintic drugs do kill the adult worms, but dead worms can block the blood vessels of the lungs. The veterinarian must keep a close eye on the dog after treatment, so dogs being treated for heartworm often stay in the hospital until the treatment is completed. Heartworms can also be removed by surgery. After the adult worms are killed, the microfilariae must be killed. Although the microfilariae have no effect on the dog, they may live for a year or more. As long as they are present, they can infect mosquitoes. Many different anthelmintics kill microfilariae.

Control of heartworm is best achieved by the use of prophylactic drugs given during the mosquito season to kill the third- and fourth-stage larvae before they get to the heart. Three prophylactics that veterinarians use for this purpose are diethylcarbamazine, ivermectin, and milbemycin. Diethylcarbamazine must be given daily and ivermectin and milbemycin once a month from 1 month after the onset of mosquito season to 1 month after the end of the season. In temperate areas, these medications can be used year-round as a preventive.

Dipetalonema reconditum

The adult *Dipetalonema* are 1 to 3 cm long and live in the skin of the dog. Like the heartworm, this worm produces microfilariae (Figure 15-9), which circulate in the blood. A flea (*Ctenocephalides*) feeding on the dog picks up the microfilariae. In the flea, the microfilariae develop to the infective stage in about 1 week. A flea with the infective larvae infects a dog by injecting the larvae into the skin as it bites. The larvae then grow to the adult stage in the dog.

FIGURE 15–9

The microfilaria of *Dipetalonema reconditum*.

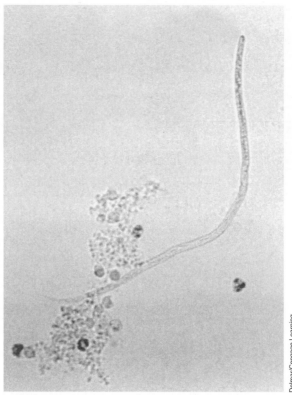

Delmar/Cengage Learning

Dipetalonema causes no problems for the dog. The only problem this parasite causes is for the veterinarian. Because the microfilariae of *Dipetalonema* live in the blood, they can be confused with the microfilariae of heartworm. It is difficult to tell which species a single microfilaria belongs to; this is a job for a parasitology expert. Veterinarians make an educated guess, however, based on the following information. Heartworm microfilariae in 1 ml of blood usually number in the hundreds of thousands, whereas *Dipetalonema* microfilariae usually number fewer than 100. *D. immitis* (heartworm) microfilariae move in place without directional motion. *D. reconditum* do have directional motion and will move across the viewing field of the slide. In addition, they have a distinguishing cephalic hook, a blunt anterior end, and in formalin fixed specimens, a "buttonhook" tail. Heartworm microfilariae are generally longer and wider than the *Dipetalonema*. Also, the serology tests for the antigen of the adult heartworm cannot detect an infection with *Dipetalonema*. Therefore, a dog with a few microfilariae in its blood and a negative heartworm antigen test falls into one of the following groups:

- It has *Dipetalonema*.

- It had heartworm, but the adults have died.

- It has a heartworm infection that is just becoming patent. If the dog retests negative for heartworm antigen about 2 weeks later, then the veterinarian knows it does not have heartworm.

There are anthelmintics that kill both *Dipetalonema* and *D. immitis* microfilariae. To prevent infection with *Dipetalonema*, the dog must be kept free of fleas, as described in the section on dog and cat fleas in Chapter 13.

Trematodes

A few **trematodes** infect dogs and cats, but these infections are rare in the United States. Some anthelmintics kill trematodes, but infection can be prevented by keeping the pet from hunting or fishing situations, where the pet may encounter the trematode's intermediate host.

The lung fluke (*Paragonimus kellicotti*) is a trematode whose adults inhabit cysts in the lungs of dogs, cats, and wild carnivores. The adults are about 1 cm long; their cysts open into the air passages of the lungs. The host coughs up and swallows the eggs, which pass out with the feces and hatch in water. Lung fluke hatchlings undergo several larval stages in snails and then a final cyst stage in crayfish. The dog or cat becomes infected by eating an infected crayfish or snail.

A pet with adult lung flukes may cough and wheeze, or have no signs at all, or die of fatal lung hemorrhage. The infection responds to anthelmintics. It is prevented by keeping pets from raw crayfish and snails.

Tapeworms

Tapeworms usually live in the small intestines of dogs and cats. Adult dog and cat tapeworms are large, anywhere from 50 to 60 cm long, but cause minimal harm to their host. Infected pets normally show no signs of infection. The first indication is usually individual proglottids on the pet's feces or around its anus.

Double Pore Tapeworm (Dipylidium caninum)

Dipylidium caninum is the common tapeworm of both dogs and cats. Fleas and occasionally dog-chewing lice are this tapeworm's intermediate hosts. Fleas become infected when the larval flea eats the tapeworm egg. Lice can become infected at any feeding stage in their life cycle. The dog or cat infects itself by swallowing the infected flea or louse.

The larval double pore tapeworm emerges from the digested insect and develops to adulthood in the host's small intestine. The adult tapeworm attaches to the intestinal wall by hooks and suckers on its scolex. Through its skin, the tapeworm soaks up its host's digested food. As the tapeworm grows, its end segments fill with eggs and break away from the worm. The eggs of *D. caninum* group together in packets of 20 to 30 (Figure 15-10). The proglottids encasing them may break and release these egg packets in the intestine (Figure 15-11). If the proglottids do not break, they pass out with the feces. The proglottids can move under the power of their own muscles, and they leave the feces. Eventually they die and release their egg packets, which larval fleas eat.

Veterinarians diagnose *D. caninum* infection by finding egg packets or proglottids in the host's feces. The egg packets each contain 20 to 30 embryos. Proglottids are white and look like cucumber seeds. A number of anthelmintics kill adult tapeworms. To prevent reinfection, the fleas and lice on the pet and in its home must be killed.

Taeniid Tapeworms (Taenia taeniaeformis *and* Taenia pisiformis)

Taenia taeniaeformis is a common tapeworm of cats. Its intermediate hosts are mice and other rodents. *T. pisiformis* is a common tapeworm of dogs. Its

FIGURE 15–10

The eggs of *Dipylidium caninum* are distinctive because they are found in packets of 20 to 30 eggs. All tapeworm eggs contain six small hooks.

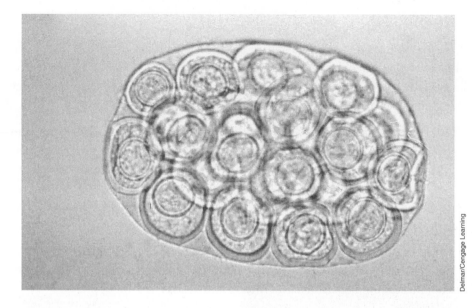

FIGURE 15–11

Proglottids of *Dipylidium caninum*. Note the two genital openings on each segment of proglottid, one on each side. These openings can be seen more easily in the specimen on the left (arrows).

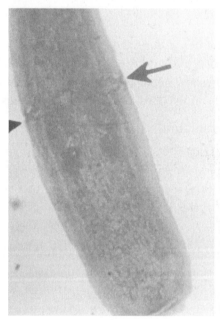
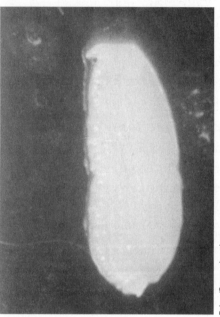

intermediate host is the rabbit. Several other taeniid tapeworms of dogs use the sheep as an intermediate host. These taeniid tapeworms are common only in dogs that are fed raw sheep or that eat sheep feces.

Veterinarians diagnose taeniid tapeworms by finding eggs (which are not in packets) or proglottids in the feces (Figures 15-12 and 15-13). The proglottids of *Taenia* are white and bell shaped. Anthelmintics kill *Taenia* adults. To prevent reinfection the pet should be kept from hunting situations.

Hydatid Tapeworms (Echinococcus granulosus and Echinococcus multilocularis)

Adult hydatid tapeworms are very small (about 5 mm long) and live in the definitive host's small intestine. *Echinococcus granulosus* adults live in dogs; the larvae live in ruminants. The section on the tapeworms of ruminants in Chapter 14 describes the life cycle of this parasite. *E. multilocularis* adults typically live

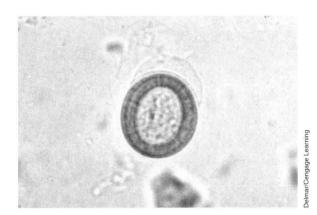

FIGURE 15–12

The egg of *Taenia* measures about 35 µm and has a thick shell that contains striations or stripes.

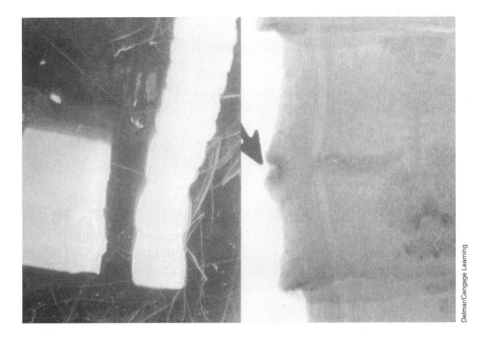

FIGURE 15–13

Proglottids of the cat tapeworm *Taenia taeniaeformis*. The segments are usually found singly, not in a chain, as seen on the left. Notice the shapes of the different proglottids. Each proglottid has only one genital opening, as shown in the stained segment on the right (arrow).

in foxes but infect cats and dogs that eat its intermediate host, the mole or other rodents. Infections with either of these two tapeworms are rare in the United States. The adult hydatid tapeworm causes dogs and cats no problems. In fact, a dog may carry thousands in its small intestine and show no clinical signs. The biggest threat this parasite poses is to the pet's owner: Either species of hydatid tapeworm can hatch and grow in the liver, lung, or other internal organs of a human who swallows its egg.

Veterinarians diagnose hydatid tapeworm infection by finding the eggs (proglottids are too small to detect) in the pet's feces. The eggs of the hydatid and taeniid tapeworms look the same, but the same drugs kill them both. Infection is prevented by not feeding dogs raw meat and keeping pets from hunting rodents and eating feces.

Protozoans

Coccidia (Isospora canis, Isospora rivolta, *and* Isospora felis)

Isospora canis is a coccidian of dogs, and *I. felis* and *I. rivolta* live in cats. These three coccidia have life cycles similar to the coccidia of ruminants. Both can infect the host as oocysts (Figure 15-14), but unlike the coccidia of ruminants, the dog and cat coccidia can use transport hosts. If a mouse or other rodent eats an oocyst, the coccidia invade its tissues and become a cyst. The dog or cat eats the rodent, and the cyst hatches to release coccidia into the cells of the pet's intestine.

FIGURE 15-14

The oocyst of *Isospora canis*, which is very similar to the oocyst of *Isospora felis*.

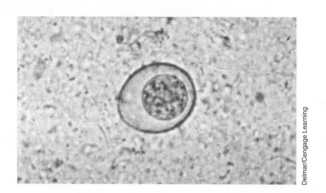

Coccidia usually cause their hosts little harm. Heavy infections (coccidiosis) in kittens and puppies, however, can cause severe diarrhea. Veterinarians diagnose coccidia infection by finding the oocysts in the suspect pet's feces. The prepatent period varies from 7 to 14 days depending on the species. The infection responds to drugs but can be prevented by keeping pets from hunting its rodent transport hosts.

Toxoplasma gondii

Toxoplasma gondii is another common coccidian of cats. Its life cycle resembles that of *Isospora felis*, except that *Toxoplasma* has various transport hosts: rodents, birds, cows, sheep, and pigs. Coccidia can invade any cell in the body of a transport host that eats their oocyst (Figure 15-15). If a pregnant woman eats an oocyst, for example, that coccidian may invade the cells of the fetus and cause birth defects. Thus, this is a zoonotic disorder. To avoid *Toxoplasma* infection, pregnant women should avoid being around cat feces and eating rare meat like

FIGURE 15-15

The oocyst of *Toxoplasma gondii* is similar to the oocyst of *Isospora*, but it is very small (10 mm) and tends to be round.

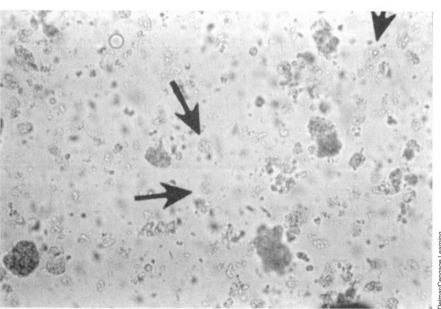

beef or pork. Gardening with gloves and washing all fruits and vegetables (especially those grown organically) are strongly recommended for prevention of *Toxoplasma* infection in women.

The cat is the only definitive host for *T. gondii* and thus can serve as a source of infection for pregnant women and their fetuses. Occasionally veterinarians are asked to euthanize the family's cat to protect the family from this potential disorder. It is thus very important that the veterinary staff be prepared to provide education to the client on how to protect the family and care for the cat. Someone other than the pregnant woman (preferably a man) should clean and empty the cat's litter box daily. This will help eliminate the exposure of the woman to the sporulated oocysts. Sandboxes should be kept covered to prevent the cats from using them as litter boxes. (This zoonotic disorder as it relates to humans is discussed more fully in Chapter 10.)

Toxoplasma causes cats few problems, but can cause extensive cell damage and eventual death to a transport host with a poor immune system. It is a grave threat to people with AIDS, cancer, or any other cause of immunosuppression. Some drugs kill *Toxoplasma* in the cat's intestine. The same drugs kill the rapidly dividing cysts, but not the "resting cysts," in the transport host, so the host's disease can be stopped but not cured. To prevent *Toxoplasma* infection, the cat should be kept from hunting and not fed raw meat.

Sarcocystis

Sarcocystis, a coccidia-like protozoan mentioned in the section on other protozoan parasites of ruminants in Chapter 14, lives in the cell lining of dog and cat intestines. Ruminants, deer, and rodents play intermediate host to different species of *Sarcocystis*, whose cyst stage lives in its intermediate host's muscles. The intermediate host infection is difficult to diagnose.

Veterinarians diagnose dog and cat *Sarcocystis* by finding the oocyst in the pet's feces. Although *Sarcocystis* can sometimes cause its intermediate host muscle pain, it causes no dog or cat disease and has no treatment. Dog and cat infections can be prevented by not feeding them raw meat. To prevent livestock infections, cat and dog feces should be kept off pastures and out of stored food. Current research indicates that *Sarcocystis* may also cause problems in immunosuppressed persons. As a precaution, individuals with a depressed immune system should follow extra precautions similar to those outlined for *T. gondii*.

Giardia lamblia

The protozoan *Giardia* also lives in the small intestine of dogs and cats. It is a microscopic, pear-shaped organism (sometimes described as looking like a tennis racket) with a sucking disk, which it uses to stick to the wall of the small intestine. It does not invade the intestinal wall but reproduces by dividing in the intestine. Large numbers can develop from one organism. *Giardia* occurs in two forms: the rarely observed trophozoite (feeding stage) and the more commonly seen cyst form, which passes out with the feces. The feces can contaminate the water with *Giardia* cysts (Figure 15-16). When an animal swallows the water, the cysts pass to a new host. Humans and many domestic and wild animals can also be infected with *Giardia lamblia*.

FIGURE 15–16

These cysts of *Giardia* can be found in both dogs and cats.

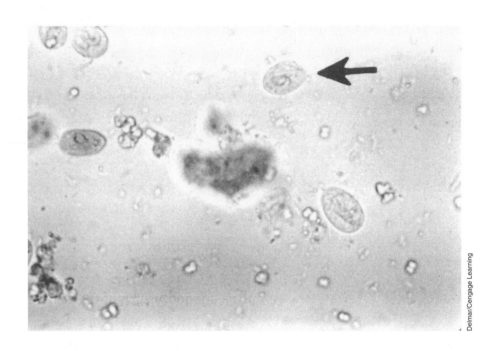

Large numbers of *Giardia* in the intestine may cause chronic diarrhea. Veterinarians diagnose infection by finding the small cyst in the host's feces (standard fecal flotation). Several drugs treat *Giardia* infections. To prevent infection, pets should be kept from drinking such potentially contaminated water sources as streams and puddles.

Endoparasites of Ferrets

Many parasites that infect cats and dogs also infect ferrets: *Toxocara cati*, *Toxascaris leonina*, hookworm, heartworm, double pore tapeworm, lung fluke, *Giardia*, *Sarcocystis*, and *Cryptosporidium*. Ferrets can be infected with ferret coccidia. The life cycle, clinical signs, diagnosis, treatment, and control measures are all the same as those for the particular parasite in dogs or cats.

Endoparasites of Rodents and Rabbits

Nematodes

The pinworms (*Syphacia* sp., *Aspiculuris tetraptera*, *Passalurus ambiguus*) are nematodes of rodents and rabbits. Like the pinworm of horses, the rodent pinworms (*Syphacia* sp., *A. tetraptera*) and the rabbit pinworm (*P. ambiguus*) live as adults in the cecum and large intestine of their hosts. The host gets the infection by accidentally eating an egg. The egg hatches, and the larvae develop to adulthood in the intestine.

Infections with pinworms are common, although these worms usually cause their hosts no problems. Veterinarians diagnose pinworm infections by finding the eggs in the feces or on the skin around the animal's anus. Anthelmintics kill the adults, but eggs in the cage make reinfection common. Therefore, the cage of an animal being treated should be cleaned thoroughly. The treatment and cage cleaning may need to be repeated several times to eliminate the infection.

Tapeworms

Wild rodents and wild rabbits can be intermediate hosts for several dog and cat tapeworms. Larval tapeworms seldom turn up in rodents and rabbits raised as pets or as laboratory animals, but a few adult tapeworms do.

Hymenolepis tapeworms live as adults (about 4–6 cm long) in their host's small intestine. A flour beetle is usually the intermediate host. Normally, these tapeworms cause their host no problem; however, very heavy infections may block food from moving through the intestine. Veterinarians diagnose *Hymenolepis* tapeworms by finding the eggs in the feces. Anthelmintics kill this tapeworm. To prevent infection, flour beetles must be kept out of the rodent's food.

Protozoans

Three protozoans live in rodent and rabbit intestines: *Giardia*, *Tritrichomonas*, and *Eimeria*. Since these parasites cause their hosts few problems, most of these infections go untreated. The *Giardia* of rodents and rabbits cannot infect humans.

The *Eimeria* coccidian is one of the few intestinal protozoan parasites that cause disease in rodents and rabbits. Like all other coccidia, the species of *Eimeria* that infect rodents and rabbits are exclusive to their host species.

The life cycle of both rodent and ruminant coccidia is the same. Several species of *Eimeria* infect the rabbit's intestine, and their life cycles are the same as those of ruminants. The rabbit, however, also has a species of coccidia (*Eimeria stiedae*) that infects its liver. The rabbit eats an oocyst of the liver coccidian, which hatches in the small intestine. The liver coccidia invade the intestinal wall and enter the blood. The bloodstream carries them to the liver, where they invade the bile duct cells. Like the intestinal coccidia, the liver coccidia multiply many times before forming oocysts. The oocysts accompany the bile to the intestine and then leave the host in the feces.

Rodents and rabbits with light to moderate coccidia infections generally suffer no problems. Heavy infections may cause diarrhea and anorexia. A rabbit severely infected with liver coccidia may die. *E. stiedae* infection is among the most common causes of rabbit death. Veterinarians diagnose it by finding the oocysts in the feces. The *Eimeria* oocysts look so much alike that only an expert can distinguish liver coccidia oocysts from those of other rabbit coccidia. Drugs (such as sulfaquinoxaline) have been used to treat rodent and rabbit coccidia, but drugs do not always work once clinical signs of infection have appeared. Veterinarians isolate a sick rodent or rabbit with oocysts in its feces, treat its infection, and prophylactically treat the other animals it may have been in contact with. Coccidia oocysts need several days outside the host to become infectious, so removing feces daily and cleaning cages frequently prevent heavy infections. Animals that recover from even a light coccidian infection are usually immune to reinfections.

Endoparasites of Reptiles

Reptiles caught in the wild have many different endoparasites; those bred in captivity generally have few. Because many reptile parasites use an intermediate host, they do not infect meat-eating snakes whose diet is restricted to mice raised away from snake feces.

Nematodes

Snake Ascarids (Ophidascaris *species*)

Adult snake ascarids live in the wall of the esophagus and stomach. Their eggs pass out with the feces. Amphibians are this parasite's intermediate hosts. When the snake eats the infected amphibian, the snake's digestion frees the larvae, and the larvae burrow into the stomach wall and develop to adulthood. Heavy ascarid infections may cause the snake to vomit. Reptile-safe anthelmintics treat this infection. Captive-bred snakes will not contract it if they don't eat amphibians.

Lungworm of Snakes and Lizards (Rhabdias *species*)

Adult snake lungworms are all females. They live in the snake's lungs, where they lay eggs that the snake coughs up and swallows. The eggs pass out with the feces and hatch on the ground. The larvae go through two free-living stages and then become infective. The snake either eats the infective larvae accidentally, or the larvae burrow through the snake's skin. The larvae then return to the snake's lungs, where they develop into adult females.

Lungworms can cause severe disease in snakes and lizards. The presence of lungworms makes it easier for bacteria to infect the lung. The animal may gape (hold its mouth wide open), a sure sign that it is not getting enough air. Veterinarians diagnose lungworm infection by finding the eggs in the feces. Lungworm infections respond to anthelmintics. Veterinarians usually give antibiotics at the same time to kill any bacteria infecting the lung. The cage of a reptile infected with lungworm should be cleaned and its feces removed daily.

Tapeworms

Snakes caught in the wild may carry a number of tapeworm larvae. No known anthelmintics kill larval tapeworms in reptiles. Several varieties of adult tapeworm infect reptiles; most cause their host no problems. Although certain drugs kill the adult tapeworms in snakes, a snake fed only animals raised away from reptile feces will not become infected.

Protozoans

The amoeba of snakes (*Entamoeba invadens*) is a protozoan that infects reptiles. The trophozoite stage of this protozoan lives in the intestine of turtles, snakes, and lizards. The cyst stage, passed in the feces, is infective. A cyst accidentally eaten by a reptile will hatch in the reptile's intestine. The trophozoite reproduces by dividing in two.

Most turtles are not severely affected by an intestinal amoeba infection, but it can kill snakes and lizards. Some of the clinical signs are anorexia and diarrhea. Veterinarians diagnose the infection by finding the cysts in the feces. Amoeba infection responds to drugs. As with lungworm infections, veterinarians treat amoeba infection with an antibiotic to kill accompanying bacteria that cause a separate disease. Turtles should never be housed with snakes or lizards. Though turtles do not get sick from this amoeba, they can still pass the cysts in their feces.

Endoparasites of Poultry and Other Birds

Nematodes

Common Ascarids of Poultry (Ascaridia *species*)

Adult ascarids are medium-sized (5–11 cm) nematodes that live in the host bird's small intestine. Eggs pass out with the feces. An infective larva develops in the egg in 5 days. The eggs enter birds in several ways:

- The bird picks up the egg while eating food from the ground.

- The egg sticks to the foot of a fly that the bird eats.

- An earthworm eats the egg, and the bird eats the earthworm.

The egg hatches in the bird's stomach and, a week later, the larva burrows into the intestinal wall. The larva eventually returns to the intestine and matures to the adult stage. The ascarids cause disease mainly in young birds. The larvae may damage the intestinal wall and cause bleeding. Light infections usually cause no problems, but heavy infections may cause unthriftiness, anemia, and diarrhea. Veterinarians diagnose ascarid infection by finding the eggs in the feces. Ascarid infections respond to anthelmintics. Poultry raised on wire have no access to the ascarid eggs unless the poultry house has a fly problem. Poultry raised on litter should have the litter changed weekly. The litter should be kept dry to slow the development of the eggs.

Cecal Worm (Heterakis gallinae)

The adult cecal worm is a small (1 cm) nematode that lives in the cecum of many different types of poultry. The eggs pass out with the feces and take about 2 weeks to become infective. The eggs get into the poultry the same ways ascarid eggs do. The larvae hatch from the eggs in the intestine and then develop to the adult stage in the cecum.

The cecal worms cause no disease in poultry but are the vector of blackhead, a disease of poultry caused by a protozoan. Veterinarians diagnose cecal worm infection by finding the eggs in the feces. The eggs look much like those of the common ascarid of poultry. Fortunately, the same anthelmintics and control measures work against both parasites.

Gapeworm (Syngamus trachea)

The adult gapeworm is a small nematode that lives in the trachea of many different birds. The male gapeworm is much smaller (5 mm) than the female (2 cm). The male remains permanently attached to the female by clasping her with his tail near the front of her body. This arrangement results in the two worms forming a Y shape. The female gapeworm lays her eggs in the trachea. The bird coughs up and swallows the eggs, which pass out in the feces. The eggs become infective in about a week and can infect the bird in two ways:

- The bird accidentally eats the egg while eating food from the ground.

- An earthworm or slug eats the egg, the egg hatches in these intermediate hosts, and the larva burrows into the tissues.

The larva can live in the intermediate host for several years. When the bird eats the intermediate host or the egg, the larva comes out in the intestine. The larva burrows into the intestinal wall and travels through the blood to the lungs. In the lungs, the larva breaks out into the air passages and moves to the trachea, where it develops to adulthood.

Heavy gapeworm infections may block the trachea, cutting off the bird's air supply. Younger birds, because they have smaller tracheas, are more prone to this problem. The gapeworm is so named for its most common clinical sign—a bird gaping while trying to get more air. Heavily infected birds may smother. Veterinarians diagnose gapeworm infection by its clinical signs and by finding the eggs in the feces. A number of anthelmintics kill the gapeworm. Infection can be prevented by raising birds on wire so they do not come into contact with the eggs or intermediate hosts. Because the larvae live so long in the intermediate hosts and because it is not possible or desirable to kill all the earthworms in an area, birds raised on the ground are always at risk. Such birds should be treated prophylactically with an anthelmintic every 3 weeks. Since wild birds carry this parasite, efforts should be made to prevent their feces from falling on the ground where poultry is being raised.

Capillaria *species*

Capillaria are small nematodes (1–2 cm) that live in the crop, intestine, or cecum of birds. Eggs pass out in the feces. In some species, a bird eats the egg, and the worms develop in the intestine or cecum. Some species develop in the intestine or cecum of a bird that has eaten the *Capillaria* egg. Other species develop to the larval stage in an earthworm and then to the adult stage in the crop or intestine of a bird that eats the worm. The predilection site (crop, intestine, or cecum) depends on the particular *Capillaria* species.

Birds with a heavy *Capillaria* infection lose weight and may have diarrhea. Veterinarians diagnose the infection by finding the eggs in the feces. Anthelmintics kill *Capillaria*; gapeworm control measures may also help prevent *Capillaria* infections.

Tapeworms

Davainea proglottina is a small tapeworm (3 mm) of the chicken. *Raillietina* are larger (12–25 cm) tapeworms that infect chickens and many other types of birds. *D. proglottina* uses a snail as the intermediate host. The *Raillietina* species use such insects as ants, flies, and beetles as intermediate hosts. The poultry tapeworms live in the small intestine of birds.

Unlike most other adult tapeworms, the adults of the poultry tapeworms can cause severe disease. Heavy *D. proglottina* infections can be fatal. The clinical signs of heavy tapeworm infection are unthriftiness and loss of weight. Veterinarians diagnose infection by finding the eggs or proglottids in the feces. Anthelmintics kill tapeworms, and good sanitation (to prevent the buildup of infected intermediate hosts) helps prevent heavy infections.

Protozoans

Coccidia (Eimeria *species*)

The coccidia are the most important parasite of poultry, but other birds can also be infected with coccidia. The coccidia of birds, like the coccidia of other animals, are host specific. The coccidia of birds live in the small intestine, large intestine, and cecum of birds. The life cycles of the bird coccidia are the same as those of ruminant coccidia. Only some bird coccidia species cause severe disease. There are, for example, six species of *Eimeria* found in chickens. Four cause severe disease, and two do not. Clinical signs of severe disease, which usually occurs in young birds and can be fatal, include bloody diarrhea and weight loss. Even those species that do not cause severe disease can cause the birds to be unthrifty. Veterinarians diagnose coccidia infection by finding the oocysts in the feces. The oocysts from different species look alike, but an expert parasitologist can tell them apart. If a bird dies, the location of the damage in the intestine may help identify the *Eimeria* species that infected the bird.

A number of drugs treat coccidia in birds, but since the coccidia have often wrought permanent damage by the time the clinical signs appear, the infected bird may never achieve the weight gain of an uninfected bird. Most poultry farmers prophylactically treat their flock by using feed that contains drugs. To ensure that no drug remains in the meat, the farmers switch to nonmedicated feed several days before slaughter. Birds raised on wire do not normally get coccidia.

Trichomonads (Trichomonas gallinae *and* Trichomonas gallinarum)

Trichomonas gallinae is a protozoan that lives in the upper portion of the digestive tract (pharynx, crop, and esophagus) of pigeons, turkeys, and chickens. In pigeons, it causes a condition known as canker. *T. gallinarum* lives in the cecum and large intestine of poultry and occasionally is found in the liver and blood. Trichomonads exist only in the trophozoite form; there is no cyst stage. An infected bird transmits *T. gallinae* when it drinks from the water dish and some of the trichomonads from its throat get into the water. An uninfected bird takes a drink from the same water and swallows the released trichomonads. An infected bird transmits *T. gallinarum* by trophozoites in its feces. These infected feces get into the water or on the food that other birds swallow.

A bird infected with *T. gallinae* looks drowsy, has foul-smelling breath, and may drool. Severe *T. gallinae* infections kill. Infected birds may be unthrifty; severely infected birds suffer depression and diarrhea. Both of these trichomonads cause more severe disease in younger birds. Veterinarians diagnose infection by finding the trophozoites in the feces or on a swab of the back of the mouth. There are drugs (such as dimetridazole) that kill *Trichomonas* in birds. This drug, however, is no longer available for this purpose in the United States. To reduce the risk of infection, wild birds should be prevented from sharing the flock's water source.

Blackhead (Histomonas meleagridis)

Blackhead is a disease caused by *Histomonas meleagridis*, a protozoan. This parasitic protozoan infects the cecum and liver of turkeys, chickens, and other fowl.

It causes the most severe disease in turkeys. Like the trichomonads, *Histomonas* has only the trophozoite stage in its life cycle. It has an unusual mode of transmission: It uses another endoparasite, the cecal worm, as a vector. The cecal worm, feeding in the cecum, takes in *Histomonas* trophozoites with its food. These trophozoites then get into the cecal worm's egg before it is laid. The cecal worm egg hatches in the bird's intestine, freeing the trophozoite, which multiplies in the cecum. The blackhead trophozoites at first feed on bacteria in the cecum, but some may invade the wall of the cecum and feed on the bird's tissues. Some of the trophozoites may travel to the bird's liver and eat liver tissue.

Turkeys (especially young turkeys) with blackhead disease become anorexic and have yellow, semisolid feces. They usually die about a week after the clinical signs appear. The name *blackhead* resulted because the combs on some turkeys' heads turn dark blue because of a lack of oxygen in the blood. This occurs in only a few birds, so the absence of a discolored comb does not rule out the disease. Chickens can be infected with blackhead, but they rarely show any clinical signs unless they are very young. Even young chickens usually recover from the infection.

Veterinarians diagnose blackhead from its clinical signs and by finding the *Histomonas* trophozoites in a dead bird's cecum. Because the trophozoites do not survive for long in the feces, they do not turn up on a fecal exam. The presence of cecal worm eggs and the symptoms of blackhead, however, aid diagnosis. Drugs kill the *Histomonas* trophozoites in the bird. Because approved drugs are changed frequently, current information on the regulatory status of anthelmintic drugs for poultry should be checked by the veterinarian and regulatory authorities.

To prevent blackhead infections, the turkeys should be given medicated feed until just before they are sent to market. Turkeys should also be regularly treated for cecal worms to prevent blackhead transmission. Moreover, chickens and turkeys should never be raised together. The chickens may carry the blackhead protozoan and the cecal worm.

Giardia

Giardia live in many different birds. The life cycle of bird *Giardia* is the same as that of dog and cat *Giardia*. The *Giardia* of birds do not infect mammals. *Giardia* infections in poultry generally do not cause the birds any problems, but these infections can cause serious disease in pet birds, such as budgerigars. *Giardia* infections are more severe, even fatal, in young birds. In these birds, the clinical signs include weight loss, depression, and diarrhea. There are drugs that kill *Giardia* in birds. Infections in overcrowded birds tend to be more severe, so birds should be kept sufficiently spaced. A bird to be added to a cage of birds should be checked for *Giardia* and treated if necessary.

| CASE STUDY | Heartworm (*Dirofilaria immitis*) |

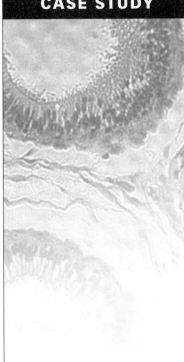

A 2-year-old cocker spaniel canine from Florida is coughing, especially after his walks in the park, and he is lethargic. The owner is concerned, especially since his dog is only 2 years old, so he takes the cocker spaniel to the veterinary hospital for an examination. The veterinarian suggests blood work and radiographs of the thorax for diagnostic procedures. The blood work (CBC/biochemistry) presents a mild anemia, and the serological test identifies adult *D. immitis* antigen. The radiographs present pulmonary artery segment enlargement and a mild lobar arterial enlargement.

The cocker spaniel is hospitalized during the adulticide administration and observed for thromboembolic complications. There are no complications after 24 hours. The owner takes the cocker spaniel home with instructions for cage confinement for 3 weeks and restriction of activity for 4 weeks.

The prognosis is good. An antigen test performed 12 weeks after the adulticide treatment is negative, and the cocker spaniel is put on heartworm prophylaxis.

CLINICAL SIGNS AND DIAGNOSIS

Heartworm disease is directly related to the amount of worms the animal is infested with, duration of the infection, and the animal's (host) response. Animals can often be asymptomatic. Clinical signs are exercise intolerance and coughing. Cachexia, syncope, and ascites can be seen in severely affected dogs. This is due to pulmonary artery embolism because of the worms. Screening tests for antigen and microfilariae are performed by taking a blood sample. Since the dog was showing clinical signs, radiographs were taken, even if the test came back negative. The radiographs might show enlarged pulmonary arteries, increased size on the right side of the heart, and possible lung disease. When a large number of heartworms are present, the veterinarian may be able to visualize them within the main pulmonary artery, or in the right side of the heart, during an ultrasound of the heart and lung area of the dog.

TREATMENT

An adulticidal treatment of melarsomine dihydrochloride (Immiticide) should be used under strict cage rest to prevent pulmonary thromboembolism. Exercise is limited to a short walk on a leash to defecate or urinate. Dogs are usually hospitalized for at least 24 hours to monitor side effects or complications, which include fainting spells called syncope, fever, coughing, and heart failure.

Microfilariae should be eliminated after the treatment of adulticidal therapy is completed, which is usually 3 to 4 weeks later. Ivermectin or milbemycin is used for this treatment. The dog should be monitored following treatment of a microfilaricide for at least 6 to 8 hours. Many veterinarians have dogs stay 24 hours for observation. Dogs may be put on fluid therapy if circulatory collapse occurs due to the rapid death of microfilariae producing systemic side effects. Three weeks after microfilaricidal treatment, a blood sample should be evaluated. The dog can be put on preventive medication for heartworm if the test is negative. If the test is positive, he should be retreated and tested again 2 weeks after.

Heartworm preventives are ivermectin or milbemycin, diethylcarbamazine, and annual retesting if the animal is at risk of infection.

Summary

The life cycle of the various endoparasites of both small and large animals is an important informational tool in the diagnosis and treatment of infestations in domestic animals. The transmission of various endoparasites between small and large animals, wildlife and domestic animals, requires the understanding of more than just small animal endoparasites by the companion animal technician. The veterinary technician must become familiar with the common diagnostic tools and procedures and be able to describe the common endoparasites to both the veterinary employer and clients. Preventive programs for the various companion and food animals need to be addressed with the client as part of an overall comprehensive health strategy.

REVIEW

MULTIPLE CHOICE

1. _____ larvae can infect kittens via transmammary transmission.

   a. Giardia

   b. Toxocara *(circled)*

   c. Sarcocystis

   d. Both b and c

2. The coccidia that infects many different animals but produces only oocysts when it infects a cat is

   a. *Toxoplasma gondii.* *(circled)*

   b. *Giardia lamblia.*

   c. *Sarcocystis.*

   d. *Dipetalonema reconditum.*

3. Control of *Davainea proglottina,* a poultry tapeworm, can be achieved by preventing the chickens from eating

   a. feces.

   b. untreated corn.

   c. snails. *(circled)*

   d. minnows.

MATCHING

Match the parasite on the right with the host site it inhabits on the left.

B 1. pulmonary artery   a. The adult of the dog hookworm

e 2. pharynx   b. The adult of the dog heartworm

d 3. trachea   c. The adult of the rabbit pinworm

a 4. small intestine   d. The adult of the gapeworm of birds

c 5. cecum   e. The trophozoite of *Trichomonas gallinae* of pigeons

FILL IN THE BLANKS

1. ___Large nematodes___ is an ascarid found in both dogs and cats.

2. The ___Prepatent___ is the time from the point of infection until a specific diagnostic stage can be identified.

3. *Dirofilaria immitis* is commonly known as ___Heartworm___.

4. The _____ Cat _____ is the only definitive host for *T. gondii*.

5. The adult _____ gapeworm _____ (*Syngamus trachea*) is a small nematode that lives in the trachea of many different birds.

6. _____ Coccidia _____ are the most important parasite of poultry.

Important Techniques for Veterinary Technicians

A compound microscope is essential for the veterinary technician and is used to evaluate urine, blood, parasites, feces, exudates, bacteria, and other body fluids.

There are two types of microscopes: a monocular microscope, with one eyepiece, and a binocular microscope, with two eyepieces. The ocular lens is located within the eyepiece(s). The eyepieces magnify the view field 10 times on most binocular microscopes, which increases the clarity. Some of the less expensive microscopes magnify the view field only 5 times. The microscope used for veterinary hospitals or laboratories should have at least three objective lenses. The 10× is used for scanning the slide. The 10× is known as the low-power lens. The 40× is used to identify fungus, parasites, and cells. The 40× is known as the high dry lens. The 100×, also known as the oil immersion lens, is used to differentiate bacteria and cytologies and confirm the morphology of cells.

The platform where the slide is placed for viewing the specimen is known as the stage. On a compound microscope, objects appear upside down and reversed. The left side of the specimen is seen as the right side. The slide, when moved by the mechanical part of the stage, is reversed. The specimen appears to be moving to the right when it is actually moving to the left. There are two knobs, one located inside the other, on the microscope. The outer (larger) knob is the coarse-focus knob, and the inner (smaller) knob is the fine-focus knob. The focus knobs bring the specimen into focus.

The microscope's light condenser delivers a concentrated beam of light to the specimen. The condenser is located under the stage. The condenser can be raised or lowered to focus the light source on the specimen being viewed and to prevent haloes and fuzziness around the specimen. The iris diaphragm controls the amount of light illuminating the specimen by manually opening or closing the diaphragm to the amount needed to view the specimen.

Operating the microscope begins by lowering the stage all the way down, switching on the light source, and placing the slide with the proper side up. If frosted tip slides are used, the frosted side should face upward. If clear slides are used, the side on which the specimen was placed must be marked. The specimen will look out of focus if the slide was placed on the wrong side, even after focusing. The lens is moved by holding on to the turret, not the lens. The turret is located above the lens. Scanning begins using low-power 10×, then moving the lens to 40× high dry. Oil must not be used on the slide with this lens. When using the 100×, focus is first on 10×. Then the lens is moved to 100× oil immersion. Prior to adjusting the 100× in place, a drop of immersion oil is added to the slide and then is wiped off the lens with lens paper. A touch of alcohol can be placed onto the lens paper if there is buildup of oil on the lens.

How to Do a Staining for Cytologic Preparations and Bacteria Examination

Following are a few examples of the many stains used in pathology and many variations in technique.

Romanowsky Stains

Romanowsky stains are used for cytologic preparations. Some of the most common Romanowsky-type stains are Diff-Quik, DipStat, and Wright's stains. Romanowsky stains are inexpensive, easy to prepare, easy to use, and have excellent quality for staining organisms and cytoplasm of the cell. That is why they are used for cytology in most practices.

After the smears and cytologies are prepared, they must be air-dried before staining with a Romanowsky stain. (Air-drying the smear helps to make the cells adhere to the slide.) Each stain has its own recommended staining procedure and techniques. There are a few variations in staining quality, but consistent use of one type is highly recommended. There are three containers in a set of Romanowsky stains: (1) the fixer solution, which is usually clear; (2) the eosin

ROMANOWSKY STAINING	
Fixer solution	Methanol
Acid dye	Eosin
Basic dye	Methylene blue
Buffer	pH 6.8 phosphate buffer (usually distilled or tap water)

1. Prepare smear or cytology and air-dry.

2. Fix by immersing in fixer solution for approximately 1 minute.

3. Transfer, without rinsing or drying, to the acid dye solution and stain for 15 to 30 seconds. Gently agitate.

4. Transfer, without rinsing or drying, to the basic dye solution and stain for 15 to 30 seconds. Gently agitate.

5. Rinse briefly in buffered water and allow to dry.

stain, which is an orange color; and (3) the methylene blue stain, an oxidative product of methylene blue (azure), which is a purple-blue color.

New Methylene Blue Stains

New methylene blue stain is a useful adjunct to Romanowsky stains and gives excellent nuclear and nucleolar detail. It can be applied directly to a slide that has been air-dried. This stain is used for bacteria, fungi, mast cells, and the presence of nucleated cells, such as reticulocytes.

Gram's Stains

Gram's stains are used to categorize bacteria. These categories are gram-positive and gram-negative, based on their cell wall structure. Once the smear has dried on the slide, it is ready to be heat-fixed. Heat-fixing is done by passing the slide through a flame two or three times with the specimen side up. Care must be taken so as not to overheat the slide. The slide should feel warm, not hot. Heat-fixing helps prevent the specimen from washing off. It also helps preserve cell morphology.

Follow these steps:

1. Place the slide on a staining rack over a sink if possible.

2. Pour the first solution, crystal violet, onto the smear (slide). Wait 30 to 60 seconds. Rinse gently with tap water. (Some cytologists and doctors prefer distilled water to rinse off the slide.)

3. Pour the second solution, Gram's iodine, onto the smear (slide). Wait 30 to 60 seconds. Rinse gently with tap water.

4. Rinse the smear with a decolorizer until the solvent flows colorlessly from the slide. This usually takes 5 to 10 seconds. Rinse again, this time with tap water.

5. The last solution to pour onto the smear (slide) is the safranin counterstain. Wait for 30 to 60 seconds. Rinse with tap water.

6. Place the slide on a rack to air-dry or blot dry.

7. Examine microscopically under 100× lens (oil immersion).

The gram-positive bacteria retain the crystal violet iodine complex and appear purple. The gram-negative bacteria lose the crystal violet iodine complex and take up the secondary stain, safranin, and appear red.

How to Prepare an Ear Cytology and Microbiological Examination

A drop of mineral oil is placed on a clean glass slide. An ear swab containing the ear debris is suspended in the mineral oil. If the debris is thick, the mixture is stirred gently to break up the debris. A cover slip is placed on top of the debris and mineral oil and examined on low power for the presence of ear mites (*Otodectes*). This method is done especially if the debris is brown and flaky.

Next, another sample is taken using a new ear swab, which is gently rolled along a clean glass slide to allow cells and organisms to adhere. Heat-fixing the slide before staining will help fix the proteins to the slide. Some of the stains used are Diff-Quik, Wright's stain, and Gram's stain. Yeast and bacteria are the most common abnormal findings.

How to Do a Fine-Needle Aspiration

Fine-needle aspiration (FNA) is a procedure used to obtain samples of cutaneous tissue masses and lymph nodes for cytology evaluation. The size of the syringe used for external masses can be 3, 6, or 12 ml (the most common size is 12 ml). The size of the syringe used is influenced by the consistency of the tissue being aspirated. The samples are collected with a 22- to 25-gauge needle. A few clean slides (dust free and no fingerprints) will be needed to place the sample on. It is best to use frosted edge slides so the samples can be labeled. Another advantage to frosted edge slides is that one side of the edge is frosted and the other side is not, so it is easy to tell which side of the slide the sample has been placed on, especially after staining.

First, the mass is swabbed with alcohol. (Some veterinarians choose not to swab the mass with alcohol.) The mass should be stabilized between the fingers to aid penetration of the skin and mass. The needle is inserted into the center of the mass. Strong negative pressure is applied by withdrawing the plunger of the syringe to about three-fourths the volume of the syringe. Several core samples are taken by redirecting the needle several times while remaining inside the mass. The syringe is not removed from the mass each time so that the sample does not become contaminated. After the sample is collected, the negative pressure is relieved from the syringe, and the needle is withdrawn from the mass. The needle is removed from the syringe, and air is drawn up into it. The needle is replaced on the syringe.

The contents in the needle are squirted onto the middle of the slide. A second slide is used to smear the contents. It is a good idea to make a few sample slides. A permanent marker should be used to label the slides so that the identification will not come off during the staining. The slides are allowed to air-dry, before fixing with the appropriate stain (Romanowsky stains). Slides for in-house cytology can be stained; slides that will be submitted to an outside laboratory should not be stained.

How to Do a Cellophane Tape Preparation for Lice or Mites

A cellophane tape preparation can be used instead of a skin scraping to collect lice or mites that live on the surface of the skin (in the keratin layer) and in the hair coat of the host—for example, *Cheyletiella* (walking dandruff), which is most commonly seen on rabbits and cats.

A piece of clear cellophane tape is applied to the skin to pick up epidermal debris, lice, or mites. A strip of mineral oil is placed on a clean slide. Then the tape with the sample is placed on the slide so that the adhesive side of the tape touches the mineral oil. A cover slip is not necessary but can be used to keep the tape from getting wrinkled. The slide is ready to be examined microscopically for ectoparasites.

How to Do a Skin Scraping

A skin scraping requires a #10 scalpel blade, three or four slides, and some mineral oil. It is best to get a sample from three or four different lesions if possible.

The lesions are not prepped. First, the lesion is squeezed to bring the mites to the surface (Figure A-1). Some are deep in the hair follicles and sebaceous glands. A small amount of mineral oil is placed on the blade to help the sample stick to it. The blade is held perpendicular to the lesion. Scraping is in one direction only, usually away from the body to avoid irritation or cutting into the skin. The periphery of the lesion is scraped until capillary blood oozing is seen. Remember, *Demodex* mites live deep in the hair follicles and sebaceous glands and *Sarcoptes* live in surface debris. One drop of mineral oil is placed on the slide, and the sample is transferred on the blade to the slide. Some veterinarians add a cover slip at this point. The slide is now ready for examining the mites under the microscope's low power.

Delmar/Cengage Learning

FIGURE A–1

Two registered veterinary technicians performing a skin scraping on a cat.

How to Do a Direct Smear for a Fecal Exam

The materials needed for a direct smear are a container of saline, an applicator stick, a microscope, slides, and a cover slip. Lugol's iodine (optional) can be used to help identify protozoal organisms, especially the internal structures of the cysts and trophozoites.

A drop of saline is placed on the slide, and an equal amount of feces is mixed in thoroughly using the applicator stick until there is a homogeneous emulsion on the slide. The smear on the slide should be thin enough to read newspaper print through the smear. Any large pieces of feces are removed. If the emulsion on the slide is not homogeneous, another drop of saline can be added and the sample remixed. The cover slip is placed over the smear. The entire smear is examined using the 10× objective lens for parasite eggs or larvae (see Figure 14-6). After examining the entire smear, the 10× objective lens is changed to a 40× objective lens for motile protozoal organisms, such as *Giardia* and *Isospora* (coccidians). The smear is examined again. More sophisticated methods using rapid assay test kits are also used to detect *Giardia*.

How to Do a Fecal Flotation

A fecal flotation is used to examine feces. The flotation is known as the concentration method based on specific gravity of the fecal debris and parasitic material. Flotation kits are common in veterinary practice. A test tube or a vial can still be used with a cover slip if a flotation kit is not available. The flotation kits have a vial with a filter and prepared zinc sulfate or sometimes sodium nitrate solution (Figure A-2).

FIGURE A-2

Preparation and steps for a fecal flotation.

About 1/2 teaspoon of feces is placed in the vial or cup, and the flotation solution (zinc sulfate) is added to the line on the cup. If a kit is not being used, 20 ml of flotation solution is added to the vial. The stick applicator is used to mix the solution and feces together, especially if the fecal sample is a little hard. The strainer is added to the cup, and more solution is added until a meniscus is formed. A cover slip is placed on top of the cup. If a kit is not being used, cheesecloth can be used as a strainer. The cheesecloth is placed over the top of the vial, and the mixture is poured through the cheesecloth into another vial until a meniscus is formed. A cover slip is placed gently on top of the second vial.

The cup or vial is allowed to sit undisturbed for 10 to 20 minutes, depending on the type of solution or kit being used. Parasite eggs in the feces flotation cup will float to the top and stick to the cover slip placed over the cup. The cover slip is removed (after the appropriate time) by picking it straight up. The cover slip is placed on a slide with the wet side adjacent to the slide, and the microscopic exam is done using the 10× objective lens.

How to Use the Modified Knott's Technique for Heartworm (*Dirofilaria immitis* and *Dipetalonema reconditum*)

The sample of blood used for this test must be uncoagulated blood. Heparin or other anticoagulants such as EDTA may be used to prevent the blood from coagulating.

1. Place 1 ml of uncoagulated blood into a 15-ml centrifuge tube and add 9 ml of 2 percent formalin.

2. Invert the sample to mix, and shake gently to lyse the red blood cells.

3. Centrifuge the sample for 5 minutes.

4. Pour off the supernatant by inverting the centrifuge tube once to let it drain.

5. Add 2 drops of methylene blue stain to the sediment on the bottom of the tube.

6. Gently tap the tube to mix the sediment with the stain.

7. Place a drop on the slide. A pipette can be used to obtain the sample from the tube and place it on the slide.

8. Place a cover slip on the slide and examine under (10X) objective lens for microfilariae.

Dirofilaria species coil and uncoil, whereas the *Dipetalonema* species may sometimes glide across the slide. By using the modified Knott's technique, the concentrates "relax" and stain the microfilariae while lysing the red blood cells, allowing for easier visibility of the microfilariae. The most accurate way to differentiate between *Dirofilaria* and *Dipetalonema microfilariae* is to measure the length and width of the body. *Dipetalonema* is shorter and more slender.

How to Use the Filter Technique for Heartworm (*Dirofilaria immitis* and *Dipetalonema reconditum*)

The filter technique is the most common in veterinary practices today. An example of the membrane filtration test kits is Di-Fil (EVSCO). The kits come with lysing solution, stain, and filters. The membrane filtration test is similar to the Knott's test, but the difference is that a millipore filter is used to concentrate the microfilariae. The filter test requires 1 ml whole blood and 10 ml of lysing solution. If the test is going to be performed immediately, an anticoagulant is not necessary. Heparin or EDTA anticoagulants should be used if the test is performed at a later time.

First, 1 ml of whole blood is drawn up into a 12-ml syringe. The needle is removed from the syringe, and the tip of the syringe is wiped. The syringe is attached to the nozzle port of the lysing solution bottle, and 10 ml of the lysing solution is drawn up (Figure A-3). The whole blood and lysing solution are mixed together by inverting the syringe a few times.

FIGURE A-3

Filter technique for heartworm. Attach the syringe to the nozzle port of the lysing solution bottle and draw up 10 ml of the lysing solution.

The cap to the filter is unscrewed and the O-ring removed from the filter. A new filter is placed on the mesh wire screen that is inside the filter. The dividing paper found between each of the membrane filters is discarded. The O-ring is placed on the filter membrane. The cap is screwed on the filter (Figure A-4).

FIGURE A-4

Filter technique for heartworm. Place a new filter onto the mesh wire screen that is inside the filter. Remove the O-ring from the filter. Place the O-ring onto the filter membrane.

A   B

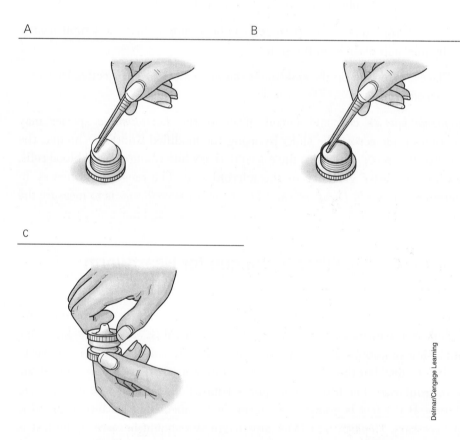

C

Next, 10 ml of water are injected into the filter holder to remove excess fragments (Figure A-5).

The filter cap is unscrewed and the filter membrane removed using thumb forceps. The filter membrane is placed on a clean glass slide. A drop of stain, which comes with the kit, is added to the center of the membrane. A cover slip is placed over the stained filter membrane (Figure A-6).

A   B

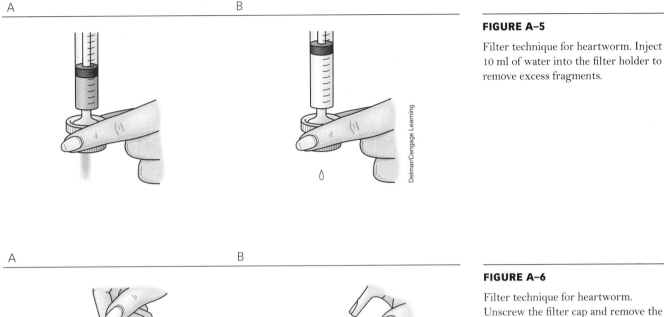

Delmar/Cengage Learning

FIGURE A–5

Filter technique for heartworm. Inject 10 ml of water into the filter holder to remove excess fragments.

A   B

C

FIGURE A–6

Filter technique for heartworm. Unscrew the filter cap and remove the filter membrane using thumb forceps. Place the filter membrane onto a clean glass slide. Add a drop of stain, which comes with the kit, to the center of the membrane. Place a cover slip over the stained filter membrane.

Delmar/Cengage Learning

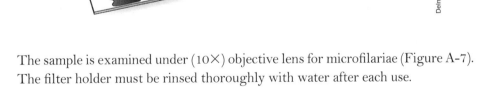

The sample is examined under (10×) objective lens for microfilariae (Figure A-7). The filter holder must be rinsed thoroughly with water after each use.

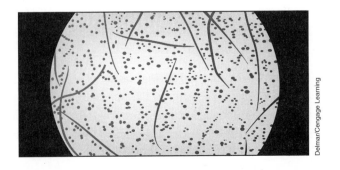

Delmar/Cengage Learning

FIGURE A–7

Filter technique for heartworm. Examine under (10×) objective lens for microfilariae. Rinse the filter holder thoroughly with water after each use.

How to Use the Buffy Coat Method for Microfilariae

This method is not as common in veterinary practices. A microhematocrit tube is used the same way as for collecting a blood sample for a PCV/TP (packed cell volume and total protein). The buffy coat method is performed after the PCV has been evaluated, but before the TP is evaluated. To mark the tube, a file or glass cutter is used to scratch the microhematocrit tube at the buffy coat level. The microhematocrit tube is snapped to break it. The section of the tube containing the buffy coat is tapped onto the slide. One drop of saline and one drop of methylene blue stain are added to the buffy coat on the slide. A cover slip is placed on top and the sample is examined under the (10×) objective lens for microfilariae.

How to Do a Fungal Culture Using Dermatophyte Testing Media

The most common cause of dermatophytosis, a fungal infection, is *Microsporum canis*. The three species of fungus that cause this condition are *Trichophyton mentagrophytes*, *Microsporum canis*, and *Microsporum gypseum*. The dermatophyte testing media is made of a Sabouraud dextrose agar, which is a more conventional medium for culture of dermatophytes. The Sabouraud dextrose agar comes in two forms: a square compact container or a round cylinder jar with a screw top. The media is amber in color prior to use.

To collect a specimen for the dermatophyte testing media (DTM), a pair of hemostatic forceps is used. The forceps are rinsed with water if they have been stored in a cold sterile solution. (The cold sterile solution has an antifungal, antibacterial agent that can interfere with the test results.) The forceps are dried so that the specimen will not stick to them.

Some veterinarians prefer to clean the area of the lesion with alcohol to kill saprophytic fungi. Allow the area to dry before collecting the specimen. This procedure should be used with pigs, since specimens from pigs are usually contaminated with saprophytic fungi. Most veterinarians do not clean the area with alcohol because they feel it could interfere with the collection of the specimen.

The hairs at the periphery of the lesion are plucked, and the hairs and scales (specimen) are pressed onto the media gently, taking care not to bury the hairs into the media because the fungus needs oxygen to grow. The cover should remain on the container loosely to allow aeration. The DTM should incubate at room temperature and be examined daily for color changes (red) and growth (white or cream fluffy). The red color media should appear simultaneously with the fluffy white or cream dermatophyte colonies. Colonies that have a green or black coloration should be noted as a contaminant. The media has a phenol red pH indicator. The fungi utilize the protein in the media, which produces alkaline metabolites that turn the media red.

The information is recorded in a DTM log daily for 10 days, unless the DTM is positive before the 10 days. To confirm the positive results, a cellophane tape slide preparation is recommended. Clear, not frosted, cellophane tape should be used. A piece of tape with the sticky side facing toward the media is pressed lightly onto the fluffy white or cream colony. The fluffy dermatophyte colony will adhere to the sticky side of the cellophane tape. (Some practitioners prefer to collect samples with forceps.) A drop of new methylene blue or lactophenol blue stain is placed onto a slide. The stain helps to identify the dermatophytes. The cellophane tape is placed on top of the stain and slide. A microscopic examination is done, first with the low-power objective lens (10×) and then the high dry power objective lens (40×).

How to Do a Blood Smear

A blood smear must be done as soon as possible before the blood clots. Cells are smeared onto the slide making the thickness a one-cell layer. (The slides must be clean and dust free.) If the slides are frosted, the frosted side must face the technician. A good way to check is to feel both sides: One side will be smooth and the other side rough to the touch. A small amount of blood is placed on the end of the slide. (The most common problem with making slides is that too much blood is applied to the slide.) The second slide (the one with no blood on it) is used to make the smear (see Figure A-8A). The slide is held at a 30-degree angle over the slide lying on a flat surface (slide with the blood). The slide is drawn back to touch the drop of blood (Figure A-8B). Now the blood should spread along the width of the spreader slide (slide without the blood). The spreader slide is moved forward (Figure A-8C), making a feather effect (Figure A-8D). If it goes off the slide, too much blood has been used. The slide must be air-dried quickly to freeze those cells as they are.

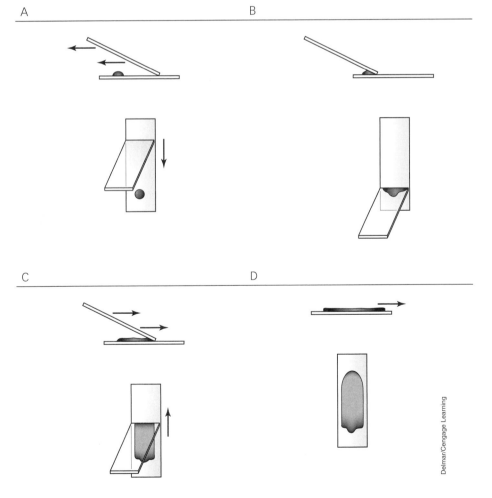

FIGURE A-8

How to do a blood smear.

Delmar/Cengage Learning

How to Use a Hemacytometer for a White Blood Cell Count

The manual Unopette hemacytometer method of counting WBCs is the leukocyte test most commonly used in veterinary hospitals. Unopette WBC test kits contain diluting reservoirs and capillary pipettes for making a 1:100 dilution of blood in ammonium oxalate or acetic acid, both of which lyse erythrocytes. Cells are counted using a hemacytometer and cover slip (Figure A-9). Unopette is an in-vitro diagnostic system for the enumeration of leukocytes in whole blood.

FIGURE A–9

How to use a hemacytometer.

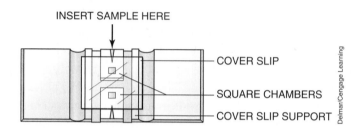

INSERT SAMPLE HERE

COVER SLIP

SQUARE CHAMBERS

COVER SLIP SUPPORT

Delmar/Cengage Learning

1. Add the appropriate amount of blood (microhematocrit tube amount) to the diluent (lysing solution).

2. After 10 minutes, shake the reservoir to mix the cells evenly and convert to a dropper assembly.

3. Carefully clean the hemacytometer and cover slip to ensure they are free of dust and fingerprints. Cover slips for counting chambers are specially made and are thicker than those for conventional microscopy, since they must be heavy enough to overcome the surface tension of a drop of liquid.

4. Apply the cover slip to the hemacytometer, and load the solution into each groove of the hemacytometer located in the middle on each side. The solution will spread beneath the cover slip. Make sure that no excess fluid goes into the grooves between chambers. If this happens or if air bubbles can be seen on the counting chambers, the filling procedure must be repeated.

5. Place the hemacytometer under the microscope using 4× objective. There is a counting grid comprising 9 primary squares with 16 tiny squares contained within each of these primary squares (Figure A-10). Count the cells in one of the 16 tiny squares using 10× objective. The number of cells counted is multiplied by the dilution and volume factors. Counts are expressed as cells per microliter (µl) of blood: total cells counted plus 10 percent of those counted, multiplied by a dilution factor of 100 (80 + 8 = 88 × 100 = 8800/µl).

How to Do a Urinalysis

A urinalysis provides information on the condition of the kidneys and the animal's ability to filter and excrete metabolites normally. When there are disturbances

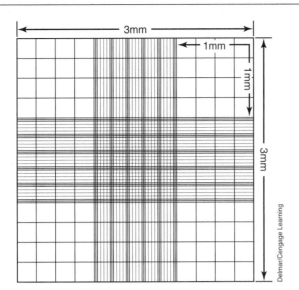

FIGURE A–10

The counting grid.

in the animal's body, such as the metabolic or endocrine system (diabetes mellitus, for example), a urinalysis shows abnormalities.

A urine sample should be analyzed as soon as possible. If the sample cannot be analyzed within 30 minutes of collection, the sample should be refrigerated. Urine that is left out at room temperature for a longer period of time will lose valid information due to degenerative changes. Urine can be refrigerated up to 6 to 12 hours. Refrigeration will help preserve urine constituents. Refrigerated urine must be warmed to room temperature for evaluation. A gross examination is an observation that can be done prior to and without the aid of a microscope or chemical reagents.

First, the sample is labeled with the animal's identification. Part of the sample is centrifuged immediately for cytologic evaluations. Next, 5 ml of urine is poured into a test tube and centrifuged for 5 minutes at a low speed (1500–2000 rpm). The supernatant is poured off, leaving only the sediment on the bottom of the test tube. The sediment is resuspended by tapping the bottom of the test tube gently. It is now ready for cytologic evaluation. A drop of urine sediment is placed on a glass slide. A sediment stain can be added to the urine if desired. A cover slip is placed on top of the drop. The sediment is examined on low power (10$\times$) first and then on high power (40$\times$) for cast, crystals, cells, and bacteria. The findings are recorded on a lab sheet with the animal's identification on it. Information on how the sample was collected is recorded (cystocentesis, catheterization, free catch, or expressing the bladder).

Specific gravity is used to evaluate the renal tubule's ability to concentrate or dilute filtrates from the glomerulus. A refractometer measures the specific gravity. A drop of urine is placed on the prism glass, and the cover is closed. To determine specific gravity, the refractometer is pointed toward the light and the scale is read on the right-hand side. The measurement at the boundary of the light to dark is read and the specific gravity is recorded on the lab sheet.

The color is recorded. The color of the urine usually correlates with specific gravity. If the urine is lighter in color, the specific gravity is lower (diluted). If the urine is darker in color, the specific gravity is higher (concentrated). Urochromes are pigments in the urine that make it yellow. Urine is normally light yellow to amber in color. Rabbits have orange to reddish brown pigments in their urine. Greenish yellow urine with colored foam indicates the presence of bile pigments.

The transparency is recorded. Normal urine is clear. Normal equine urine is cloudy due to calcium carbonate crystals. The terms *cloudy*, *clear*, and *flocculent* (particles seen with the naked eye) are used to describe the transparency of the urine.

Odor is not really diagnostic but does have an importance if the urine has a sweet-fruity odor. This indicates ketones, which are commonly found in animals that have diabetes mellitus, acetonemia in cows, and pregnancy disease in ewes. An ammonia odor can indicate cystitis, caused by bacteria, which produces urease that has metabolized urea to ammonia.

The biochemical evaluation is conducted using a reagent strip. The strip is dipped into the urine sample quickly and the strip is tapped on the edge, removing excess urine. (The urine must not be allowed to run in between the pads or touch any objects.) Each pad is read at the appropriate time, comparing color changes to the chart provided with the reagent strips. The information is recorded on the lab sheet. Reagent strips are commonly used to determine pH, ketones, and glucose. The pH expresses the hydrogen ion concentration of the urine (assess the body's acid-base balance). Ketones, which are produced during fat metabolism, are an important source of energy. Excessive ketones found in the urine indicate that the body is using more fat and fewer carbohydrates to metabolize into energy, which results in catabolism and ketonuria. The ketone concentration in the urine is proportional to the color intensity on the reagent pad. A few of the other pads on the reagent strip test for protein, bilirubin, and blood in the urine sample. Nitrite, urobilinogen, leukocytes, and SG (specific gravity) pads are also available. Care must be used in the evaluation of urobilinogen. The urobilinogen test pad must not be relied on to screen patients for hemolytic disorders, hepatic disorders, or patency of bile ducts. The specific gravity test, if present, should be ignored because it never matches with the refractometer.

The microscopic examination of the sediment is usually performed by taking a drop of the sediment, also known as supernatant, left at the bottom of the tube after decanting the urine in the tube that was centrifuged for 5 minutes at 1500 to 2000 rpm. The drop of sediment is placed on the slide, and a cover slip is added. The sediment is examined on low-power objective (10✕) assessing for casts, cells, and crystals. It is examined on high-power objective (40✕) to identify morphology of cells and elements and to detect bacteria. The results are recorded.

How to Do a Quantitative Fecal Egg Count

Two common quantitative fecal egg count techniques are the McMaster and Baermann. Each has its own qualities and limitations in calculating parasite eggs per gram (EPG) of a feces sample. It is increasingly important to be able to identify controlling populations of endoparasites on the pasture and rangelands where our equine and food-producing animals reside.

Modified McMaster Technique

The McMaster technique is used for demonstrating and counting helminth eggs in fecal samples. The McMaster chamber has two chambers, each with a grid etched onto the upper surface. When filled with a suspension of feces in flotation fluid, much of the debris will sink while eggs float to the surface, where they can easily be seen and those under the grid counted. There are several commercially available kits. The Paracount-EPG is a fecal analysis kit designed to assist the veterinary staff in making recommendations for parasite control within a herd or flock of grazing livestock/horses and is also routinely used in monitoring the parasite control of individual horses. The McMaster technique is the most widely employed method for this purpose. Its touted capabilities include the following:

1. Determining the presence/absence of strongyle and ascarid eggs expressed as EPG

2. Determining the efficacy of a dewormer drug

3. Tracking the required interval between dewormings

4. Determining the relative contaminative potential of an individual animal

The veterinary staff's goal should be to monitor and maintain an EPG count at low levels and to deworm when appropriate to maintain acceptable sanitation of the animals' pastures, water sources, and corrals (Figure A-11).

FIGURE A-11

Students preparing the McMaster test.

Delmar/Cengage Learning

Equipment Needed (in addition to the Modified McMaster kit)

1. Microscope (total 100× power) with mechanical stage

2. Flotation solution (use a commercial brand of sodium nitrate with a specific gravity 1.2) or zinc sulfate (1.8 sp.gr.)

3. Paracount-EPG kit which includes:

   a. Two McMaster slides with standard grid

   b. Two mixing vials

   c. Two transfer syringes

Collection of Sample

Fresh samples are preferred (best results if samples are taken within a few hours at room temperature). The storing of samples should be in a sealed container in the refrigerator.

Procedures (Figures A-12a, A-12b, A-13, A-14)

1. Add 26 ml of flotation solution to calibrated vial (provided with kit).

2. Add feces to bring volume up to 30 ml level.

3. Stir/mix thoroughly.

4. Charge one McMaster slide chamber with first sample.

5. Charge the second chamber with a different sample.

6. Count the eggs under each grid promptly, utilizing grid lines.

7. Calculate EPG. Add the numbers of eggs for both grids. Multiply their sum by 25. The result is equal to the EPG for this sample, using the 26.4 ml ratio of flotation solution to feces. There may be a slight variance in the above formula for species other than horses or cattle. For sheep or goats check with the manufacturer of the particular kit purchased for modifications based on species differences in EPG calculations and initial fecal sample dilutions.

FIGURE A–12a

Adding feces to the flotation solution.

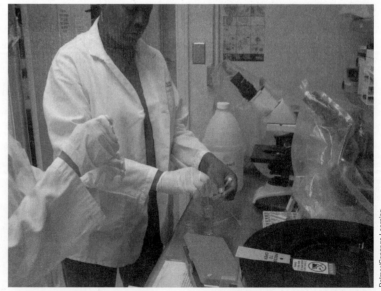

Delmar/Cengage Learning

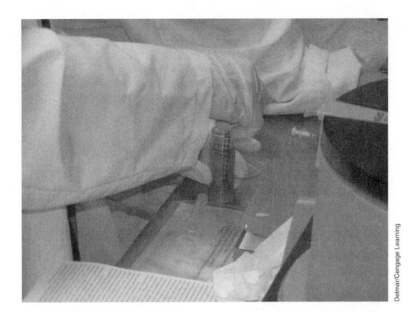

FIGURE A–12b

Adding enough fecal sample to total 30 ml.

FIGURE A–13

Charging the chamber.

When to Test

Sheep and goats are generally tested 7 to 10 days after treatment; horses and cattle are tested 10 to 14 days posttreatment. Adjustments to dewormer can be made after determining efficacy based on this test.

Baermann Technique

The Baermann technique is used to separate nematode larvae from fecal material. It requires more time, equipment, and finesse than the McMaster. It is especially important when diagnosing suspected lungworm (*Dicytocaulus, Aelurostrongylus, Filaroides, Crenosoma, Muellerius, Protostrongylus*) or threadworm (*Strongyloides* spp.) infections. *Strongyloides stercoralis* is a zoonotic parasite between human and canine species. Use extra precaution when handling feces of suspected animals.

FIGURE A–14

Counting the EPG.

Although there is no standard equipment for performing the Baermann technique, some commonly used equipment can be used as each laboratory modifies its procedures to fit its own specific needs.

Equipment Needed

Microscope (total 100× power) with mechanical stage
Centrifuge
Funnel and ring stand
Rubber or plastic tubing
Pinch clamp or clip
Cheesecloth
Metal rod or thin stick
Strainer or wire screen
15-ml centrifuge tubes
Scissors
Spatula
Rubber band
Flask
Iodine
Pasteur pipette

Setting Up Equipment

Attach 3 to 4 inches of rubber tubing to the narrow portion of the funnel. Place the funnel in the ring stand. Make a tight seal using the pinch clamp at the end of the rubber tubing.

Procedures

1. Measure approximately 5 to 10 grams of fecal material and place in the center of the cheesecloth (using several layers). Only use enough cheesecloth to make a small pouch containing the sample.

2. Bind the stick or metal rod to the cheesecloth pouch with the rubber band.

3. Suspend the pouch containing the fecal sample in the funnel with the metal rod.

4. Fill the funnel with lukewarm water covering the entire sample and let stand for 24 hours.

5. Draw off 3 to 4 ml of fluid from the stem of the funnel and place into a test tube.

6. Centrifuge at 1000 rpm for 2 minutes.

7. Use a Pasteur pipette to transfer a small droplet of the sedimented fluid to a microscope slide.

8. Add one drop of iodine to fix the larvae to the slide, and gently cover with a cover slip.

9. Examine under the microscope at 10× magnification and identify larvae.

Common Diseases

This appendix outlines the more common diseases of certain select species. It is by no means a complete list. The diseases are arranged according to the portion of the body that is affected and specific symptoms.

Mice and Rats

Signs of ill health are frequently reflected in the external appearance of mice and rats. The coat becomes ruffled, and the animals often remain immobile even when handled. Soiling of the perineum, the area between the genital opening and the anus, may occur if diarrhea is present. In addition, nasal or ocular discharges are frequently seen in cases of respiratory infection. Poor husbandry is the most common cause of ill health.

Skin Disorders

1. Skin wounds/abscesses—often occur in males that are housed together

    a. Lesions: small puncture wounds or abscesses

    b. Areas of hair loss and ulceration: often self-inflicted due to stress

2. Ringworm—scaly areas of hair loss; reddened and pruritic patches

3. Mammary glands

    a. Mastitis: infrequent inflammation of the mammary gland

    b. Neoplasia: usually malignant in mice; frequently benign in rats

Gastrointestinal Disorders

1. Malocclusions—overgrowth of incisors resulting in inappetence, or loss of appetite, and weight loss

2. Diarrhea—common clinical sign of ill health caused by a wide range of organisms:

   a. *Protozoa:* large numbers may cause inflammation of the gastrointestinal tract

   b. *Entamoeba murisii, Trichomonas murisii, Giardia, Eimeria* (sp.): common coincidental finding, but can cause disease if animals are kept in poor conditions

   c. Helminthes: cestodes (*Hymenolepis*) and nematodes (*Syphacia obvelata, Aspiculuris* sp.) can result in loss of condition and diarrhea. *Syphacia* can cause rectal prolapse in mice.

   d. Bacteria: may lead to generalized loss of condition or sudden death in addition to diarrhea

   e. Tyzzer's disease: caused by *Bacillus piliformis*; also results in generalized illness and death

Respiratory Disorders

These constitute the most common problems in mice and rats. Respiratory disorders often produce extensive pneumonic lesions at necropsy; however, the affected animal may be asymptomatic.

1. Signs

   a. Dyspnea

   b. Weight loss

   c. Rhinitis

   d. Abnormal respiratory sounds

   e. Hunched appearance

2. Causes

   a. Viral

      i. *Sendai* virus

      ii. Pneumonia virus of mice

   b. Bacterial

      i. *Mycoplasma pulmonis*

      ii. *Pasteurella pneumotropica*

      iii. *Bordetella bronchiseptica*

   c. Environmental

      i. Stress

      ii. Overcrowding

      iii. Unclean housing

Neurological Disorders

1. Central nervous disease

   a. Viral: can cause paralysis in mice; occurs rarely

      i. Lymphocytic choriomeningitis: zoonotic and usually asymptomatic, but death with or without paralysis can occur

      ii. Mouse (Theiler's disease) encephalitis virus: rare progressive paralysis of hindquarters; produces general signs of ill health

   b. Bacterial: rolling disease (*Mycoplasma neurolyticum*) due to exotoxin; causes roll in 1 to 2 hours and death within 4 hours

2. Middle (inner) ear disease

   a. Signs

      i. Head tilt

      ii. Circling

      iii. Loss of balance

   b. Causes

      i. Bacterial

      ii. Mycoplasmal

Guinea Pigs

Skin Disorders

1. Skin wounds/abscesses—localized abscesses, scratches, and bite wounds; common if several animals are housed together (damaged food bowls or caging may damage skin and cause localized infection)

   a. Hair loss (without obvious itching)

      i. Hair chewing, either by other guinea pigs or self-inflicted

      ii. Alopecia: deficiency of hair growth; frequently found in females during late pregnancy

   b. Ectoparasites

      i. Mites (commonly *Sarcoptes*): cause extensive pruritus, or itching; diagnosed by skin scraping

      ii. Lice: rarely cause pruritus unless infestation is severe

2. Ringworm—fairly common, producing patches of hair loss; affected skin usually scaly, but may be markedly inflamed and pruritic (example: zoonotic *Trichophyton mentagrophytes*)

3. Mammary glands—conditions such as mastitis can be avoided with improved hygiene and frequent cage cleaning

4. Cervical lymphadenitis, or "lumps"—lymph nodes of neck often burst and discharge yellow-white pus

   a. Often results in *Streptococcus zooepidemicus*, a secondary infection

   b. Can become enzootic (the disease is sporadic but in the population)

   c. Chronic abscessation most common; some animals die suddenly with septicemia

5. Pododermatitis—bacterial infection of the feet, producing swelling and ulceration of foot pads

   a. Often occurs with *Staphylococcus aureus*

   b. Associated with rough, infrequently cleaned flooring

Gastrointestinal Disorders

1. Malocclusions—can occur in the incisor or cheek teeth; once one group of teeth has become affected, the abnormal jaw position tends to allow the other group of teeth to overgrow; results in anorexia and excessive drool

2. Diarrhea—relatively common problem in guinea pigs; often of unknown etiology

   a. Protozoa: often a secondary problem, developing as a sequel to initial intestinal disorder

   b. Helminthes (of little clinical significance)

   c. Bacteria

      i. *Salmonella:* can cause acute enteritis and sudden death, abortion, or generalized loss of weight and condition

      ii. *Yersinia pseudotuberculosis:* causes diarrhea and weight loss; death after 3 to 4 weeks (some animals die acutely in 24–48 hours or develop nonfatal infections restricted to cervical lymph nodes). The condition is zoonotic; chronic disease may cause enlarged mesenteric and abdominal lymph nodes.

   d. Nonspecific: follows stress or dietary change

   e. Antibiotic-induced: may kill normal gut flora, allowing overgrowth of *Clostridium* (sp.) that produce fatal enterotoxin

Respiratory System Disorders

Respiratory disease is relatively common in guinea pigs and may be caused by a variety of agents.

1. Signs

   a. Dyspnea, or shortness of breath

   b. Abnormal respiratory sounds

   c. Sneezing

   d. Nasal discharge

   e. Weight loss

   f. Inappetence

   g. Depression

   h. Death

2. Causes (*Bordetella bronchiseptica* is the most common cause)

Musculoskeletal System

1. Hypovitaminosis C (scurvy)—signs develop within 2 to 3 weeks of restricted vitamin C intake

   a. Joints swollen, painful; animals refuse to move; loss of condition; hair scruffy, teeth loose

   b. Multiple hemorrhages in muscles and around joints

Miscellaneous Conditions

1. Pregnancy toxemia—occurs during last week or two of pregnancy or during first 4 days postpartum; abnormal carbohydrate metabolites develop in blood, causing depression, anorexia, and often dyspnea (labored breathing); death usually rapid; obese animals more susceptible

2. Conjunctivitis—may be associated with respiratory infection but usually appears secondary to local trauma or irritation

Rabbits

Skin Disorders

1. Skin wounds/abscesses—result from cage mates and fights with other animals

   a. Hair loss (without obvious itching)

   b. Ectoparasites

   c. Mites: can cause severe pruritus

      i. *Psoroptes cuniculi:* usually involves the ear canal, but can also include the pinnae (the external ear); canal becomes reddened, ulcerated, and plugged with brownish exudate

      ii. *Cheyletiella parasitivorax:* causes loss of hair in tufts with an accumulation of skin flakes and debris at base of hair; underlying skin reddened and sore

      iii. Sarcoptic and notoedric mange

2. Ringworm—produces areas of hair loss with scaling of the skin; severe pruritus may result, and the lesion may be encrusted with exudate

3. Mammary glands—mastitis can develop when female's litter is abruptly weaned; poor husbandry increases incidence

4. Necrobacillosis—necrotic skin lesions; condition progresses gradually, rabbit losing condition, eventually becoming severely emaciated and dying

5. Pododermatitis (sore hocks)—hair loss and ulceration of the skin on the bottom surface of the hind feet; due to trauma or constant exposure to damp and dirty solid floors

6. Bacterial dermatitis—areas of reddened and ulcerated skin on the ventral abdomen and groin region of rabbits housed under poor conditions

   a. *Staphylococcus* (sp.)

   b. *Pseudomonas* (sp.)

Gastrointestinal Disorders

1. Malocclusions—result in impaired appetite and drooling

2. Gastric hairball—excessive grooming and ingestion of hair may lead to the production of a large hairball; animals are inappetent and often develop diarrhea due to secondary digestive tract disturbance

3. Diarrhea—a range of different factors can produce diarrhea or enteritis:

   a. Protozoa: *Eimeria magna*, *E. media*, *E. perforans*, and *E. irresidua* may be primary causes of diarrhea

   b. Helminthes: in large numbers, the nematode *Passalurus ambiguus* may cause diarrhea

   c. Bacteria

      i. *Salmonella typhimurium* may cause enteritis with high mortality in young rabbits (zoonotic)

      ii. *Bacillus piliformis*, also known as *Clostridium piliforme* (Tyzzer's disease), results in anorexia and diarrhea with high mortality or a chronic form with progressive loss of condition

      iii. Enteritis complex: caused by a wide range of factors, including poor husbandry, diet, and bacterial, parasitic, and fungal agents; causes depression, fecal staining of perineum, and a watery-feeling abdomen

4. Pseudotuberculosis—leads to loss of condition and development of caseous nodules in the abdomen; zoonotic

5. Hepatic coccidiosis—caused by *Eimeria stiedae*; parasitizes the bile ducts and produces severe liver damage; signs are absent or nonspecific (poor growth, loss of condition)

Respiratory Disorders

1. Pasteurellosis—probably most frequent cause of illness in pet rabbits; *Pasteurella multocida*, found in the respiratory tract, produces a clinical

disease triggered by stress, pregnancy, and lactation; can be chronic and result in sudden death due to the rapid onset of severe bronchopneumonia

    a.   Secondary conditions, rhinitis and conjunctivitis; purulent discharge most common sign of either

    b.   Remission can occur following treatment

  2. *Bordetella bronchiseptica*—causes acute bronchopneumonia

Neurological Disorders

1. Middle and inner ear disease—most frequently caused by *Pasteurella multocida*

2. *Encephalitozoon (Nosema) cuniculi*—parasite that affects the CNS and kidneys; rarely causes clinical illness, but generalized CNS disturbances or renal dysfunction (interstitial nephritis) may be seen

3. *Toxoplasma gondii*—causes toxoplasmosis, paralysis, or convulsions following a period of depression and inappetence, followed by death

4. Posterior paralysis—damage to the lumbar spine

Miscellaneous Conditions

1. Myxomatosis—viral disease transmitted via fleas and mosquitoes; leads to development of purulent conjunctivitis with swelling of the eyelids and periorbital tissue, often causing partial closure of the eye; subcutaneous swelling then occurs on the head, neck, and anogenital region, with death in 11 to 18 days following onset of signs

2. Rabbit syphilis—caused by *Treponema cuniculi*; produces inflammation and ulceration of the genital region and possible secondary lesions on the face and paws; usually self-limiting

3. Pregnancy toxemia—rare condition that results in collapse, dyspnea, and sudden death during last few days of pregnancy

4. Pox—found in naturally occurring outbreaks; highly fatal; produces febrile, profuse nasal discharge; crusty nodules form 5 days after initial infection

5. Botfly—*Cuterebra* (rabbit botfly) occasionally infests both the dog and the cat

Cats

Skin Disorders

1. Bite wounds and abscesses

2. Eosinophilic granuloma complex

3. Ringworm—significant endemic problem, occurs in 70 percent of all kittens (adults serve as inapparent carriers); prompts facial dermatitis with circular areas of hair loss, scaly patches, or raised and reddened plaques with protruding stubbled hair; condition may affect feet as well

a. *Trichophyton mentagrophytes*

b. *Microsporum canis*

4. Mammary glands

a. Mastitis: not common in the queen

b. Metritis: follows parturition; may be due to retained placental membranes or lack of aseptic technique during assisted delivery

5. External parasites

a. *Otodectes cynotis:* ear mites

b. *Ctenocephalides canis/felis:* dog and cat fleas

Gastrointestinal Disorders

1. Panleukopenia (feline infectious enteritis)—a parvoviral illness found especially in kittens; may cause death or progressive anorexia, dehydration, diarrhea, vomiting, and panleukopenia; infected pregnant queens may experience abortion, stillbirth, and neonatal death; cerebellar hypoplasia can occur in kittens exposed in utero

2. Salmonella

3. Enteric parasites

a. *Toxocara cati* and *Toxascaris leonina:* roundworm

b. *Dipylidium caninum:* tapeworm

c. *Toxoplasma gondii:* toxoplasmosis

Respiratory System

1. Feline viral rhinotracheitis (FVR)—known as herpesvirus and responsible for 40 to 45 percent of upper respiratory infections; can promote secondary invasion; sneezing, coughing, fever, and hypersalivation are first signs, followed by photophobia, chemosis, serous ocular and nasal discharge, and depression; ocular and nasal discharge may become purulent and form crusts

2. Feline calicivirus (FCV) of the *Caliciviridae* family—responsible for 40 to 45 percent of upper respiratory infections; generally milder than FVR, but wide differences in pathogenicity of strains; conjunctivitis, rhinitis, and tracheitis not usually seen

a. Mildest: causes fever and ulcers on tongue, hard palate, and nasal commissure

b. Most virulent: causes severe pneumonia, complicated by secondary bacterial pneumonia

3. Feline pneumonitis (FPn)—caused by *Chlamydia psittaci* and responsible for 4 percent of upper respiratory infections; characterized by sneezing, coughing, serous and mucopurulent ocular and nasal discharge, and fever

4. Reovirus—usually mild, consisting of mild conjunctivitis and possible pharyngitis

Neurological Disorders

1. Rabies—*Rhabdoviridae* family; *Lyssavirus* genus

2. "Spring neuralgia"—vestibular disease (*vestibular* means "relating to the vestibule of the ear")

Miscellaneous Conditions

1. Feline infectious peritonitis (FIP)—infects cats of all ages; initial infection asymptomatic or manifest as mild upper respiratory disease; characterized by chronic weight loss, depression, and fever that is nonresponsive to antibiotics; causes inflammatory lesions of the blood vessels involving the liver, kidney, pancreas, eyes, brain, heart, and lungs

   a. Peritoneal exudate: found in most all clinical cases

   b. Pleural and peritoneal exudate: found in some clinical cases

2. Feline leukemia virus, or FeLV (Retroviridae)—a type C oncovirus, can cause fetal resorption and abortion; kittens may develop thymic atrophy and immunological incompetence; chronic FeLV causes a variety of bone and bone marrow disorders, inflammation of the kidney, anemia, and neoplasia; immunosuppression leads to increased susceptibility to other pathogens

3. Kitten mortality complex (KMC)—etiology unknown, though FeLV or FIP may be implicated; causes high frequency of reproductive failures (fetal resorption, abortion, stillbirths) and kitten deaths in successive pregnancies; kittens born weak and die soon after birth or, after seeming healthy for a few weeks, become anorectic and depressed, lose weight, and die; some kittens and adults die of acute congestive cardiomyopathy

4. FIV (Retroviridae)—lentivirus discovered in 1986; initial clinical phase of lymphadenopathy includes fever, malaise, and leukopenia intervening with normal clinical phase; terminal phase due to secondary opportunistic infection and FIV-associated neurological involvement; chronic unresponsive stomatitis, inflammation of the upper gastrointestinal tract found in 50 percent of all cats; respiratory disease, dermatitis, and chronic enteritis found in remaining percentage

Dogs

Skin Disorders

1. Tooth root abscesses

2. Flea bite dermatitis

3. Ringworm—*Microsporum canis, M. gypsum, Trichophyton mentagrophytes*

4. Parasites

   a. Fleas: *Ctenocephalides canis/felis*

   b. *Sarcoptes scabiei:* sarcoptic mange

   c. *Demodex canis:* demodectic mange

5. Mammary glands and related structures

    a. Metritis: an inflammation of the uterus that usually occurs in the first week postpartum; toxemic bitch is depressed, anorectic, febrile, with decreased milk production; excessive or abnormal vaginal discharge, usually fetid, watery, and red (green to black if placenta is retained)

    b. Mastitis: occurs when nursing ceases abruptly and the mammary glands become engorged; purulent milk can be threatening to puppies' health

    c. Neoplasia: usually benign

6. Pustular dermatitis—occurs in pups before weaning, usually affecting entire litter; staphylococci usually colonize skin on account of high humidity and poor sanitation; lesions found predominantly on head and neck

Gastrointestinal Disorders

1. Parvoviral enteritis (CPV)—serious threat to young puppies, mortality high in 5- to 12-week-old puppies; symptoms include lethargy, anorexia, depression, diarrhea, and vomiting with rapid dehydration; panleukopenia or lymphopenia may be present in early course of disease; produces villous atrophy and crypt epithelial necrosis in the small intestine (crypts lined with large, irregularly shaped epithelial cells); also causes lymphoid necrosis of the spleen and lymph nodes and Peyer's patches; acute death brought on by nonsuppurative myocarditis

2. Coronavirus enteritis—similar to parvovirus, but without panleukopenia or lymphopenia

3. Diarrhea and vomiting—common following weaning; most likely related to dietary change and stress

4. Parasites

    a. *Toxocara canis:* roundworms

    b. *Ancylostoma caninum, Uncinaria stenocephala:* hookworms

    c. *Trichuris vulpis:* whipworms

    d. *Dipylidium caninum:* tapeworms

    e. *Giardia*

Respiratory Disorders

Approximately 70 percent of newly received dogs show clinical signs of respiratory disease, and a mortality rate of up to 20 percent is common. Affected dogs exhibit mucopurulent nasal discharge, conjunctivitis, coughing, dyspnea, depression, and anorexia. The condition can progress to bronchopneumonia.

1. Canine distemper—Morbillivirus; major pathogen; most infections fatal

2. Parainfluenza—paramyxovirus; one of the most prevalent pathogens but not a major one

3. Canine adenovirus I (infectious canine hepatitis)—also causes liver, lymphatic tissue, and vascular endothelium damage

4. Canine adenovirus II—one of the most prevalent pathogens but not a major one

5. Tracheobronchitis (kennel cough)—synergy of *Bordetella bronchiseptica* and canine parainfluenza; although usually mild in older dogs, can be severe in pups; secondary invaders may cause life-threatening disease; produces a dry, hacking cough that can progress to a productive cough, pneumonia, and nasal or ocular discharge

6. *Streptococcus zooepidemicus*—acute necrotizing pneumonia; clinical course often peracute, resulting in death without clinical signs; less severely affected dogs exhibit coughing and moist rales, often with purulent nasal discharge and tonsillitis; produces diffuse hemorrhagic pneumonia and septic thrombi in kidneys, lymph nodes, spleen, brain, and adrenal glands

7. Other

   a. Reovirus type I, feline reovirus types 1 and 3—causes conjunctivitis, photophobia, and lacrimation. The virus spreads rapidly by contact. Reovirus type I infection in susceptible mice causes type I diabetes and growth failure.

   b. Canine herpesvirus—acute fulminant disease found in puppies 1 to 3 weeks old with sudden onset, constant crying, cessation of nursing, and death within several hours; adults remain unaffected or show mild upper respiratory signs

   c. Minute virus of canines—canine parvovirus is closely related to feline panleukopenia virus (FPV). Parvovirus affects rapidly dividing cells such as intestinal crypt epithelium.

   d. Canine coronavirus—infects mature villus epithelium

   e. Reovirus type II and III—reovirus type II is primarily bovine in nature. Reovirus type III is the most pathogenic reovirus of laboratory rodents. The primary importance of reovirus type III is as a contaminant of transplantable tumors and cell lines.

   f. *Filaroides hirthi:* filarid lung worm; coprophagy responsible for transmission

Neurological Disorders

Rabies *Lyssavirus* (Rhabdoviridae)—affects dogs, cats, and other carnivores, but all warm-blooded mammals are susceptible to the virus. Opossums are the least susceptible. The virus can be spread by inhalation as well as from bite wounds.

Miscellaneous Conditions

1. Brucellosis (*Brucella canis*)—suspected when history of abortion, infertility, testicular abnormalities, or lymphadenopathy exists; may provoke prolonged bacteremia (2 years) without fever; symptoms in males include

epididymitis, orchitis, prostatitis, testicular atrophy, and lymphadenopathy; females abort or have early embryonic death, along with prolonged vaginal discharge; pups can weaken and die or become bacteremic with lymphadenopathy; condition is zoonotic

2. Toxemia and septicemia ("toxic milk syndrome")—associated with uterine subinvolution; may cause crying and bloating in 4- to 14-day-old pups and red, edematous rectums; bitch may exhibit signs of metritis; septicemia characterized by crying, bloating, cessation of nursing with resultant hypothermia, hypoglycemia, and dehydration; visceral congestion and intestinal distention with gas can occur

Primates

Skin Disorders

1. Dermatitis—self-inflicted by bored or stressed animals or brought on by deficiency of fatty acids in diet

2. External parasites

   a. Sarcoptic mange

   b. Pediculus longiceps: sucking lice

3. Ringworm—causes lesions in debilitated animals

   a. *Microsporum* (sp.)

   b. *Trichophyton* (sp.)

Gastrointestinal Disorders

1. Loose stool—can be caused by fright, stress, or incorrect feeding

2. Diarrhea—changes in environment can turn a latent carrier of enteric organisms into a clinical case

   a. Protozoa: several species are commonly found and are frequently the cause of persistent or recurring diarrhea

      i. *Entamoeba histolytica (Balantidium coli.)*: produces a variety of helminthes (nematodes, filariae, cestodes) and can cause enteritis

      ii. *Strongyloides* and *Trichostrongylus (Prosthenorchis)*: damage intestine

      iii. *Enterobius vermicularis* (peri): causes anal irritation

   b. Bacteria

3. *Herpesvirus simiae* (B virus)—primarily seen in Rhesus monkey; produces vesicles on tongue and lips that progress to ulcers; rarely causes systemic illness; zoonotic

4. Hepatitis—highly zoonotic; easily transmitted to humans from recently caught asymptomatic chimpanzees and several species of monkeys

Respiratory Disorders

1. Tuberculosis—*Mycobacterium tuberculosis* occurs in both apes and monkeys, running a course of up to 6 months; usually remains occult until it becomes terminal

    a. Signs

        i. Chronic cough

        ii. Weight loss

        iii. Intractable pneumonia

    b. Lesions: found on lungs, liver, and spleen

2. Herpesvirus T—exists subclinically in squirrel monkeys and certain other species but may be fatal in marmosets and owl monkeys; produces rhinitis, pneumonia, and hepatitis, with death occurring in 2 to 7 days

3. Measles—often reported in monkeys (though a rash is not always present); some cases are fatal; transmitted from humans; common signs are rhinitis, pneumonia, facial edema, rash, anorexia, and diarrhea

4. Influenza—young chimps and other species are susceptible to acute rhinitis, which is self-limiting; may produce a cough that leads to fatal bronchopneumonia

Neurological Disorders

Rabies—rhabdovirus; acute viral encephalomyelitis

Musculoskeletal Disorders

1. Fractures—may occur as a result of nutritional bone disease or trauma

2. Dental conditions

    a. Facial abscesses: due to tooth-root infection, usually of the canine tooth

    b. Dental caries: common if fed sugary, man-made food

    c. Abnormal tooth eruption and "floating teeth": usually secondary to nutritional bone disease

3. Nutritional bone disease—a common condition; signs include lethargy, pain, "cage paralysis," inability to jump, and vertebral damage

    a. Rickets or osteomalacia: often present, dependent on age

    b. Pathological fractures

    c. Rarefaction of bone shafts; old, healed fractures; and distorted bones

    d. Predisposition to secondary infection

Miscellaneous Conditions

1. Pseudotuberculosis (*Yersinia pseudotuberculosis*)—involves sudden loss of condition, anorexia, collapse, and possible enlarged abdominal lymph

nodes; lesions produced by focal necrosis in spleen and liver; stress can cause condition to flare up

2. Marburg disease—comparatively rare, febrile condition that is usually fatal; few clinical signs show until 24 to 48 hours before death; zoonotic

3. Soft tissue injury—caused by deep, gaping wounds sustained during fighting with other primates; best to leave wounds on hands and feet open

4. Cheek pouch food accumulations—normal in some primates (macaques)

5. Edematous swelling/bright coloration of the perineal region—normal signs of estrus in some species

Cattle, Sheep, Goats, and Horses

Skin Disorders

1. Clinical signs in ruminants include primary lesions (papules, vesicles, pustules, nodules) and secondary lesions (scales, crusts, and alopecia)

2. Secondary lesions often result from self-trauma or superimposed bacterial infections

3. Demodicosis—common pustular disease in the goat. Greatest significance in canine. Horses, cattle, pigs, sheep can also carry *Demodex* species but rarely cause disease.

   a. Itch mites—contagious but slow spreading in the surface layers of the skin of wool sheep

   b. Chorioptes mites—cause foot, rump, tail mange in horses, sheep, dairy cows

   c. Sarcoptes mites—cause skin reddening and pustules. Found in the pig, sheep, horses, cattle. In cattle, found predominantly in the head, neck, and udder regions. Is contagious to dogs, cats, and humans

   d. Psoroptes mites—cause pustules, yellowish sticky scabs and crusts. Found in sheep, cattle, horses

4. Pox lesions—acute viral diseases acquired by either inhalation or through the skin, or mechanically (arthropod bites)

   a. Contagious ecthyma (orf in humans), capripox in the goat, sheeppox—widespread skin eruption in sheep (related to lumpy skin disease in cattle)

   b. Swinepox—papules, pustules on face, ears, inside of legs, abdomen; transmitted primarily by biting louse

   c. Lumpy skin disease— nodules on the skin of cattle; related to sheeppox virus

   d. Pseudocowpox, cowpox, bovine papular stomatitis

      i. Pseudocowpox—related to contagious ecthyma; small red papules on the teats or udder, scabbing; characteristic horseshoe or circular ring of small scabs

    ii. Cowpox—a benign, contagious disease of cows characterized by pox lesions, pustules, and scabs on teat and udder skin

    iii. Bovine papular stomatitis—reddish, raised, ulcerative lesions on the lips, muzzle, mouth. Clinically disease is mild, importance lies in the need to differentiate it from foot-and-mouth disease. Been reported in sheep, goats, and hands of milkers (humans)

5. Pruritus—common with lice, mange, zinc deficiency, and photosensitization. Also common symptom of scrapies or TSE (sheep), BSE (cattle), insect bites (ticks, lice, fleas, and blackflies) in donkeys, cattle, horses, sheep, goats.

6. Alopecia—endocrine disorders, bacterial skin diseases, burns, immune-mediated, inflammatory, nutritional deficiencies (protein).

    a. Horses, sheep—temporary alopecia during pregnancy, lactation

    b. Cattle—alopecia anemia, a congenital disorder in polled Hereford cattle

    c. Goats, sheep—zinc or selenium deficiency

    d. Wool eating disease in sheep—deficiency of calcium, phosphorus, sodium chloride, copper, zinc, manganese, cobalt, vitamin E, and/or a protein deficiency

Gastrointestinal Disorders

1. Malocclusions—faulty bite

    a. In horses, incisor and/or cheek teeth malocclusions—interfere with grasping or chewing food and premature loss of teeth. Require removal or smoothing of sharp points on the outer edge on upper molars and inner edge on lower molars (floating the teeth). Unbalanced incisors result in poor function and wear of the molars.

    b. As horses age, an ongoing wear of teeth causing animal to not properly masticate (chew) hay or grain

    c. Homozygous dwarfism in beef cattle—malocclusions common with this disorder

    d. Brachygnathism (abnormal shortness in mandible) and pragmatism (where either mandible or maxillae project forward)—occur frequently in goats

2. Scours (diarrhea)

    a. Protozoa—*Coccidiosis (Eimeria zurnii, Eimeria bovis)* affects goats, swine, horses, cattle, sheep, poultry, rabbits

    b. Bacteria—*Escherichia coli (Colibacillosis), Salmonella, Enterotoxemia (Clostridium perfringens), Johne's disease (M. paratuberculosis), Equine monocytic ehrlichiosis (Potomac horse fever)*

    c. Viruses—*Rotavirus* Scours, *Coronavirus* Scours, Bovine Virus Diarrhea, Bovine kobuviruses

    d. Nutritional—anything that disrupts normal nursing habit (strong wind, storm, cow going onto new grass), overconsumption (more common with bottle/bucket feeding), bottle feeding of milk too cold (especially during first 3 weeks of life)

    e. Helminthes (cestodes, nematodes)

        i. Cattle—*Ostertagia ostertagi*, *Trichostrongylus axei*, and *Haemonchus* spp. Diarrhea more common in calves than in adult cattle.

        ii. Horses—*Cyathastomes*, *Strongyles*

        iii. Sheep—*Trichostrongylus axei*

        iv. Pigs—Stomach worms (*Hyostrongylus* sp.)

        v. Goats—*Haemonchus contortus*, *Teladorsagia circumcincta*

    f. Miscellaneous

        i. Horses—gastric ulceration; more common in horses fed fewer times/day

Respiratory Disorders

1. Signs

    a. Horses—dyspnea, flared nostrils, coughing, and strange noises during exercise

    b. Cattle—noisy, open mouth breathing, frothing, infrequent coughing, separation from herd, anxiety, death

    c. Sheep, goats—dyspnea, tachypnea, rapid tiring (especially with exercise), cyanosis, abnormal sounds with breathing, nasal discharge, coughing

2. Causes

    a. Viral—caprine arthritis, encephalitis virus (goats), chronic progressive pneumonia (sheep, goats), Parainfluenza virus-3 (PI-3) (cattle, sheep, goats), bovine respiratory syncytial virus (BRSV), equine herpesvirus 1 (EHV-1) and EHV-4

    b. Bacterial—contagious caprine pleuropneumonia (goats), chlamydial pneumonia (calves, lambs, goats), shipping fever pneumonia (cattle), *Streptococcus equi* (strangles in horses), atrophic rhinitis (*B. bronchiseptica*) in pigs

    c. Environmental

        i. Airborne particles—dust, soot, mold and fungal spores, grain and hay dust, ammonia (from urine), ozone

        ii. Exercise-induced pulmonary hemorrhage (EIPH) in racehorses

        iii. Heatstroke

    d. Nutritional

        i. Cattle— "fog fever" (acute bovine pulmonary edema and emphysema)

        ii. Sheep, goats—pregnancy toxemia (ketosis)

    e. Parasitic—lungworms (*Dictyocaulus*), nose bots (*Oestrus ovis*)

Musculoskeletal System

Lameness is the third most important problem on many modern dairy farms after mastitis and reproductive failure. In swine, musculoskeletal problems are the second highest cause of culling. In horses, pain (both weight-bearing and non-weight-bearing related) lameness needs to be separated from mechanical lameness (restriction lameness).

Causes include:

Cattle—sole ulcer, sandcracks, heel erosion, footrot, interdigital dermatitis, fractures, dislocations, arthritis

Goats—mineral imbalances (Ca:P), lack of vitamin D, excess protein, excess Fe, Cu deficiency, Se or vitamin E deficiency
      Bacterial—joint ill, footrot, *Mycoplasmosis*
      Viral—caprine arthritis and encephalitis
      Trauma—fractures (jumping, getting caught in fences)

Pigs—footrot and joint-ill (related to environment), abnormal bone growth (nutritional), exudative epidermitis and polyarthritis (bacterial)

Sheep—Lambs: joint-ill, tetanus, white muscle disease, enzootic ataxia (copper deficiency), polyarthritis (chlamydial), rickets, poisonous plant intoxication (e.g., sneezeweed), and contagious ecthyma. Adults: mastitis, epididymitis, and mineral and trace element imbalances, erysipelas, laminitis, bluetongue, ulcerative dermatosis, foot-and-mouth disease, and dermatophilosis

Horses—fibrotic myopathy (stringhalt), stress fractures, fatigue, bone cysts, bog and bone spavin, bruised soles, navicular disease, osselets, puncture wounds, ringbone, sandcrack, sidebone, thrush, osteochondrosis dessicans, tying-up syndrome (*recurrent exertional rhabdomyolysis*, or Monday morning disease), degenerative joint disease

Neurological Disorders

Signs of neurological disorders are sometimes confounded with other systems (musculoskeletal). In goats signs include excessive bleating, resistance or over-reaction to touch, fear or aggression, compulsive walking or running, grinding of teeth, and/or fluttering of the eyelids. In most animals these disorders are indicative of inflammation of the brain (encephalitis). Livestock will tend to stagger, have personality changes, appear disoriented, and make vocalization (distressed sounds).

Causes include:

Goats—nutritional: thiamine deficiency (polioencephalomalacia), pregnancy toxemia, grass tetany (hypomagnesemic tetany)

Microbial—rabies, pseudorabies, Borna disease, cowdriosis, and bacterial meningoencephalitis

Toxic agents—organophosphates, chlorinated hydrocarbons, cyanide, nitrates, urea, nitrofurans, and the coyotillo (buckthorn, or tullidora) plant (*Karwinskia humboldtiana*)

Cattle, sheep—considerable degree of overlap between the neurological diseases of sheep and cattle. Intoxications are more common in cattle, and parasitic and microbial are more common in sheep. *Hypoderma bovis* (cattle) can cause serious problems if they die when migrating near the spinal cord. There are numerous genetic disorders in cattle and few in sheep. Copper deficiency and swayback are more of a problem in lambs as is the slow virus disease of sheep (Maedi-visna and caprine-arthritis-encephalitis virus and scrapie) whereas bovine spongiform encephalopathy is found in cattle

Horses—equine encephalitides (arboviruses, *Sarcocystis neurona*, *Neospora* sp., West Nile virus). Both cervical vertebral myelopathy (CVM, or wobblers) and osteochondrosis dessicans are anatomical deformities that can either pinch or cause inflammation of the spinal cord affecting the signals from the brain to the nerves controlling the legs.

Narrative Guide for the Image Library

Note: In this supplement the genus and species names are given for many of these parasites. These names are given in italics, with the species name following the genus name. A genus name should always be followed by a species name. However, when an unidentified species is referred to, this is designated by the abbreviation sp. following the genus name. When no species name is given with the genus, all species within that genus are being referred to.

Contents

Internal Parasites of Domestic Animals

How to Identify the Diagnostic Stage

Parasites of Small Animals

Parasites of Large Animals

Identification of External Parasites

External Parasites—Cause and Carriers of Disease

Types of Animal Diseases

Detecting Illness

Canine Diseases

Feline Diseases

Large Animal and Avian Diseases

Internal Parasites of Domestic Animals

How to Identify the Diagnostic Stage

The purpose of the Internal Parasite Photo Cards included with this supplement is to familiarize the reader with how veterinarians can visually recognize the diagnostic stage of the common internal parasites of domestic animals, both small and large. This diagnostic procedure is difficult, as many of the parasite eggs look alike. In many cases, the treatment for different parasites will vary; therefore, the proper diagnosis must be made. For this reason, the diagnosis can be performed only by a licensed veterinarian. A veterinary assistant would under no circumstances be involved in identifying the diagnostic stages of internal parasites. Nevertheless, a veterinary assistant who has visually examined the various diagnostic stages should be better able to understand the effect that parasites have on the health of an animal.

The diagnostic stage of a parasite is the life cycle stage that the veterinarian uses to identify the parasite that has infected an animal. In the case of parasites that live in the gastrointestinal tract, the diagnostic stages are the eggs of the worms and the cysts of the protozoa found in the feces. For parasites that live in the circulatory system, the stages found in the blood are used for diagnosis.

Photos of parasites also show a lot of plant debris that may resemble parasite eggs. There are a few general rules to help determine what is being viewed. Parasite eggs and cysts have a regular outline and contain an embryo or cells. Nematode larvae have a head and tail and contain an intestine as well as other organs. Hairs and plant spines do not show the internal detail seen in a larva.

Most of the pictures in this collection were taken using a microscope. Microscopes have various magnifying capabilities. The various magnification settings on a microscope are referred to as the microscope objective. For example, a 40× objective would magnify the object 40 times its normal size.

The microscope objective used to view an object is dependent on the size of the object. The smaller the object, the higher the microscope objective must be to view the object. Because many eggs and cysts are colorless and hard to see, iodine is often used to stain the fecal smear to provide some contrast. Iodine stains many eggs and cysts and colors them from a yellow to a dark brown, making them easier to view. The descriptions that follow indicate whether the object seen in the photo has been stained. Several stains in addition to iodine are used to color the diagnostic stages of parasites, thus highlighting features that may aid in the identification of the parasite. Most of these stains are a combination of red and blue dyes.

Parasites of Small Animals

Some of the eggs of the nematodes of small animals are similar in appearance, even though the worms are very different. The first four eggs covered fall into this group.

Internal Parasite 1

On the left is an egg of *Toxocara canis* (*T. canis*), an ascarid or roundworm of dogs. This egg measures about 80 μm (micrometers). The egg of *T. cati*, the ascarid of cats, looks exactly like that of *T. canis*, except that it is a little smaller (70 μm). *Toxocara* eggs are round (as in this photo) to slightly oval in shape. They have a rough shell and a brown central mass that fills most of the shell.

On the right is the egg of *Toxascaris leonina*, an ascarid that can be found in both dogs and cats. This egg has the same shape as that of *Toxocara*, but the surface is smooth and the central mass does not fill all of the space in the shell. The dark brown color of the egg on the right is due to staining with iodine. If it was unstained, it would be the same color as the egg on the left (40× objective).

Internal Parasite 2

The egg on the left is *Toxocara canis*. Here the focus is on the surface of the egg's shell. The texture is rough and pitted. On the right is the smoother surface of the *Toxascaris leonina* egg (40× objective).

Internal Parasite 3

There are two different eggs in this photo that look very much alike. The egg on the left is that of the whipworm of dogs, *Trichuris vulpis*. It is a symmetrical lemon-shaped egg with a plug at either end. The whipworm egg has a smooth shell and is about 80 μm long. On the right is the egg of the lungworm *Capillaria aerophila*, which is found in both dogs and cats. This egg is more oval and less symmetrical than that of the whipworm. It is also slightly smaller (65 μm long) and has a rough, pitted shell. It is important to be able to tell these two eggs apart, because the treatment is different for each infection (40× objective, unstained).

Internal Parasite 4

This photo focuses on the surface of the shells of the whipworm (left) and the lungworm (right) eggs. The whipworm egg has a smooth shell, while that of the lungworm is pitted. This difference is most easily seen on the area of the shell between the plug and the central mass (40× objective).

Internal Parasite 5

This is the egg of the hookworm of dogs and cats (*Ancylostoma*). The shell is very thin, and a mass of cells is in the center. Hookworm eggs measure 55 to 75 μm in length. This egg has been stained with iodine; the cell mass of an unstained egg would appear colorless (40× objective). Occasionally when a cat or dog vomits, a nematode will be found in this discharge. In order to treat the animal, the veterinarian must identify the worm.

Internal Parasite 6

This photo shows the heads of the most common nematodes that may be found in the stomach contents of small animals. On the left is the anterior (front) end of *Toxocara cati*. Wide "wings" run along the side of the head. The other ascarids of dogs and cats also have these wings, but they are not as wide. On the right is the anterior end of *Physaloptera*, the stomach worm of dogs and cats. This worm is not as common as the ascarids. The head looks as if it has been pulled back into the body, forming a collar. The stomach worm is the only worm that has this collar. Both worms in this photo have been decolorized to take this picture. Normally, they are a light pink or cream color. These worms were photographed using the 4× and 10× objectives, respectively.

Larval nematodes are sometimes found during fecal exams of dogs and cats. These larvae, if found in fresh feces, can be of *Strongyloides stercoralis* or one of the lungworms. In feces over 24 hours old, hookworm larvae may have hatched from their eggs. Therefore, it is important to examine fresh feces if at all possible.

Internal Parasite 7

This is the first-stage larva of *Strongyloides stercoralis*, a nematode that lives in the wall of the small intestine of dogs and cats. The tail is tapered, and an oval-shaped group of cells is near the center of the worm (arrow). These two characteristics are enough to tell this larva from that of the lungworm (40× objective, unstained).

Internal Parasite 8

This is the first-stage larva of the dog lungworm *Filaroides*. This larva is similar to the larva of the cat lungworm *Aleurostrongylus*. Both have a small notch in the tail (arrow), which differentiates them from *Strongyloides* (40× objective, unstained).

There are two genera of tapeworms commonly found in dogs and cats. *Dipylidium caninum* is very common in dogs and may also be seen in cats. Tapeworms of the genus *Taenia* are common in cats and are often seen in dogs as well. In both *Taenia* and *Dipylidium* infections, the terminal segment of the worm (proglottid), which contains the eggs, drops off and passes out with the feces. Unless this segment breaks apart while in the feces, no eggs will be seen during a fecal exam. Therefore, it is important to be able to identify not only the tapeworm eggs but also the proglottids.

Internal Parasite 9

The eggs of *Dipylidium caninum* are distinctive because they are found in packets containing 10 to 30 eggs. The entire packet measures 100 to 200 μm, depending on the number of eggs in it. Individual eggs measure 30 to 50 μm in width. All tapeworm eggs contain six small hooks (40× objective, unstained).

Internal Parasite 10

These are proglottids of *Dipylidium caninum*. The proglottid on the left was recovered from the surface of a stool sample. The white color and the curved sides make this segment resemble a cucumber seed. These proglottids range in size from 5 to 15 mm. There are two genital openings on each segment, one on each side. These openings can be seen more easily in the stained specimen on the right (arrows). The identity of the tapeworm can be confirmed by crushing the proglottid between a cover slip and a slide. This will release any egg packets it may contain, and these eggs are easy to identify. A proglottid that has dried out should be placed in some water and allowed to soften before crushing it (unstained proglottid, 2× objective; stained proglottid, 4× objective).

Internal Parasite 11

This is the egg of *Taenia* sp. The eggs of *Taenia taeniaeformis*, a tapeworm of the cat, and those of *Taenia* sp., *Multiceps* sp., and *Echinococcus* sp., which are found in the dog, all look like this egg. *Taenia*-type eggs measure about 35 μm wide and have a thick shell that contains striations or stripes. The embryo within the shell contains six hooks. The hooks are not visible in this photograph (40× objective, unstained).

Internal Parasite 12

These are proglottids of the cat tapeworm *Taenia taeniaeformis*. *Taenia* segments are usually found singly, not in a chain, as seen on the left. (The segments on the left are still attached to the worm.) The segments of a *Taenia*-type tapeworm are as wide (or wider) as they are long (just the opposite of those of *Dipylidium*). There is only one genital opening on each proglottid, as can be seen in the stained segment on the right (arrow). (The stained segment was photographed using a 4× objective, the unstained using a 2× objective.)

The cysts of protozoa that infect the gastrointestinal tract of dogs and cats are much smaller than the eggs of parasitic nematodes. Many of these protozoan cysts are too small to see with the 10× objective; the 40× objective must be used to find them.

Internal Parasite 13

These are cysts of *Giardia*, which are found in both dogs and cats. These cysts are small (about 10 μm long) and oval in shape. In the cyst to which the arrow is pointing, two nuclei (just below the arrow), as well as some other structures, can be discerned. This photograph was taken with the 100× oil objective, and the cysts were stained with iodine. Unstained cysts are colorless.

Internal Parasite 14

This is the oocyst of *Isospora canis*, a coccidian parasite of dogs. There are many species of *Isospora* that infect cats and dogs. All of the oocysts of the different species of *Isospora* look very much alike, although they vary in size. The oocyst of *I. canis* is one of the largest, measuring about 40 μm long. *Isospora* oocysts range in size from 10 to 40 μm long. Its egg shape, thin cyst wall, and central mass are the typical appearance of a coccidian oocyst in fresh feces. Older cysts may have two central masses (40× objective, stained with iodine).

Internal Parasite 15

This is the oocyst of *Toxoplasma gondii*, a coccidian parasite of cats. It is similar to the oocyst of *Isospora* but very small (10 μm) and tends to be round. Like *Isospora*, there is only a single central mass in a young cyst and two masses in older (24–48 hours) cysts (40× objective, unstained).

Internal Parasite 16

Trichomonas sp., seen in this photo, is a protozoan that lives in the large intestine of cats and dogs. This parasite has no cyst stage. The only way to diagnose this protozoan is to make a direct smear of diarrhea. *Trichomonas* sp. measures 15 to 20 μm long and is pear shaped. It has several flagella (hairlike projections) coming out of its front end and an undulating membrane running the length of the body (arrow). When this parasite is seen alive, the flickering of this membrane identifies it as *Trichomonas* sp. (100× oil objective, stained).

Some protozoan parasites are found in the blood of small animals. These may be seen when a blood smear is being examined for white blood cells. Two of these protozoan blood parasites are seen in photo 17.

Internal Parasite 17

On the left are two red blood cells infected with *Babesia canis* (arrow). This organism, which lives inside the red blood cells of the dog, is very small (1–2 μm). They usually occur in pairs within the red blood cell. On the right are blood cells of a bird, some of which are infected with a protozoan parasite (*Haemoproteus* sp.). All bird red blood cells have a nucleus. The protozoa (arrows) lie in the cytoplasm of the red blood cells, next to the nucleus. Both of these parasites were stained with Giemsa stain and photographed with the 100× oil objective.

Other parasites that can be found in the blood of the dog are the heartworm (*Dirofilaria immitis*) and the filarial worm *Dipetalonema reconditum*. The microfilariae of *D. immitis* and *D. reconditum* are found free in the blood; they are not in the blood cells. Both of these worms are nematodes.

Internal Parasite 18

The microfilaria of the dog heartworm, *D. immitis*, can often be seen in a direct smear of the blood of an infected dog. Although the numerous red blood cells may hide part of the worm, the movement of the microfilaria is easy to see. As seen in this photo of a direct blood smear, the microfilaria is fairly large (300–320 μm long) (arrow) (10× objective, unstained).

Internal Parasite 19

Because the heartworm microfilaria may be present in numbers too few to be seen in a direct smear, the blood is usually concentrated before being examined. There may be two different microfilariae in smears of the concentrated blood from which the red cells have been removed. *Dirofilaria immitis* lives in the hearts of dogs and causes a serious disease that is hard to treat. *Dipetalonema reconditum* lives in the skin of dogs and causes no disease. Therefore, being able to distinguish these two parasites is important.

The microfilaria on the left is that of *Dirofilaria immitis*, the dog heartworm. When killed in formalin (part of the concentration process), these microfilariae have a straight or slightly curved body. The tail is also straight and the head tapers slightly (arrow). The heartworm microfilaria and the white blood cells around it have been stained with methylene blue (40× objective).

On the right is the microfilaria of *D. reconditum*. Formalin-killed microfilariae of this worm tend to have a curved body. Their tail may also curve, and the head is blunt (arrow). Also, *Dipetalonema* microfilariae are shorter than those of *D. immitis* (275 μm versus 300 μm) and narrower (6 μm versus 7 μm) (40× objective, unstained).

Because these microfilariae look so much alike, several worms should be examined before making a decision. *D. reconditum* is not as common as *D. immitis*.

Parasites of Large Animals

Many of the nematode parasites of large animals have eggs that are similar to those found in small animals. Whipworms of pigs, sheep, and cows have eggs that resemble those of *Trichuris vulpis* from the dog (Internal Parasite 3). *Strongyloides* sp. of the horse, *Ostertagia* sp. of the cow, cattle hookworm, and many other gastrointestinal nematodes of large animals all have eggs that look very much like those of the dog hookworm *Ancylostoma caninum* (Internal Parasite 5). The ascarids of horses and pigs have eggs that are very similar to those of the ascarids of small animals (Internal Parasite 1). Similarly, the oocysts of the coccidian parasites of large animals look like those of *Isospora canis* (Internal Parasite 14). The *Giardia* of large animals produce cysts that are similar to those seen in small animals (Internal Parasite 13).

Internal Parasite 20

The eggs of the horse parasite *Strongylus vulgaris*, and any nematode egg that resembles it, are known as strongyle-type eggs. The hookworms of dogs and cats have a strongyle-type egg. In the horse, both large and small strongyles have this type of egg. In cattle and sheep, *Ostertagia*, *Haemonchus*, *Trichostrongylus*, *Nematodirus*, and hookworms all have strongyle-type eggs.

The egg on the left is that of *Strongylus vulgaris*, but it is difficult to tell it apart from the dog hookworm egg (Internal Parasite 5) or any other strongyle-type egg. These eggs have an oval shape, are 60 to 90 μm long, and have a thin shell and a central mass of cells (40× objective, unstained).

The egg on the right is that of *Nematodirus*, a nematode parasite of sheep and cattle. Although this is a strongyle-type egg, it can usually be identified as that of *Nematodirus*. It is a large egg (greater than 130 μm in length), much longer than that of the other strongyle-type eggs. Also, the ends of the egg tend to be more pointed than those of the other worms (40× objective, unstained).

Internal Parasite 21

The eggs of the ascarid of pigs (*Ascaris suum*) and the ascarid of horses (*Parascaris equorum*) are very similar.

On the left in the photo is the egg of *Parascaris equorum*. It is round, about 95 μm across, and usually brown in color. It contains a central mass that can just be made out in this photograph (10× objective, unstained).

The egg of *Ascaris suum* is on the right. This egg is smaller than that of the horse ascarid (about 60 μm) and may be either round (top) or slightly oval (bottom). It too is brown and contains a central mass (more easily seen at higher magnification, bottom) (top: 10× objective; bottom: 40× objective; both unstained).

Internal Parasite 22

The pinworm of horses lays its eggs on the skin around the anus. Eggs may be obtained for examination from the perianal skin or the feces. *Oxyuris equi* eggs are about 90 μm long, they are an elongated oval in shape, and they have a plug at one end (arrow) (40× objective, unstained).

Internal Parasite 23

Unlike *Strongyloides stercoralis* of small animals, the *Strongyloides* sp. found in horses, sheep, cattle, and pigs lay eggs that pass out in the feces. These eggs resemble strongyle-type eggs, except that they contain a larva rather than a cell mass. This egg measures about 55 μm in length. Lungworms of horses and pigs also have a strongyle-type egg that contains a larva (40× objective, unstained).

Internal Parasite 24

The nematode *Trichinella spiralis* of pigs does not lay eggs. Instead, the female worm gives birth to first-stage larvae, which quickly penetrate a muscle cell and encyst. In order to diagnose a *Trichinella* infection in a pig, a piece of muscle must be crushed between two slides and examined for larvae. At slaughter, a piece of the diaphragm is most often used for this purpose. This photo shows a crushed piece of muscle tissue in which two larvae are encysted (arrows). When the larvae encyst, they coil up as seen here (10× objective, unstained).

Internal Parasite 25

The trematode (flatworm) *Fasciola hepatica* lives in the liver and bile ducts of cattle and sheep. Occasionally, it may be found in dogs. The egg is large (about 140 μm long) and spindle shaped and has a lid at one end (arrow). The thin shell is smooth, and the contents of the egg fill most or all of the shell (40× objective, unstained).

Internal Parasite 26

The tapeworm of horses (*Anoplocephala*) and that of sheep and cattle (*Moniezia*) have square- to triangular-shaped eggs. The eggs in this photo are those of the horse tapeworm, but the eggs of *Moniezia* would look similar. They have a triangular shape with rounded corners. These eggs measure 50 to 80 μm. The embryo appears as a circle in the bottom egg (arrow). This embryo contains six hooks, although they cannot be seen in this photograph (40× objective, stained with iodine).

Internal Parasite 27

Cryptosporidium is a protozoan parasite of the intestine of large animals and causes severe diarrhea. This parasite is related to the coccidia and produces an oocyst that passes out in the feces. The oocyst of *Cryptosporidium* is very small (about 4 μm) and is just visible when viewed with the 40× objective of the microscope. Even with the oil (100×) objective, it is difficult to distinguish these oocysts from yeast. The left side of this photo shows an unstained oocyst (arrow) (100× oil objective).

When a special staining procedure is used (acid-fast staining), these oocysts will stain red, while yeast will stain blue. The right side of this photo shows the oocysts (red) that have been stained with the acid-fast stain (100× oil objective).

Internal Parasite 28

Balantidium coli is a common protozoan parasite of the large intestine of pigs. The cyst of this parasite is very large (about 50 μm), and the distinctive nucleus helps to identify it (40× objective, stained).

There are many things in a fecal sample that may resemble parasite eggs and cysts but are actually plant material. Also, eggs of free-living organisms may enter an animal with its food and pass out, undamaged, in the feces.

Internal Parasite 29

The eggs on the left in this photo were seen during a fecal exam of a dog. Although they resemble hookworm eggs, they are about twice the size (110 μm) of hookworm eggs. These are the eggs of a free-living mite that the dog ingested with its food. The legs of the developing larval mite can be seen in one of the eggs (arrow) (10× objective, stained with iodine).

The cyst on the right was found in cat feces and is that of a parasitic amoeba (protozoa) of mice. This parasite does not infect cats: The cat had eaten an infected mouse, and the amoeba cysts had passed through the cat unharmed. The eggs or cysts of parasites that do not infect the animal being examined are known as spurious parasites. This cyst is about 10 μm wide (100× oil objective, stained with iodine).

Internal Parasite 30

In this photo are some common plant materials that may be mistaken for parasites during a fecal exam.

On the top left is a plant spore that is in the size range of nematode eggs. However, the nick at one end, where the spore once was attached to the plant, indicates that this is not an egg. These spores are common in the feces of large animals (40× objective, stained with iodine).

The object on the bottom left is a plant spine. It resembles the head end of a nematode larva. Nematode larvae rarely break apart, and thus the abrupt ending of this spine indicates it is not part of a worm (40× objective, stained with iodine).

The two objects on the top right that resemble strongyle-type eggs are actually plant cells. Their cell walls have an irregular outline, whereas the shells of strongyle-type eggs have a smooth, oval outline (10× objective, stained with iodine).

On the bottom right are two different types of objects that may be confused with parasite eggs. The two round objects at the top (arrows) are fat globules, and the object at the bottom (arrow) is pine pollen. There is nothing inside the fat globules—parasite eggs would have cells or an embryo inside. Pine pollen is easy to recognize; it is made up of a circular central part to which two half circles are attached (40× objective, stained with iodine).

Identification of External Parasites

External Parasites—Cause and Carriers of Disease

A parasite is an animal that lives on or in another organism and causes some effect on or disease in that organism. The organism on or in which the parasite lives is the host. External parasites live on their host and may also be called ectoparasites.

External parasites, like internal parasites, live in a particular place on the host, which is called a site or a predilection site. Some external parasites spend their entire life on the host—on the skin, in the pores of the skin, attached to the hair, or in the ears—whereas other external parasites spend only a short time on their hosts. The presence of an external parasite on a host is referred to as an infestation.

All external parasites affect their hosts in some way. Often, the level of infestation of the host will determine the severity of the effect. When there are only a few parasites on the host, they may go unnoticed; however, the presence of many may cause disease. Not only may these external parasites cause direct injury to their hosts, but they may also serve as vectors for other diseases, such as those caused by viruses, bacteria, or internal parasites. A vector carries the microorganism, or internal parasite, from one host to another. When the internal parasite, or microorganism, must spend time developing within the external parasite before it can infect an animal or a human, the external parasite is termed an intermediate host. The veterinarian identifies the presence of external parasites so that the disease they cause or transmit may be prevented or cured.

These photo cards cover the external parasites that are of major importance to veterinary medicine, that is, those that cause problems for the domesticated animals that people protect and raise. In all cases, the scientific name of the parasite, its common name, the host it infests, the type of disease that it causes, and how the parasite is identified are noted.

External Parasite 1

Fleas are ectoparasites of both dogs and cats. This photo shows the dog flea *Ctenocephalides canis*. Fleas are insects that are small (4 mm long), narrow, and dark in color, with powerful legs for jumping. They spend most of their life off the host, but when on the host, they are found most commonly around the tail and hind end of the dog.

Fleas of dogs may infest cats and vice versa. Some ectoparasites, however, may infest only one host; they are host specific. Others, such as the fleas of dogs and cats, may infest more than one host and are nonhost specific.

External Parasite 2

This cat flea is very similar to the dog flea in appearance. Identification is made by analyzing similar features. Although it is usually unnecessary to differentiate between the dog flea and the cat flea, it may be done based on the size and shape of the head and the size of the spines on the mouth comb.

External Parasite 3

This photo shows a male mosquito of the genus *Anopheles*. Mosquitoes are classified in the order Diptera, which contains all of the true flies, mosquitoes, and gnats. Members of this group have one pair of wings and reproduce. The stages of this group's life cycle are eggs, larvae, pupae, and adults. The eggs, larvae, and pupae are found in still or running water or moist soil. The eggs are laid in groups or singly, while the larvae have a breathing tube and hang from the water's surface. It is the adults that cause great annoyance to humans and animals alike. The adult mosquito is small (3–4 mm in length) and quite delicate in appearance. The mouthparts form a slender tube through which blood is taken directly from blood vessels. The thorax contains two wings; the abdomen is slender and may be banded. Blood may often be seen in the abdomen of an engorged female after a blood meal. The male feeds on plant juices and may be distinguished from the female by the shape of its antennae. The male antennae are very fluffy and are termed *plumose*; those of the female are slender and less fluffy.

The adult female prefers to suck blood from certain hosts at specific times of the day. Therefore, the mosquito makes a very good vector for the transmission of disease from one host to the next. Anopheline mosquitoes transmit the dog heartworm *Dirofilaria immitis.* In this case, the mosquito serves as an intermediate host, since the parasite develops from a first-stage to a second-stage larva in the gut of the mosquito. When the mosquito bites a dog and sucks blood, the infective third-stage larvae of the dog heartworm are injected into the skin and begin migrating to the heart of the dog. The bite alone of the mosquito can cause great annoyance to its host. When the mosquito bites, it injects saliva into the area around the bite to allow the blood to flow more readily. This saliva is very irritating and can cause large bumps on the skin, which is a form of allergic response.

External Parasite 4

This photo shows the *Culex* mosquito. Although not clear in this photo, the *Culex* mosquito has horizontal bands across the abdomen like those of the anopheline mosquito. This mosquito can transmit *Dirofilaria immitis* to dogs, as well as western and eastern encephalitis viruses to horses.

External Parasite 5

This photo shows a *Cuterebra* sp. fly. These flies are large (20 mm or more in length), with beelike bodies and tiny mouthparts. Their larvae cause problems in animals. These flies lay their eggs in the soil, usually close to rodent or rabbit burrows. The larvae hatch and penetrate the skin of the baby animals. Dogs and cats near the area may be infested as well. The larvae are large (25 mm in length), stout, dark in color, and covered with bands of spines. The larvae feed on the tissues of their host and cause tissue damage and skin irritation. The puppy, kitten, or rabbit that is bitten will have a large swelling on the skin around the larvae. This condition is most often seen in the warm summer months. The best treatment is surgical removal, and animals usually do well.

Infestation with *Cuterebra* larvae is a form of myiasis, a condition in which the larvae of flies enter the host through a wound in the skin, through the nose, or by direct penetration of the skin and feed on the tissues of the host while they develop to the next stage.

External Parasite 6

The stable fly, *Stomoxys calcitrans*, looks very much like the common housefly. It is gray in color and approximately 6 to 8 mm in length. Its mouthparts form a stout needle that is thrust through the skin. The end of these mouthparts has a rasping apparatus that lacerates the tissues, causing a pool of blood to form. The fly then sucks in this blood. The common housefly *Musca domestica* can be differentiated from the stable fly because it does not have these piercing mouthparts. The housefly prefers instead to sponge up fluids. The stable fly can cause annoyance to all domesticated animals and may transmit worm parasites of the eye and stomach (such as *Thelazia* and *Habronema*) in the horse.

External Parasite 7

Musca autumnalis looks identical to the common housefly (*Musca domestica*). This fly feeds in the same way as the common housefly, by sponging up blood and tissue juices from its host. It causes most problems to the host because it prefers to feed from the secretions in and around the face of the host. These flies are often seen crawling around the eyes of cattle, horses, and dogs, where they can produce swelling and irritation. They frequently carry bacteria and viruses from the eyes of one animal to another, because they repeatedly go from one host to another. They are also associated with dirty conditions around the barnyard, close to areas where manure is stored.

External Parasite 8

The horsefly, or *Tabanus* sp., is a large, sturdy fly with powerful wings and large eyes. Tabanids have clear wings with characteristic veins that may be used to tell them apart from other related flies. The female is known for the deep, painful wounds she inflicts in the skin of a horse when she feeds. These flies have bladelike mouthparts that they use to slash the skin of the host, causing a pool of blood to form. Later, other flies may be attracted to these large wounds in the skin, gathering there and causing further pain and annoyance to the horse. These flies may mechanically transmit bacteria and viruses from a horse or other animal to another one in the area. When blood from the large pool sticks to the legs or mouthparts of these flies, it is then spread on or injected into the next host on which the flies feed.

External Parasite 9

The *Chrysomere* sp. fly is a medium-sized fly with banded wings that are at an angle. The antennae are made up of three segments of equal length. The female fly feeds on many animals and causes much damage to the skin due to the biting mouthparts. Flies of these and other species may be controlled to some extent by the application of chemicals to the skin of the host. These either repel or kill the flies when they come to feed. Pyrethrins are chemicals that have often been used to protect horses and cows. They usually come in an oily spray that is applied to all areas of each animal's skin except around openings such as the eyes, nose, and genital organs.

External Parasite 10

The flesh fly, *Sarcophaga* sp., is a medium-sized fly with a gray checkerboard pattern on its abdomen. It feeds and flies around during the day and can cover long distances because of its strong flying ability. Larvae and pupae of this fly are most often found in decaying matter in the environment, but occasionally the flesh fly infests wounds of animals at pasture and lays eggs in the wounds. When this happens, the larvae and pupae that develop from the eggs can cause a lot of tissue destruction. Also, because these flies frequent dirty areas where there is dead material present, they can carry any bacteria and other pathogens mechanically on their bodies and deposit them into the wounds when they lay their eggs.

External Parasite 11

The *Callitroga hominivorax* flies are often called metallic flies or blowflies. They have a metallic greenish blue color, with an orange face and three dark stripes on the abdomen. As adults, they feed by sponging up fluids from the host and therefore have mouthparts that are adapted for this purpose. The reason *Callitroga hominivorax* flies are important in veterinary medicine is that they develop only in living tissues. The females lay their eggs on the edge of a fresh wound; then the eggs hatch, and the larvae feed from the wound. They fall to the ground to pupate and complete their life cycle. The animals most likely to be parasitized by these flies are cattle, horses, and pigs; but dogs, humans, and birds may also be affected. Because of these metallic flies, it is advisable to postpone elective surgeries on farm animals to the colder months when such flies are not present.

External Parasite 12

The ox warble, or *Hypoderma bovis*, fly is a beelike fly. These flies are from the family Oestridae and characteristically look like bees. They are hairy, with eyes that are set wide apart and mouthparts that are not used to bite the host. Instead, the females lay their eggs on the hair shafts of a cow. Larvae, as shown in this photo, develop from these eggs, penetrate the skin, and migrate into the connective tissue, where they grow. They escape by creating holes in the skin of the cow's back and fall on the ground to pupate. The major problem caused by these flies is not disease, since they are well tolerated by the cow. Rather, they create an economic problem: The larvae make many holes in the hide, destroying its value as leather.

External Parasite 13

The sheep nasal fly, or *Oestrus ovis*, is another of the beelike flies. The adult fly has a dark gray color, with small black spots on the thorax, and it is covered with light brown hair. It also has wide-set eyes and mouthparts that are nonfunctional. The female fly lays her eggs in the nasal passage of sheep, where they develop in the nose and sinuses. The larvae cause a great deal of damage to these areas, making it very difficult for the sheep to breathe. This photo shows two larvae that might be found in the nose of a sheep. Frequently, the larvae grow so large in the sinuses that they are unable to escape. In such a case, they may die, causing infections in the nasal and sinus passages of the sheep. In the parts of the country where this fly exists, sheep farmers have a hard time protecting their flocks from this pest.

External Parasite 14

The horse botfly, or *Gastrophilus* sp., is a beelike fly. The adult flies are brown and hairy and have nonfunctional mouthparts. The only stages of this ectoparasite that feed are the larval stages. The adult female fly lives for only a short time, during which she deposits her eggs on the hairs of a horse. The horse may ingest these eggs in grooming or scratching itself. Then the eggs hatch, and the larvae penetrate the tissues of the mouth. They finally develop and are swallowed, attaching themselves to the wall of the stomach and developing further. They remain inside the horse for 10 to 12 months and may cause irritation to the stomach.

External Parasite 15

This photo shows the larvae of the horse botfly *Gastrophilus* sp. They are parasitic maggots made up of 12 segments, of which the first two are fused together. By means of mouth hooks, which they usually have, the larvae attach to the stomach lining. Many of the species have rows of spines on each segment. Organophosphate worm medication is usually effective in getting rid of these larvae from the stomach of the horse, but the infection may be prevented by removing the eggs from the hair shafts as soon as they are laid.

External Parasite 16

The black fly, or *Simulium* sp., is a small fly (1 mm–5 mm in size), dark gray to black in color, with a humped thorax, short hornlike antennae, broad, clear wings, and short legs. The female fly sucks blood from many animals, especially cattle and horses, but dogs, humans, and birds may be attacked also. The bites are very painful and cause toxic and allergic reactions in the skin, as well as considerable blood loss.

In particular these flies are frequently found in the ears of horses, where they cause an annoying dermatitis, especially in the spring. The eggs of these flies are laid in fast-moving streams, frequently near rapids, thus becoming particularly annoying to animals in these areas.

External Parasite 17

Culicoides sp. flies are frequently called no-see-ums because of their very small size. They are so small that they can pass through ordinary window screen. These flies may be recognized by their small size, their humped thorax, and hairy wings. The mouthparts are adapted for sucking blood; they have a short proboscis (mouthpart), with jaws that act like scissors to break the skin of the animal they are attacking. The females of this group attack humans and animals to get blood and can be very annoying. Their bites are irritating, causing swelling and itching that often must be treated. This type of allergic reaction is frequently seen in the horse, where fly bites from *Culicoides* sp. can cause swelling and itching of the horse's belly during the summer months.

External Parasite 18

The sheep ked, *Melophagus ovinus*, which is found in most parts of the world, is wingless and hairy, with a leathery appearance, and is about the size of a tick (4–6 mm long). The head is short and broad, the thorax brown, and the abdomen grayish brown. It has heavy claws on stout legs that it uses to hold on to the sheep. This ectoparasite may be easily spotted on white sheep due to the dark color. It causes damage to the sheep with its piercing mouthparts, which can open wounds in the skin. These wounds may make the sheep susceptible to attack by flies, or they may become infected with bacteria or protozoa. The sheep ked may also cause great economic loss to the sheep industry because it damages the sheep's wool.

External Parasite 19

Lice are flat and wide and have no wings. They are pale in color and about 2 to 3 mm long. The sucking louse, *Linognathus setosus*, has a head that is more slender than the thorax, with mouthparts designed to suck the host's tissue, body fluids, and blood. This louse has stout legs that are used for grasping the hairs of the host; it has no eyes. The louse pictured here is an ectoparasite of dogs and foxes. It causes the host a great deal of irritation and itching and is transferred by direct contact with an infested animal. It can also be transferred by combing an infested animal and then using the same comb on another animal.

External Parasite 20

Trichodectes canis is a biting louse. This type of louse looks very much like the sucking louse. It has no eyes, and its body is broad and flat. However, the head of the biting louse is wider than the thorax and as wide as the abdomen. The mouthparts of the biting louse are very different from those of the sucking louse. They are designed to bite and chew, and they feed on debris on the surface of the skin rather than suck like the other group of lice. *Trichodectes canis* has been shown to transmit the internal parasite *Dipylidium caninum*, the dog tapeworm, from one dog to another.

External Parasite 21

Ticks are divided into two groups: hard ticks (Ixodidae) and soft ticks (Argasidae), depending on whether they have a shield on their backs. The hard ticks have the dorsal shield. There are many complex characteristics that are used to identify each species of tick. This information is grouped into a table called a key. Ticks cannot be identified based on their size, because their size changes greatly, depending on whether they have sucked any blood. Also, the young nymph ticks look much like the adults, but they are very small.

Ticks spend a great deal of their life off the host but they are dependent on the host for a blood meal, which allows them to reproduce. Each species of tick has different feeding habits. Some feed on only one host, others feed on two hosts, and still others feed on three hosts. The greater the number of hosts from which the tick takes blood, the greater the chance for ticks to transmit disease from one host to the next.

The *Rhipicephalus sanguineus* tick is an Arachnida tick. This brown dog tick is a three-host tick, which means that it feeds on three hosts during its life cycle, but it is found mainly on the dog. This is a hard tick, having a hard dorsal shield and a rather plain appearance. This tick has eyes and festoons, which are small, squared-off grooves on the back edge of its shell. The hypostomes, or mouthparts, are short, as are the palps, which the tick uses to feel where it is in space. This tick can spread many diseases from one host to the next, such as, for example, the virus Rocky Mountain spotted fever.

External Parasite 22

The deer tick is a hard tick from the group Ixodidae. This tick does not have eyes or festoons. The most important characteristic that distinguishes this tick from other hard ticks is the ring that surrounds the front of its anus. The anus of the tick is found by turning the tick over and looking for a small opening in the middle of the abdomen. A distinct groove surrounds the opening. *Ixodes dammini* is very important to veterinarians because it can transmit the *Rickettsia* that causes Lyme disease in humans and dogs. This is a disease that has become common on the East Coast of the United States. The germ that causes Lyme disease is *Borrelia burgdorferi*. This disease is serious in both dogs and humans, causing fever and pain in the joints. Dogs and humans are at risk for getting the disease when they go into wooded areas where the deer tick lives.

External Parasite 23

Dermacentor variabilis, the American dog tick, is very common in dogs in the United States. This tick is ornate and has eyes, plus festoons on the back of its shell. As in *Rhipicephalus*, the palps are short. This tick is most numerous in the spring and summer months, since it needs moisture to survive in the environment. However, all stages of this tick are able to survive the cold temperatures of winter. The dog tick is also a three-host tick and can spread Rocky Mountain spotted fever.

External Parasite 24

Otobius megnini is a soft tick that gets its name, *spinose ear tick*, from the site where it is most likely to be found on its hosts. This tick is found on many domestic animals, including dogs, horses, cows, goats, pigs, and cats. The larval and nymph stages of these ticks are ectoparasites of these animals. The adults are not parasitic and spend all their time in the environment. This is a one-host tick, meaning that whatever host it chooses to take blood from, it will remain on that host and not move to another host.

The main problem caused by this tick is one of irritation. They enter the ear to feed, which causes great pain and annoyance to the host. Dogs that have these ticks will shake their heads and scratch their ears. The veterinarian will usually find a waxy or bloody material in the ear canal. The ticks must be removed by hand from the ear of the animal.

External Parasite 25

Mites are the second most important group of arachnids that causes problems for the domestic animals. They are so small that they cannot be seen with the naked eye. The mange mite, *Sarcoptes scabiei*, is one of the most common of the mites. It is small (0.2–0.5 mm long) and has a round body. The legs of the mange mite are short, and spines stick out from the side. The female mites burrow into the skin to lay their eggs. The eggs hatch, and the larvae come out to the top of the skin and wander around. All of this activity causes intense itching in the skin of the dog. The skin becomes very red, and the dog will bite and scratch until much of the hair has been removed from the body. The best way a veterinarian can tell if the dog is infected with this mite is by taking a scalpel blade and scraping the skin until a small amount of blood is seen. This scraping is then placed on a slide under the microscope to examine it for the mite.

External Parasite 26

Demodex canis is a mite that lives in the hair follicles and sweat glands in the skin of the dog. It belongs to a group of mites that is very specialized in its selection of a site in which to live. This means that a *Demodex* mite from a dog cannot live in the skin of a horse and vice versa. The body has a long cigar shape. It is small, about 0.25 mm long, with short, stumpy legs located in the front half of the abdomen. This mite also causes mange in the dog. It is found in the skin of healthy dogs, however, and causes disease only when the dog's immune system is compromised. Dogs that have this mange are not as itchy as dogs with sarcoptic mange, but there is thickening of the skin, redness, scaling, and loss of hair. This mite and other skin mites in the dog may be killed with organophosphate dips applied to the surface of the skin. The veterinarian must find the underlying cause of the mite infestation so that it will not return when the treatment is ended.

External Parasite 27

Notoedres cati is the mite that infests the skin of the cat. It causes mange-like skin irritations that are usually seen around the ears and neck but may extend to the face and feet of the cat. It is very small and looks similar to the sarcoptic mite in the dog, with a round body, short, stumpy legs, and bristles that extend from the rear of the body. This mite can also infest rabbits and foxes and causes very itchy skin in all of these. If a cat is infected, the treatment a veterinarian will probably recommend is to apply lime sulfur dip with a sponge to the affected area, since the cat cannot tolerate many of the organophosphate compounds that are used in other species.

External Parasite 28

The ear mite, *Otodectes cynotis*, infests both the dog and the cat. It burrows into the skin of the ear canal, causing intense irritation, redness, and swelling. The ear responds by making lots of dark black wax. Because of the intense irritation, animals react by scratching and shaking their heads. The mite itself is about the same size as the sarcoptic mite, with an oval body, but the four pairs of legs are much longer. Identification may be made by taking a swab of the material in the ear canal, smearing it on a slide, and looking at it under the microscope on low power. To treat the animal for this mite infestation, the veterinarian first cleans the ear with a mild antiseptic solution. Then the veterinarian recommends to the owner to place a solution of oil with the drug thiabendazole and an antibiotic in the ear canal for several days in a row. Since this condition is very contagious, spreading from one animal to another when they touch each other, all animals living together in one household at the same time must be treated.

External Parasite 29

Cheyletiella is a mite that causes a condition called walking dandruff in rabbits, cats, and dogs. It is so called because it looks like dandruff that is moving, especially along the fur on the back of the host. This ectoparasite may infest people who come in contact with infested animals. The mite itself has a small, round body, less than 1 mm in size. It has a hooklike structure on the head and characteristic bristles and combs on the legs. These mites do not burrow into the skin, but the hook on their head pierces the skin while they are feeding on tissue fluids. They can cause dandruff and scaling of the skin, but they mainly cause itching.

Types of Animal Diseases

Detecting Illness

This part, in conjunction with the Animal Disease Photo Cards, shows what a veterinarian sees when diagnosing canine and feline diseases, including tumors. When diseases are not discovered, the resulting health problems can cause serious consequences for the animals—severe illness and even death.

Canine Diseases

Animal Disease 1

Malignant melanoma is the most common malignant tumor of the canine oral cavity. Disease 1 shows a dark brown to black lobulated mass involving the front of the jaw of a dog. This melanoma arises in the gums and invades locally, sometimes destroying bone. The tumor often spreads to the nearby lymph nodes and lungs.

Animal Disease 2

The skin of this dog contains numerous scaly, firm, raised patches. This is one form of cutaneous lymphosarcoma, a tumor made up of sheets of lymphoid cells. Cutaneous lymphosarcoma almost always spreads to internal organs.

Animal Disease 3

The histiocytoma shown here is an extremely common benign tumor, usually arising in young dogs. This photo shows its usual appearance—hence the common name of "button tumor." These tumors usually regress on their own.

Animal Disease 4

The kidneys of this neonatal dog are speckled with numerous pinpoint hemorrhages (petechiae), caused by canine herpesvirus infection. Herpesvirus typically affects dogs less than 2 weeks old, resulting in necrosis (death of cells) and hemorrhage in many organs, including lungs, liver, and kidneys. It is almost invariably fatal. Dogs in groups, such as in breeding colonies or kennels, are at most risk.

Animal Disease 5

This dog's liver has many irregular nodules separated by fibrous bands. This is the appearance of cirrhosis, which is the final result of many insults to the liver, including toxic necrosis or chronic active hepatitis. Damaged hepatocytes are replaced by fibrous connective tissue, while undamaged hepatocytes multiply, leading to new nodules.

Animal Disease 6

In parvoviral enteritis, the parvovirus attacks rapidly dividing cells. The cells lining the crypts of the intestine are especially vulnerable, leading to the necrosis and red discoloration present in this photo. Lesions in the intestine are often segmental.

Animal Disease 7

This photo shows the "punched-out" Peyer's patches typical of this form of parvoviral enteritis of the intestine. Peyer's patches are localized lymphoid tissues of the intestine that become depleted when infected with parvovirus.

Animal Disease 8

Calcinosis circumscripta is a nodular lesion seen in the skin and tongues of young, large-breed dogs. The chalky white material, which is primarily mineral, has the consistency of granular toothpaste. The cause is unknown. The cure is to completely remove, or cut out, the lesion.

Animal Disease 9

The heart base tumor is a tumor that arises from the aortic body (a small organ at the base of the heart that senses the amount of oxygen and carbon dioxide in the blood). Brachycephalic breeds, such as Boston terriers and boxers, are most prone to developing this tumor, which often results in hydropericardium (excess fluid in the pericardial sac). This condition could result in heart failure.

Animal Disease 10

The coronary vessels in this photo show the yellow plaques characteristic of atherosclerosis. This accumulation of fat within the vessel walls is rare in the dog, but when it occurs, it is almost always associated with hypothyroidism (deficient activity of the thyroid gland).

Animal Disease 11

The stifle joint of this dog is replaced by a large tumor. Osteosarcoma, a malignant tumor of osteocytes, is the most common bone tumor in the dog. It almost invariably spreads to the lungs.

Animal Disease 12

The interstitial cell tumor, the most common testicular tumor of the dog, is encapsulated. It is colored yellow and red. It is benign and cured by castration.

Animal Disease 13

Hemangiosarcoma is a common tumor of German shepherds, golden retrievers, and Irish setters. It is a malignant tumor of blood vessels, and it often arises in the spleen, right atrium of the heart, liver, and lung. Dogs may die from severe hemorrhage associated with the bloody masses.

Animal Disease 14

Squamous cell carcinomas of the tonsils of dogs are extremely aggressive tumors, which spread to many lymph nodes and internal organs. These carcinomas appear as firm, gray-tan irregular masses in the oropharyngeal area.

Animal Disease 15

This photo shows an opened colon heavily infested with *Trichuris vulpis*, the so-called whipworm of dogs. The disease is known as trichuriasis. Heavy infestations may result in death due to intestinal bleeding and dehydration from uncontrollable diarrhea.

Animal Disease 16

Lung lobe torsion is a rare condition in which one or more lobes of the lung may become twisted at the base, cutting off blood flow into or out of the lobe. The affected region remains engorged, and necrosis occurs due to lack of oxygen.

Animal Disease 17

This photo shows a skin tag, an elevated piece of excess skin often found in older dogs. It is of unknown cause. The cure is to excise it, that is, to cut it out.

Animal Disease 18

This photo of the opened left ventricle of a dog's heart shows irregular, warty growths on the leaflets of the mitral valve. The growths are composed of inflammatory cells, bacteria, and fibrin. In many instances, these "vegetations" are the result of bacteria circulating in the blood from an infection elsewhere in the body. The disease is called vegetative endocarditis.

Animal Disease 19

The circular regions in this kidney are regions of infarction—localized necrosis due to vascular blockage. In this case, the blockage is due to bits of the "vegetations" on the mitral valve that have broken off, circulated in the blood, and eventually lodged in renal arterioles. If these septic infarcts damage enough of the renal tissue, renal failure may occur.

Animal Disease 20

This photo is of an opened urinary bladder with attached ureters and transected kidneys. There is a gray, plaquelike growth at the neck (trigone) of the bladder. Transitional cell carcinoma usually occurs in older female dogs. This malignant tumor may block the ureters, preventing urine outflow. The result is dilated ureters and dilated renal pelvis.

Animal Disease 21

As shown in this photo, there are numerous well-defined, often wedge-shaped, gray-brown areas rimmed in black along the edge of the spleen. These are infarcts, localized areas of necrosis associated with loss of blood to these areas. Many older dogs have splenic infarcts. Most are of unknown cause; however, some are associated with tumors that may fill or disrupt blood vessels.

Animal Disease 22

This photo shows two cross sections of a brain with a pituitary adenoma, which compresses the hypothalamus and thalamus. Although this type of tumor is benign, the compression of structures in the area may lead to a variety of clinical signs. Brachycephalic breeds, such as boxers and Boston terriers, are the most commonly affected.

Animal Disease 23

Two ovaries with attached uterine horns are shown in this photo. Each ovary has numerous thin-walled cysts, called ovarian cysts, which contain clear fluid. Some cysts may secrete excess estrogen or progesterone.

Animal Disease 24

This photo shows the consequences of some ovarian cysts. Excess hormones from the cysts lead to excess growth of uterine glands, often with cyst formation. The opened uterus shown here has cystic endometrial hyperplasia. This uterine change often predisposes the animal to pyometra (bacterial infection of the uterus with pus formation).

Animal Disease 25

In this photo are shown a normal fetus and a littermate with anasarca, generalized edema. The excess fluid gives the animal its swollen, wet look. The cause of fetal anasarca is unknown, but it is incompatible with life. The animal dies late in gestation.

Animal Disease 26

The tan-red lobulated mass shown in this photo surrounds the molars of a dog's jaw. This mass, or acanthomatous epulis, is a tumor of the tissues that make up the teeth and their surrounding tissues. It is locally invasive and often results in bone destruction. Removal of a large portion of the jaw is usually necessary to prevent recurrence.

Animal Disease 27

The mitral valve of this dog's heart is diffusely thickened and nodular, resulting in endocardiosis, a degenerative change seen commonly in old dogs. When severe, endocardiosis may lead to valvular insufficiency (failure of the valve to completely close) and heart failure.

Animal Disease 28

This photo shows a cross section of a dog's eye. At the lower edge is a black mass extending from the iris to the back of the eye. This tumor, known as a uveal melanoma, has arisen from melanocytes of the anterior uvea (iris and/or ciliary body). The majority of melanomas arising in this site are only locally invasive and necessitate removal of the eye. A few spread to internal organs.

Animal Disease 29

This photo gives a close-up view of the inside of a dog's rib cage, showing the pleural surface. There are parallel stacks of minerals just underneath the pleura between the ribs. This uremic mineralization is common in animals with chronic renal disease and is thought to result from the toxins associated with uremia (urine in the blood).

Feline Diseases

Animal Disease 30

This photo shows the organs from the thorax of a cat with dyspnea (a condition characterized by difficulty in breathing). There is a large, many-lobed pink mass involving the cranial mediastinum and surrounding the base of the heart. This is a common tumor of lymphocytes in the cat, known as cranial mediastinal lymphosarcoma.

Animal Disease 31

The tip of one lobe of the liver shown here has a well-rounded, many-lobed mass with numerous small, thin-walled cysts. The biliary cystadenoma is a common benign tumor of old cats, arising from bile duct epithelium. Surgical excision is the cure.

Animal Disease 32

Two thyroids from a cat are shown, each containing markedly enlarged white-tan parathyroids. (In the cat, normal parathyroids are not usually visible.) Bilateral parathyroid hyperplasia is seen in all vertebrates and is reflective of calcium-phosphorus imbalances associated with chronic renal disease or improper diet.

Animal Disease 33

This photo shows cross sections of a normal feline heart (above) and a heart with hypertrophic cardiomyopathy (below). The increased thickness of the left ventricular wall is evident, as is the obliteration of the lumen. The cause of this lesion is unknown, but it leads to cardiac failure and may result in sudden death.

Animal Disease 34

Shown in this photo is an opened distal aorta at the branch into the external iliac arteries. A cylindrical red-purple embolus has lodged at this site, cutting off the blood supply to the distal extremities. (An embolus is a piece of a fixed blood clot that has broken off and circulated in the blood.)

Animal Disease 35

This cat's trachea is exposed, showing a large, irregular, many-lobed thyroid (the larger brown bulb) next to a smaller-than-normal thyroid (the smaller brown bulb). The large thyroid has undergone adenomatous hyperplasia, with production of excessive thyroid hormones. The resultant hyperthyroidism is manifested by weight loss, polyphagia, and increased heart rate. This lesion is of unknown cause and may involve one or two thyroids.

Animal Disease 36

In this photo, the cat's kidneys are infected with a coronavirus, leading to the disease feline infectious peritonitis (FIP). Many organs may be affected, with the formation of nodules composed of inflammatory cells. This disease is most often seen in cats less than 2 years of age and is usually fatal.

Animal Disease 37

Below this cat's tongue is a firm mass. Squamous cell carcinoma is the most common malignant tumor of the feline oral cavity. Its growth beneath the tongue makes it difficult to remove surgically. Tumor cells often spread to lymph nodes and lungs.

Animal Disease 38

The liver in this photo is discolored yellow and is greasy to the touch due to hepatic lipidosis. Sections of this liver float in water. The individual liver cells are swollen because of excessive amounts of fat in the cytoplasm. This fatty change is of uncertain cause; however, this syndrome is often seen in obese cats that go off feed for extended periods of time. Many cats die but some survive with long-term supportive therapy.

Animal Disease 39

This colonic mesentery of a cat contains numerous smooth, oval structures that can be mistaken for parasites. These structures are normal pressure receptors of the cat, called pacinian corpuscles.

Animal Disease 40

These kidneys contain several soft, bulging, pink-gray masses. This is a typical presentation of renal lymphosarcoma in the cat. Other organs may be similarly affected. Impression smears would show sheets of neoplastic lymphocytes.

Animal Disease 41

This photo shows that the thoracic cavity of this cat is filled with a thick brown fluid and the lungs are covered by a creamy tan exudate. This is an example of pyothorax, or pus in the thoracic cavity, usually caused by *Actinomyces* sp. or *Nocardia* sp. These bacteria are often carried into the chest by the penetration or inhalation of foreign bodies.

Animal Disease 42

Feline infectious peritonitis (FIP) may affect neural tissue as well as other organs. One classical lesion is FIP ventriculitis, shown here as a bluish accumulation of fibrin-rich fluid in the ventricles of the brain.

Animal Disease 43

Portions of the abdominal and subcutaneous (beneath the skin) fatty tissue of this cat are discolored. There is also an associated "fishy" odor. This disease, called steatitis, is an inflammation of the fat associated with increased polyunsaturated fats in the diet, resulting in a relative vitamin E/selenium deficiency. There is resultant necrosis and inflammation of the fat. This disease is usually very painful.

Large Animal and Avian Diseases

Animal Disease 44

This photo shows the opened abdominal cavity of a calf. A region of small intestine has twisted 360 degrees around its mesenteric root, resulting in lack of blood flow into or out of the area. As in this case, intestinal volvulus leads to necrosis of the involved portion of gut and to the death of the animal due to the toxins released from the damaged gut wall.

Animal Disease 45

This mammary gland from an adult cow has areas of red and green-brown discoloration. There are necrosis and hemorrhage due to invasion by virulent bacteria. *Staphylococcus* sp. is the most common cause of gangrenous mastitis, the disease shown in this photo.

Animal Disease 46

Contagious ecthyma, also called orf, is a disease of sheep and goats caused by a poxvirus. The lips of the sheep in this photo are covered by scabby growths. Similar lesions may be seen on the face and feet. The viral infection is not life threatening; however, young animals that are affected will not suckle or graze, leading to possible body deterioration.

Animal Disease 47

Chronic obstructive pulmonary disease, also called heaves, is a serious disease of horses because of the increased respiratory effort seen in affected animals. This photo shows the cut surface of the lung with numerous distended airways (bronchiectasis). The cause is uncertain, but allergic responses to inhaled substances are considered to be an important component in the development of this disease.

Animal Disease 48

The skin and muscles have been removed from the jawbone of this cow to reveal the moth-eaten appearance of the skeleton. The animal became infected with *Actinomyces bovis*, which led to the inflammation of the bone and the disease actinomycosis. The bone is often deformed, hence the common name given to this disease—lumpy jaw. Affected animals often have difficulty eating and are destroyed.

Animal Disease 49

This horse kidney has a single, well-defined white mass, or renal adenoma, within the cortex. Renal adenomas are not uncommon in horses. They are benign tumors of no clinical significance and are usually an incidental finding at postmortem.

Animal Disease 50

This photo presents the opened colon from a pig showing a marked narrowing of the rectum a few centimeters from the anus. This lesion is preceded by inflammation of the area, caused by the bacteria *Salmonella typhimurium*. Because of poor vascular supply to this area, healing leads to rectal stricture. Intestinal obstruction and death result.

Animal Disease 51

The leg of this calf shows the regional white discoloration typical of nutritional myopathy, more commonly called white muscle disease. Animals grazing pastures deficient in vitamin E and selenium will have multifocal necrosis of skeletal and cardiac muscle. Severe cardiac involvement may result in death.

Animal Disease 52

This lesion involving the vulvar lips of a horse is called coital exanthema and is caused by equine herpes virus type 3. Watery blisters and pustules are found on the genital organs of mares and stallions. The lesions usually heal by themselves within a few weeks after infection, leaving areas of pigment loss.

Animal Disease 53

This photo shows a normal intestine (above) and an intestine from a calf with Johne's disease (below). Infection with *Mycobacterium paratuberculosis* results in severe inflammation, leading to the widespread thickening evident here. Young animals are most commonly affected and have profuse watery diarrhea, which may lead to dehydration and death.

Animal Disease 54

The cross sections of the noses of three pigs are shown here to demonstrate the progression of lesions seen in atrophic rhinitis. With time, there is moderate to severe atrophy and loss of the cone-shaped bones in the nasal cavity. This disease occurs commonly in young pigs and is a major cause of economic loss in pig-raising areas of the world. The cause is uncertain, but it appears to be related to infection by *Bordetella bronchiseptica* and *Pasteurella multocida*. Facial deformity and upper respiratory signs are usually evident by 6 weeks of age.

Animal Disease 55

This chicken's trachea contains a white plug of dead tissue and inflammatory cells. Infectious bronchitis is caused by a coronavirus and results in mild to moderate upper respiratory infections. The infections are not usually life threatening but result in decreased egg production.

Animal Disease 56

This opened horse's stomach reveals two common lesions. Bots is the lay term for the larvae of the botfly of the genus *Gasterophilus*. These larvae spend a portion of their life cycle in the stomach of the horse. Severe infestations may result in chronic inflammation and ulceration. The raised area with four burrow marks is caused by a *Habronema* nematode. These nematodes usually cause only mild gastric inflammation.

Animal Disease 57

Ventricular septal defects are the most common heart defects in the newborn bovine. The calf in this photo had a defect (the large, open hole) just below the aortic valve. Most of these defects are clinically insignificant and are incidental findings at postmortem.

Animal Disease 58

The muzzle of this cow is ulcerated and reddened. Although nonspecific, this change is typically seen in malignant catarrhal fever, a disease of worldwide importance. In Africa, the disease is caused by a herpes virus. The virus results in vascular necrosis, resulting in lesions involving the skin and many other organs.

Animal Disease 59

This cockatoo has marked feather loss, which in this case is due to self-mutilation. Feather picking may be associated with boredom, or it may be a reflection of internal diseases such as endocrine imbalances.

Animal Disease 60

This opened jejunum from a calf reveals the classic plug of tissue, or "cast," seen in intestines infected with *Salmonella* sp. The disease is intestinal salmonellosis, and the cast is composed of dead intestinal epithelial cells, fibrin, and inflammatory cells. Severe infections often lead to death.

GLOSSARY

A

abomasal worms (*Ostertagia,*
Haemonchus, and *Trichostrongylus*):
Important endoparasites of the
stomach of cattle, sheep, and goats.

abomasum: Fourth compartment of the
stomach of a cow, sheep, goat, or deer
(ruminant).

abscess: A localized collection of pus in
tissue.

acaricides: Pesticides that kill ticks and
mites.

acquired immunity: Exposure to an
antigen that results in antibody
production in response to that
antigen; also known as *active*
immunity.

active immunity: Exposure to an
antigen that results in antibody
production in response to that antigen;
also known as *acquired immunity.*

acute infection: Type of infection
that occurs when resistance is
overwhelmed. Also known as a
peracute infection.

acute inflammation: An inflammatory
episode occurring 4 to 6 hours
following the stimulus and remaining
fairly constant in appearance.

acute inflammatory event: The
reaction of living tissue to a local
injury in which the reaction has a
fairly rapid onset and a clear and
distinct end.

adaptation tolerance: The characteristic
of a parasite that makes it
immunologically inert, or invisible to
the host.

adhesions: Inflammatory bands that
connect opposing serous tissues.

adverse drug reaction: A response to
a drug that is serious, undesirable,
and possibly life threatening. These
reactions occur at dosages normally
used for prevention, diagnosis, or
treatment.

agenesis: The absence or imperfect
development of a part of the body due
to deficient growth.

American dog tick (*Dermacentor*
variabilis): A three-host hard tick. The
larvae and the nymphs prefer to feed
on rodents and rabbits. The adults
feed on dogs, cats, and such wild
carnivores as foxes and coyotes.

amyloidosis: An accumulation of
amyloid between cells and fibers of
tissues and organs. Also known as
amyloid degeneration.

anaplasia: The failure within a cell to
differentiate parts of the body.

anastomosis: A joining of the capillaries.

anemia: A disease in which the blood
cannot carry enough oxygen to
support the tissues of the body.

aneuploid: Possessing an abnormal
number of chromosomes that is not a
multiple of 23 (such as 24).

This glossary gives the
definitions of all important
key words in this book. Take a
moment to study the terms you
have just learned. Then be sure
to review these words and their
definitions before taking an
examination or quiz.

Anoplura: An order of insects that includes bloodsucking lice.

anorexia: A condition in which an animal will not eat due to a loss in appetite.

anthelmintic: A substance destructive to worms.

antigen: Chemical that has the ability to cause an immune reaction within the body.

antimicrobial agent: A therapeutic agent that tends to destroy microbes or prevents their growth or pathologic action.

aplasia: A type of agenesis, referring specifically to the total failure of an organ to develop.

apoptosis: Endogenous programmed cell death.

arachnid: An animal in which the adult has eight legs and two body parts (head and abdomen); an arthropod such as a spider, scorpion, mite, or tick.

arrested larvae: Larvae with slowed development. Also known as *larval quiescence*.

ascarid: Roundworm.

ascites: Fluid in the peritoneal cavity.

asymptomatic carrier: An animal that has eliminated a virulent infection but that was never stricken with the disease.

atrophy: The reduction in size of a fully developed tissue.

attenuation: A dilution process in the manufacture of vaccine, used to diminish the virulence of an organism but retaining its antigenicity.

autoimmune hemolytic anemia: The destruction of red blood cells.

autoimmune reaction: Process by which normal cells provoke the formation of autoantibodies, which results in the destruction of normal tissues.

autoinfection: A process by which first-stage larvae of threadworms develop to the infective stage while they are in the host's large intestine and then burrow through the intestinal wall and find their way back to the small intestine, where they become adult females.

autosomal recessive: A type of Mendelian genetic disorder in which the disease does not generally affect the parents; instead, one in four offspring are affected upon inheriting two copies of the gene—one from each parent.

B

babesiosis: A disease caused by infection with the protozoan parasite *Babesiidae*.

bacteremia: A condition by which the body is unable to keep bacteria from circulating into the blood.

bacterial toxin: A chemical produced by bacteria that can cause disease and even death.

bacteriophage: A virus that infects bacteria.

basophil: White blood cell; respond to allergic reactions.

Baylisascaris procyonis: A large roundworm found in the raccoon and one of the causative agents of human VLM (visceral larva migrans).

beef tapeworm of humans (*Taenia saginata*): A taeniid tapeworm whose larvae reside in cattle muscles. Adult taeniid tapeworms can live in humans as well as dogs.

benign: Lacking the ability to invade and grow at sites away from the original tumor.

bilirubin: A substance found in the blood, an excess of which causes the skin and mucous membranes to appear yellow or jaundiced.

biopsy: The surgical removal of lesions within organs and tissues, primarily for the purpose of examination.

black disease: Infectious necrotic hepatitis, a bacterial liver infection.

blackhead: A disease of poultry caused by a protozoan (*Histomonas meleagridis*).

blacklegged tick: See *deer tick*.

blowflies: Metallic black or green flies that feed on moist, decaying matter.

blushing: Sudden and brief redness of the face and neck usually caused by overactivity in the sympathetic nervous system; physiological hyperemia.

bot: Larvae of botfly which infest the stomach and intestines of the horse.

bovine anaplasmosis (*Anaplasma marginale*): A bacterial infection characterized by fever, jaundice, and emaciation and commonly transmitted by ticks.

bovine babesiosis: A protozoal disease of cattle. Also called *tick fever*.

bronchi: Air passages in the lungs.

brucellosis: Infectious and highly contagious abortive disorder affecting cattle, dogs, goats, pigs, and humans.

bubonic plague: An acute infectious disease caused by the bacterium *Yersinia pestis*. Transmitted to humans by fleas infesting rodents.

C

calor: Heat of the skin and extremities caused by movement of blood from the body's core to those areas.

calcification: Abnormal deposition of calcium salts together with small amounts of iron, magnesium, and other mineral salts.

cancer: A general term frequently used to indicate any of the various types of malignant neoplasms.

canine ehrlichiosis: A bacterial disease of dogs that may cause tick paralysis.

canine piroplasmosis (babesiosis): A protozoal disease of dogs.

Capillaria: Small nematodes that live in the crop, intestine, or cecum of birds.

capsule: A well-defined, discrete, fibrous tissue layer that surrounds most benign tumors.

carcinogens: Physical agents, viruses, or chemicals that can cause changes in cells, resulting in tumor formation.

carcinoma: Malignant tumor from epithelial tissue.

carrier: An infected animal that is capable of transmitting the infectious agent but shows no apparent signs of the disease.

caseous necrosis: A type of necrosis involving the slow, progressive destruction of cells that prevents a high response of white blood cells.

cattle grub (*Hypoderma* species): A fly that looks like a large bee but belongs to a group of flies known as botflies.

cattle tick (*Boophilus annulatus*): A one-host hard tick, which has been eliminated from much of the United States but sometimes turns up on cattle in Texas. Common on cattle in Mexico and travels to Texas on deer and stray cattle.

cecal worm (*Heterakis gallinarum*): A worm whose adult is a small nematode that lives in the cecum of many different types of poultry.

cell degeneration: An altering of a cell.

cell membrane: The structure that keeps the environment inside and outside a cell balanced and separate.

cell proliferation: An increase in cell size or number.

cellular morphology: The outward appearance of a cell.

cellulitis: An inflammation of cellular or connective tissue.

cestode: Tapeworm.

chemotaxis: The chemical process by which cells orient themselves and home in on a target.

chicken body louse (*Menacanthus stramineus*): A chewing louse common on chickens, which transmit it by direct contact with each other.

chicken mite (*Dermanyssus gallinae*): A bloodsucking mite that attacks poultry and many different wild birds.

chigger: A larval mite that feeds on animals.

chlamydiosis: Disorder caused by the bacterium *Chlamydia* (of which there are several genera).

chorioptic mange (*Chorioptes bovis*): Common mange in cattle and horses. Also called *leg mange* as it usually causes scabs and irritation in that area of the body.

chromosome mutations: Changes that are the result of a rearrangement of genetic material.

chronic infection: The persistent presence of an organism and the body's persistent attempt to destroy or contain it.

chronic inflammation: A clinical concept referring to the existence of a persistent inflammatory stimulus.

clotting factors: Chemicals in the blood that allow clotting to occur.

coagulation necrosis: A type of necrosis in which the cytoplasm of the cell coagulates (thickens or gels) and details within the cell are lost.

Coccidia: A protozoa that lives in the cells of the host's intestine.

colic: Severe pains in the abdominal region.

collagen: A type of connective tissue.

colostrum: The first milk secreted following parturition.

compartmentalized fluids: Fluids that are restricted to certain spaces within tissues, organs, or cells.

complete blood count (CBC): Test that evaluates red blood cell count, platelet count, hemoglobin concentration, and white blood cell count.

congenital: Born with abnormalities.

congestion: A passive process in which venules engorge with blood.

contact carriers: Animals that may harbor and eliminate dangerous organisms that they picked up from contact with other animals.

contagion: The transmission of any infectious organism from one body to another by direct or indirect contact.

convalescent carrier: An animal that has had a recognized disease and has not rid itself of the infecting agent.

cranial mesenteric artery: The artery that supplies the large intestine with blood.

cross-reactivity: A phenomenon by which normal cells appear indistinguishable from infected cells.

crutching: Shaving the wool away from the area below a sheep's tail; a ranching practice aimed at removing the site most likely to become soiled with blowflies.

Cryptosporidium: Microorganism responsible for diarrhea in humans. Usually spread through fecal-contaminated water. This protozoan also infects ruminants.

cuticle: The tough skin that covers nematodes.

cyst: A closed sac containing fluid.

cytogenetic: A type of genetic disorder that results from chromosomal disorders.

D

deer liver fluke: A fluke that makes its home in deer, cattle, and sheep, and behaves differently in each.

deer tick (*Ixodes scapularis*): Blacklegged tick; a three-host hard tick that feeds on birds and small mammals as larvae and nymphs. Adults feed on large mammals such as cattle, horses, dogs, and deer. It may transmit anaplasmosis and tularemia.

definitive host: The host that harbors the adult, sexual, or mature stages of a parasite.

degeneration: A sometimes reversible retrogressive pathological change in a cell or tissue that results in its function often being impaired or destroyed.

dehydration: Loss of water from the body.

deletion: The loss of a portion of a chromosome.

demodectic mange (demodicosis): A disease characterized by the formation of small nodules in the skin (large animals) and alopecia in small animals.

demodectic mange mite: A common mite that lives in the hair follicles and sweat glands in the skin of the dog.

demodectic mange mite of gerbils: A rarely found mite that occasionally attacks gerbils and hamsters.

demodicosis: Condition caused by demodectic mites (mange).

Descriptive Classification System: A system that classifies tumors according to their gross or microscopic appearance.

diagnosis: Identification of the cause of a disease.

dichlorvos: An insecticide.

differential diagnoses: Systematic method of diagnosing a disease that lacks unique clinical signs.

dipping: A practice of submerging an animal in a bath containing a recommended pesticide to control mange.

docking: Removing the tail of a sheep; a ranching practice aimed at removing the site most likely to become soiled with blowflies.

dolor: Pain.

dominant: A gene that hides or masks the effect of another gene in the same allelic series.

double pore tapeworm (*Dipylidium caninum*): The common tapeworm of both dogs and cats. Also called *cucumber seed tapeworm.*

dysplasia: The abnormal differentiation or development of a tissue; may be the precursor of a neoplasia.

dystrophic calcification: When calcium is deposited in injured, degenerating, or dead tissue.

E

ear mange mites of rodents: Small mites that live in the skin of a rat's or hamster's ears, nose, and tail.

ear mite (*Otodectes cynotis*): A mite that lives deep in the ears of dogs and cats. Its feeding causes skin irritation and an increase in ear wax, and a crusty material builds up in the ear.

ear mite of rabbits (*Psoroptes cuniculi*): The most common ectoparasite of rabbits. Lives mainly in the rabbit's ear, but in heavy infestations may travel to the skin of the head and neck.

ecchymosis: Blood in tissue that is larger and more spread out.

ectoparasites: Parasites that live on their hosts; external parasites.

ectotherms: Cold-blooded animals.

edema: An excess of fluid in the interstitial fluid compartment of the body.

ELISA: Enzyme-linked immunosorbent assay test used for detecting human exposure to roundworms.

embolism: An impaction in some part of the vascular system of an embolus too large to complete the circulatory route.

embolus: A thrombus that breaks loose from the blood vessel in which it is formed in order to circulate throughout the bloodstream.

emigration: The movement of blood cells from the blood vessels and into tissue.

endemic typhus: An acute infectious disease carried by rat and mice lice.

endo: A prefix meaning "inside."

endoparasites: Parasites that live in their hosts; internal parasites.

endoplasmic reticulum: A cellular substance that creates proteins.

endothelium: Cells that line the blood vessels.

endotoxin: A poison made within the bacteria; may contain structural components of the organism.

entrapment: A malignant growth causing a pathology where an organism, organ, or tissue becomes abnormally trapped.

eosinophils: A type of white blood cell often associated with parasites and inflammation.

epistaxis: Bleeding from the nose.

equine encephalitis: A severe viral brain disease of horses.

erosion: Shallow ulcer usually limited to the mucosa with no penetration of the muscularis mucosa.

erythema: An abnormal redness of the skin caused by dilation of blood vessels.

erythrocyte: Red blood cell.

estrus: A condition in which hormones cause the cells lining the uterus to divide.

etiology: The study of the cause of disease.

euploid: Any chromosome number that is a multiple of the haploid number of chromosomes.

exotoxin: Proteins produced within some bacteria.

exudation: The presence of fluids and the migration of inflammatory cells into an inflammatory lesion.

F

face fly (*Musca autumnalis*): A parasite that looks very much like a housefly; it laps up fluids, especially those around the host's eyes, nose, and mouth.

fat necrosis: A type of necrosis that usually occurs as a result of injury to the pancreas.

fatty degeneration: The accumulation of fat in cells.

fibrin: A chemical that solidifies and sticks to platelets and tissues, enabling blood to clot.

fibrinopurulent exudate: A fluid containing the clotting protein fibrin and pus.

fibrinous exudate: A liquid containing fibrin but lacking pus because of a severe necrotizing infection.

first intention: The healing of a primary union.

fleeceworms: A term ranchers use for the larvae blowflies in a sheep's decaying wool; wool maggots.

flesh flies: Large flies that feed on decaying meat.

fluke: A parasitic flatworm (trematode).

fly worry: A condition caused by many horn flies on a cow; so named because the cow spends its time shooing off the flies instead of grazing.

fomite: An inanimate object that carries infections from one animal to another.

fox maggot fly: A flesh fly that lays larvae on the intact skin of young foxes and mink.

free-living stage: That period of a parasite's life when it lives apart from a host.

free radicals: Chemicals that help induce neoplasia.

fur mite of rabbits (*Cheyletiella* species): A mite that spends its entire life cycle on its host and travels by direct contact between hosts.

fur mites of mice (*Myobia musculi* and *Radfordia affinis*): Common parasites of both wild and pet mice. These mites live on the skin and in the fur and feed by sucking fluids from the skin.

G

gadding: A behavior in cattle in which they try to run away from an adult cattle grub buzzing around their legs.

gangrene: Localized death of living cells or tissues.

gangrenous necrosis: A type of necrosis that combines ischemia with a superimposed bacterial infection.

gape: To hold one's mouth wide open; a condition caused by lungworms.

gapeworm (*Syngamus trachea*): A worm whose adult stage is a small nematode that lives in the trachea of many different bird species.

gene mutations: Partial or complete deletions of genes on the chromosomes.

gene tracking: A process that helps determine whether members of the same family have inherited a defective gene.

generalized congestion: A condition in which the heart is diseased, causing blood buildup throughout multiple organs or the entire body.

genetics: Biology of heredity and variation in organisms.

genome: The complete DNA sequence of an animal.

genome mutations: Changes involving the loss or gain of an entire chromosome.

Giardia: A flagellate protozoan responsible for causing giardiasis.

global warming: Belief that an increase in atmospheric CO_2 levels is causing air temperatures throughout the world to rise.

gopher tortoise tick (*Amblyomma tuberculatum*): One of the largest hard ticks; lives on mammals and birds as a larva and on the gopher tortoise as a nymph and adult.

granulation tissue: Vascular connective tissue.

granuloma: A tumor composed of granulation tissue.

granulomatous inflammation: A type of chronic inflammation, consisting of focal inflammatory lesions in which the tissue reaction is primarily of chronic inflammatory cells.

grub: Maggot of the heel fly (*H. bovis* and *H. lineatum*).

guinea pig fur mite (*Chirodiscoides caviae*): A mite that spends its entire life on its host. Infestations usually cause the guinea pig no problems.

guinea pig louse (*Gliricola porcelli* **and** *Gyropus ovalis*): An insect that feeds on skin debris. Generally causes its host no problems.

Gulf Coast tick (*Amblyomma maculatum*): A three-host hard tick that lives in the United States along the Gulf of Mexico. The larva and nymph prefer to feed on such ground-dwelling birds as quail. The adult ticks feed on cattle, horses, and other large mammals.

H

heartworm (*Dirofilaria immitis*): One of the most dangerous nematode parasites of dogs, associated with heart failure and pulmonary artery disease with hypertension.

hematemesis: The vomiting of blood.

hematin: A rare golden brown, granular pigment associated with massive red blood cell destruction or transfusion reactions.

hematoidin: A golden brown pigment probably composed of the same chemical as bilirubin, differing only in its site of origin.

hematoma: A collection of blood within soft tissues resulting in swelling.

hematuria: Blood in the urine.

hemodynamic disturbances: Alterations in fluid flow.

hemoglobin: A pigment within red blood cells responsible for the transport of oxygen.

hemopericardium: Presence of blood in the heart sac.

hemoperitoneum: Presence of blood in the abdominal cavity.

hemoptysis: The coughing up of blood from the lungs.

hemorrhage: The escape of blood from the cardiovascular system.

hemorrhagic exudate: A liquid rich in red blood cells.

hemosiderin: A golden brown pigment found inside cells of the spleen and bone marrow.

hemostasis: Clotting.

hemothorax: Presence of blood in the chest.

heparin: An anticoagulant.

hereditary: A trait derived from one's parents.

hermaphrodite: An animal that has both ovarian and testicular tissue.

heterotopic bone formation: The formation of bone where there is normally cartilage.

Hippocratic method: A philosophy of care that has become the basis of modern medicine. The Hippocratic method consists of several important points: observing all, studying the patient (not the disease), evaluating honestly, and assisting nature.

histamine: A naturally occurring cellular chemical released during an allergic reaction.

hog louse (*Haematopinus suis*): A large (6 mm) sucking louse; the only louse that feeds on swine.

hookworms (*Ancylostoma caninum, A. tubaeforme, A. braziliense, Uncinaria stenocephala*): Gastrointestinal roundworm parasites of dogs and cats.

horn fly (*Haematobia irritans*): A gray bloodsucking fly about half the size of a housefly.

host: The organism that gives a parasite a home; the animal being parasitized in a parasite-host relationship.

host reaction: The presence of inflammatory cells and vascular phenomena at the site of inflammation.

host specific: A parasite that infects only one type of host.

humors: Body fluids such as blood, phlegm, yellow bile, and black bile.

hydatid tapeworm (*Echinococcus granulosus* and *Echinococcus multilocularis*): A tapeworm whose definitive host is the dog and intermediate hosts are ruminants, horses, and humans that accidentally eat the hydatid tapeworm's eggs. The adult tapeworm lives in the definitive host's small intestine.

hydropic degeneration: A severe form of cell swelling.

hypercalcemia: An elevation of blood calcium levels.

hyperemia: An active process in which the arterioles fill with blood.

hyperplasia: An increase in cell number.

hypersensitivity reaction: Inflammatory response of immunological origin.

hyperthermia: Abnormally high body temperatures caused by factors such as heat stroke and burns.

hypertrophy: An increase in cell size.

hypoplasia: Deficiency of growth and a diminution of size, occurs during development.

hypotension: A decrease in blood pressure.

hypothermia: Abnormally low body temperatures.

hypovolemic shock: Shock that occurs from a loss of blood.

I

iatrogenic: A negative response to therapy or medicine.

iguana tick: A hard tick that attacks only cold-blooded animals.

immune carrier: An animal that has eliminated a virulent infection, though it has never actually suffered from the disease.

immunity: The ability to resist disease.

immunosuppression: A method by which a parasite can cause the host to tolerate its presence. During immunosuppression, parasites can kill immune system cells, stimulate cells until the supply is exhausted, or even release immunosuppressive chemicals to inhibit the immune system from mounting a defense.

immunosuppressive agent: Therapeutic agent; can cause immune system problems.

incision: Cut into the body.

inert: Parasite invisible to the host.

infarction: The result of an embolism blocking off the blood supply or venous drainage of a tissue.

infection: A disturbance in the function of any part of an animal's body, caused by the entering of living agents into that body; the presence of endoparasites in a host.

infestation: The presence of ectoparasites on a host.

inflammation: Swelling with heat and pain caused by congestion of blood vessels.

inheritance: Transmission and reception of genetics from one generation to the next.

insect: An animal with three pairs of legs and three body parts (head, thorax, abdomen); includes lice, fleas, ants, bees, wasps, yellow jackets, beetles, and cockroaches.

intermediate host: The host that harbors the larval, juvenile, immature, or asexual stages of a parasite.

interstitial: Between cells.

intestinal strongyles: Nematodes that lay eggs in the gut. Adults live in the host's small intestine.

intracellular: Within a cell.

intraintestinal: Fluid-filled compartment in the digestive system.

intravascular: Within the blood vessels.

inversion: A genetic phenomenon that occurs when the chromosome breaks into three pieces and one piece reverses its position before joining back with the remaining pieces.

ion: A charged particle.

ischemic: A type of necrosis that involves a deficiency of blood to an area and therefore a lack of oxygen.

isochromosome: Formations that occur when one chromosome arm is lost and the remaining arm duplicates itself to make up for the missing piece.

ivermectin: A broad-spectrum anthelmintic (a substance destructive to worms).

K

karyolysis: A nuclear cellular change in which the nucleus dissolves.

karyorrhexis: A nuclear cellular change in which the nucleus breaks into small fragments.

karyotype: The chromosomal makeup of a somatic or body cell. It is the photomicrograph of an animal's chromosomes.

karyotypic changes: Changes in chromosome number or shape.

keloid: An abnormally large scar.

kidney worm of pigs (*Stephanurus dentatus*): A worm whose adult stage lives in the kidney and walls of the ureters of pigs.

knockouts: Animals that have been engineered to lack a gene that they normally have.

L

labile cells: Cells that divide frequently to replace lost cells.

larva: An immature free-living invertebrate post-hatching and prior to metamorphosing into an adult.

leptospirosis: An infectious disease caused by *Leptospira* and transmitted to humans, livestock, and other domestic animals by infected urine.

lesion: A morphological alteration in tissue that occurs with disease.

lethal: Deadly or fatal.

leukocyte: White blood cell.

leukocytosis: An increase in the total white blood cell count of blood.

leukopenia: A decrease in the total number of circulating white blood cells.

life cycle: The developmental stages of a parasite's life.

liquefaction necrosis: A type of necrosis more commonly known as *pus*.

lithiasis: The formation of pigment or protein "stones" in hollow organs, such as gallstones and kidney stones.

liver fluke of ruminants (*Fasciola hepatica*): A large trematode that resides in the liver, specifically the bile ducts, of a ruminant, horse, or human.

local congestion: Problems with the heart's outflow.

lone star tick (*Amblyomma americanum*): A three-host hard tick that feeds on small mammals and birds as a larva and a nymph.

lung fluke (*Paragonimus kellicotti*): A trematode whose adults inhabit cysts in the lungs of dogs, cats, and wild carnivores.

lungworms (*Filaroides osleri, F. hirthi, F. milksi*): Nematodes whose adults are found in the air passages of their hosts.

Lyme disease: A zoonotic bacterial disease transmitted by ticks.

lymphocytes: The immune cells that make antibodies.

lymphocytosis: An increase in lymphocytes and mononuclear cells.

lymphokines: A cytokine involved with cell-mediated immunity.

lysosome: A chemical body that is able to dissolve food particles, foreign invaders, and even cells.

M

maggot: The legless, soft-bodied larva of an insect.

malignant: The ability of a neoplasm to invade and spread to remote sites in the body.

Mallophaga: Order which includes biting lice.

mange: Contagious skin disease causing inflammation, itching, and hair loss. Causative organism is the mite.

mange mite of guinea pigs (*Trixacarus caviae*): A mite that lives in the skin of guinea pigs and may cause hair loss, crusty skin, and itching.

mast cell: Large connective tissue cell that contains heparin, histamine, and serotonin, which are released in response to allergies, injury, or inflammation.

mediators: Components that help control inflammatory reactions.

melanin: A brown-black pigment normally found in the skin, adrenal glands, and parts of the brain.

melanoma: Skin cancer.

melena: Blood in the feces.

Mendelian disorder: A type of genetic disorder in which mutant genes play a large part.

metaplasia: The change of one fully differentiated tissue into another type of tissue.

metastasis: A growth away from the primary site in which a tumor originates.

metastatic calcification: The deposition of calcium in soft tissues that are not the site of previous damage.

microfilaria: Embryos released by female heartworms.

microhepatia: A condition characterized by a small liver.

mite: A minute arachnid, some of which are responsible for causing mange.

mitochondria: The portion of a cell that produces adenosine triphosphate (ATP) for energy.

mixed infection: An infection involving more than one organism.

molting: The shedding of an outer layer, as when a nematode sheds its cuticle.

monoclonal: Derived from a single clone.

monocyte: Cells with one nucleus.

mouse mange mite (*Myocoptes musculinus*): A mite common on both wild and pet mice that feeds on the mouse's skin tissue.

mucoid degeneration: A conversion of any of the connective tissues into a gelatinous or mucoid substance (such as a mucopolysaccharide).

mucosa: The lining of the stomach.

multifactorial: A type of genetic disorder in which multiple genes and the environment exert a large influence.

murine: Pertaining to or affecting mice or rats.

mutation: Permanent changes in DNA.

mycotoxins: Chemicals produced by fungi.

myiasis: Disease caused by larval flies.

myocardial infarction: The inability of oxygen to reach the myocardial or muscular layer of the heart; more commonly known as a heart attack.

N

necropsy: The examination of an animal's body after death.

necrosis: Cell death.

nematode: Roundworm; small, wormlike organism covered by a tough skin called a cuticle.

neoplasia: A pathological formation or growth of an abnormal tissue. This term is sometimes used interchangeably with *cancer* when the neoplasm is malignant.

neoplasm: An abnormal tissue that grows by cellular proliferation more rapidly than normal and continues to grow after the stimuli that initiated the new growth cease.

neutrophils: Type of white blood cell that serves as a first line of defense against infection.

nit: The eggs of a female louse.

nodular worm of pigs (*Oesophagostomum dentatum*): A small nematode that lives in the large intestine of a pig.

nodules: Growths.

nonclassical inheritance: A type of genetic disorder consisting of a single-gene disorder with a nonclassical pattern of inheritance.

nonpathogenic parasite: A parasite that does not destroy its host and can continue to reproduce.

normovolemic shock: A type of shock in which blood volume is normal; however, oxygen is lacking.

northern fowl mite (*Ornithonyssus sylviarum*): The most common ectoparasite of U.S. poultry; feeds on domestic fowl and wild birds.

nosocomial: Disease acquired at the hospital or clinic.

notoedric mange mite (*Notoedres cati*): A mite that causes mange in cats.

nucleus: The portion of a cell in which DNA is stored and converted to RNA.

O

obesity: State of being fat.

one-host tick: A parasite that spends all three of its feeding stages on one host.

opsonization: A process by which the immune system coats antigens with chemicals, such as complement or antibody, in order to make it easier for immune cells to phagocytize (eat) them.

ornithosis: Parrot fever, psittacosis.

OSHA: Occupational Safety and Health Administration. Part of the federal government (Department of Labor) responsible for ensuring worker safety and health within the United States.

osteoma: A benign tumor of the bone.

Ostertagia: Roundworms or nematode genus occurring in the ruminant digestive tract.

P

parasite: An organism that lives on or in another organism, getting its nourishment from that other organism.

parathyroid: The endocrine gland that controls blood calcium levels.

parenchyma: Tumor tissue representative of the functional tissue of organs.

passive immunity: Produced by the transfer of living lymphoid cells from an immune animal; doesn't stimulate antibody production and thus is short termed.

pathogenesis: The study of mechanisms in the development of disease.

pathological mineralization: The abnormal deposition of calcium salts together with small amounts of iron, magnesium, and other mineral salts. Also called *calcification*.

pathological pigmentation: An accumulation or deposition of abnormal amounts of pigment in tissue cells or tissue fluids.

pathology: The scientific study of disease.

pediculosis: Condition or disorder caused by lice infestation.

peracute inflammation: An inflammatory episode usually caused by a potent stimulus and becoming apparent rapidly.

peripheral pooling: The process by which overwhelming bacterial infections cause blood to collect in an infected area, effectively removing the blood from circulation and mimicking hypovolemic shock.

permanent cells: Cells that lose the ability to divide after birth.

petechia: Small, pinpoint hemorrhages.

phagocytosis: The ingestion of such things as dead cells or bacteria by white blood cells.

phlegmon: A suppurative inflammation of subcutaneous connective tissue. Also known as *discharge pus*.

phytotoxin: A toxin produced by plants.

pig whipworm (*Trichuris suis*): A common parasite of livestock and pet pigs raised outside.

pinworm (*Oxyuris equi*): A nematode that lives in a horse's large intestine and cecum.

plasma: One half of the blood volume, 90 percent water.

plasma cells: B-cell lymphocytes that are "turned on" to produce antibodies.

platelet: Particle that makes up the composition of blood.

platelet plug: Platelets sticking together at the site of vascular injury.

porphyria: A congenital disorder involving abnormal hemoglobin synthesis and resulting in hypersensitivity to light.

prenatal transmission: A condition in which the second-stage larvae migrate into the uterus of a pregnant dog, invade the lungs of the fetus, and molt into third-stage larvae just before the pup's birth. They complete their life cycle in the newborn puppy and become adults in the small intestine.

prepatent: Evident, latent, or prior to the shedding of an infectious agent.

primary union: A surgical incision that has been sutured.

proglottid: A body segment of a tapeworm.

proliferation: An increase in the number of cells as a result of cell division and cell growth.

proliferative response: Cell injury and tissue adaptation that results in hyperplasia and growth abnormalities.

prostaglandin: Hormone-like substance found in several body tissues and produced in response to trauma or stress.

proteolytic enzymes: Chemicals that break down proteins.

protozoa: A single-cell animal.

pruritus: Itching.

pseudohermaphrodite: A female who has ovaries but the external appearance of a male; a male who has testicular tissue but externally appears to be a female.

pseudomembranous exudate: A liquid found in the respiratory and gastrointestinal tract, consisting of areas of tissue that mat together and appear membranous.

psoroptic mange: Sheep scab (*Psoroptes ovis*); the most important mange of sheep. Also found in cattle (*Psoroptes bovis*), where it is characterized by vesicles on the skin that cause intense itching, and horses (*Psoroptes equi*), where it affects the base of the mane and tail.

purpura: Blood in the tissue that results when ecchymoses combine with one another.

purulent (suppurative) exudate: A cloudy yellow fluid rich in white blood cells. Often referred to as *pus*.

pustules: Pimple-like growths containing pus.

pyknosis: A nuclear cellular change in which the nucleus shrinks and becomes very dense.

pyrethrins: Flower-based insecticides.

pyrogen: A fever-inducing agent.

R

rabbit louse (*Haemodipsus ventricosus*): An insect in which the adult usually feeds on a rabbit's back, sides, and groin.

rat and mouse louse (*Polyplax serrata* and *Polyplax spinulosa*): An insect that feeds on rats and mice. Infested animals appear restless, have ruffled fur, and may constantly scratch.

rat fur mite (*Radfordia ensifera*): A mite that spends its entire life on its host.

recessive: Gene whose expression is hidden by a dominant gene and is only expressed when in the homozygous state.

recombinant DNA: DNA that is made in the laboratory when new DNA sequences are inserted by chemical or biological means.

regeneration: The restructuring of tissue by the cells that belong there.

Regional Classification System: A system that classifies tumors according to where they occur within or on the body.

repair: The replacement of destroyed tissue by connective tissue.

resolution: The absorption or breaking down and removal of the products of inflammation.

retrovirus: Virus that contains two single strand linear RNA molecules per virion, such as HIV, FeLV, and SIV.

reverse genetics: Marker genes are placed to localize mutant genes.

ring chromosome: A type of chromosome that results when the ends of the chromosome fuse with each other.

ringworm: Fungal infection of the skin caused by dermatophytes.

Rocky Mountain wood tick (*Dermacentor andersoni*): A three-host hard tick that lives in the western United States. The larva and nymph feed on wild rodents, while the adult prefers large animals such as cattle and horses.

rodent botfly (*Cuterebra* spp.): A large (1 inch), hairy fly that looks like a bumblebee. The female fly lays her eggs near a rodent's burrow.

roundworm: Small, wormlike organisms with a tough skin; nematodes.

rubor: Redness.

ruminant tapeworm (*Thysanosoma actinoides*): A tapeworm. The adult lives in the small intestine, bile ducts, and pancreatic ducts of cattle, sheep, and goats.

S

salmonellosis: Disease caused by ingestion of salmonella bacteria.

sarcoptic mange (*Sarcoptes scabiei* var. *bovis*): A potentially serious disease of cattle characterized by intense itching and a large loss of hair in the affected areas.

sarcoptic mange mite (*Sarcoptes scabiei*): A small, round mite that can cause mange in dogs and other animals. The variety *canis* is zoonotic.

scabies: Sarcoptic mange.

scaly-leg mite (*Cnemidocoptes pilae*): A mite that lives on poultry and wild birds.

scar: Mark left by the healing of injured tissue.

scolex: The head of a tapeworm.

screwworm fly (*Cochliomyia hominivorax*): Blue-green flies twice the size of a housefly. The adult screwworm fly is not parasitic but causes myiasis.

second intention: The healing of an extensive dermal wound with complete loss of the epithelium.

secondary infection: A type of mixed infection in which the first organism makes a favorable environment for a second organism.

septicemia: Bacteria and their toxins in the blood.

serological test: A test performed on blood serums.

serosanguineous exudate: A fluid rich in serum and red blood cells.

serotonin: Vasoconstrictor.

serous exudate: A fluid low in dissolved protein; it has no clotting factor and a low number of cells.

shaft louse (*Menopon gallinae*): A common ectoparasite of chickens and other fowl.

sheep ked (*Melophagus ovinus*): A wingless fly that spends its entire life cycle on the sheep.

sheep nasal botfly (*Oestrus ovis*): A parasite of sheep. The female nasal botfly squirts a stream of fluid into a sheep's or goat's nasal opening. This fluid contains up to 25 small larvae, which migrate deeper into the nasal passages, feeding on mucus as they go.

sheep scab (*Psoroptes ovis*): Psoroptic mange; the most important mange of sheep.

shock: A condition caused by acute generalized circulatory failure of the capillary bed.

side effect: Predictable reactions to a drug that are not life threatening.

sodium pump: A mechanism that forces sodium ions out of a cell to maintain water balance.

species specific: An organism that infects only one species and does not infect any host from another species.

spinose ear tick (*Otobius megnini*): A one-host soft tick that spends its larval and nymphal stages in a cow's ear.

stable cells: Fully differentiated cells that do not divide often unless induced to do so.

stable fly (*Stomoxys calcitrans*): A blood-feeding fly that looks like a housefly.

stick-tight flea (*Echidnophaga gallinacea*): A flea found on chickens and turkeys in the southern United States, usually in older, poorly managed poultry houses.

stomach worm of pigs (*Hyostrongylus rubidus*): A stomach worm whose adult stage lives in a pig's stomach, where it lays eggs that pass out with the feces.

stroma: A normal body tissue that acts as a framework and develops the blood vessels that nourish a tumor.

strongyles: Common nematode infecting the cecum and colon of horses.

subacute inflammation: An inflammatory reaction characterized by a decrease in the redness of tissue and fluid accumulation.

sublethal: An injury that alters cell or tissue function without destroying the cell.

superinfection: A fresh infection or reinfection.

synergism: A phenomenon that occurs when the action of two or more agents produces a result that neither could bring about alone.

systemic: Occurring throughout the entire body.

T

tabanids: Large flies commonly called horseflies (*Tabanus* species), deer flies (*Chrysops* species), or green-head flies (*Tabanus americanus*).

taeniid tapeworms, canine (*Taenia pisiformis*): A common tapeworm of dogs and cats.

taeniid tapeworms, cattle (*Taenia saginata*): Tapeworms that use ruminants as intermediate hosts, living in the ruminant's internal organ or muscle.

tapeworm: Cestode; flat, ribbonlike worm with no digestive tract.

tetanus: An acute disease caused by toxins produced by the bacterium *Clostridium tetani*.

thermolabile: Destroyable by heat.

threadworm: Nematode whose adult females live in the small intestine of dogs, humans, and occasionally cats.

threadworm of horses (*Strongyloides westeri*): A parasite with a life cycle much like that of the threadworm of pigs. This threadworm does not infect the fetus. Rather, the major problems threadworms cause happen to foals infected through the milk.

threadworm of pigs (*Strongyloides ransomi*): A parasite whose adult stage lives in the small intestine of pigs.

three-host tick: A parasite that feeds on three different hosts—one for each life stage.

thrombocytopathy: Abnormality or disease of the platelets.

thrombocytopenia: A deficiency of platelets in the blood.

thrombus: The proper term for a blood clot within a vessel.

tick paralysis: A spreading paralysis that may eventually impair the muscles involved in breathing.

Tissue Classification System: The most widely used method for categorizing tumors. The principal basis of the system is the type of tissue from which the tumor arises.

toxemia: Toxins in the blood.

toxin: Poison.

toxoid: An antigenic substance that is not toxic.

trachea: Windpipe.

transgenics: A new technology in which recombinant DNA is placed within an animal which can either produce proteins the animal needs or get rid of proteins the animal doesn't need.

translocation: Occurs when a segment of one chromosome is transferred to another.

transmammary transmission: A condition in which second-stage larvae get into the mammary glands of a nursing mother, molt to third-stage larvae, and then pass with the milk into the suckling pup or kitten.

trematode: Fluke; flat, leaf-shaped worm that has a mouth and gut but no anus.

trophozoite: A term used for the feeding stage of a protozoan.

tropical rat mite (*Ornithonyssus bacoti*): A mite that lives on rats, mice, hamsters, and many other mammals and some birds too.

true gid: The circling and lack of balance that results when a taeniid tapeworm of sheep travels to the sheep's brain and destroys brain tissue as it grows.

tularemia: A bacterial disease (*Francisella tularensis*, formerly known as *Pasteurella tularensis*) of rabbits and humans, carried by ticks and sometimes the rabbit louse.

tumor: A swollen tissue.

two-host tick: A parasite that stays on one host for two of its life stages and then feeds on a second host during its third feeding life stage.

U

ulcer: The loss of a superficial layer of an organ or tissue, with acute inflammation at the base of the lesion.

Undernutrition: Not getting enough nutrients from one's diet to sustain normal life functions.

unthrifty: Not putting on weight.

ureters: The vessels that drain the urine from the kidney.

V

vascular permeability: "Leakiness" of the blood vessels.

vasoactive amines: Combinations of nitrogen and other chemicals that can constrict or dilate blood vessels.

vasoconstriction: Decrease in the diameter of blood vessels.

vasodilatation: Dilation of the blood vessels.

vasodilation: Dilation of the blood vessels.

vector: An organism (usually an arthropod) that carries pathogens or infective agents from one host to another.

vesicles: Small blisters.

virology: The study of viruses.

virulence: The power or malignancy of an infectious organism.

W

warble: A lumpy abscess under the hide of cattle caused by a grub.

water intoxication: Intake of an excessive amount of water resulting from salt loss or limited salt intake.

whipworm (*Trichuris vulpis*): A major gastrointestinal nematode of dogs.

winter tick (*Dermacentor albipictus*): A one-host hard tick that feeds on large mammals like cattle, horses, moose, and deer. Also called *moose tick*.

wool maggots: A term ranchers use for the larvae of blowflies in a sheep's decaying wool; fleeceworms.

X

X-linked disorder: A type of genetic disorder in which daughters are carriers of the disease, while sons suffer the disease's effects.

Z

zoonoses: Diseases shared or transmitted between animals and humans.

zootoxin: Poisons from animals.

INDEX

A

Abomasal worms, 178–179
Abscesses, 27
Acarexx, 160
Acaricides, 156
Acetaminophen toxicity, 50
Acidophils, 40
Acquired etiology, 3
Acquired immunity, 68
Active immunity, 68
Acute infection, 79
Acute inflammation, 25–28, 70
Acute local active hyperemia, 44
Acute systemic anaphylaxis, 69
Adenoma, 61
Adenosine tri-phosphate (ATP), 9
Adhesions, 28
Adverse drug reaction, 102, 103
Aeromonas hydrophilia, 168
Agenesis, 13
Agranulocytes, 40–41
Air pollution, 102
Amblyomma americanum, 128
Amblyomma dissimile, 167
Amblyomma maculatum, 128
Amblyomma tuberculatum, 167
American Animal Hospital Association (AAHA), 110
American dog tick, 133, 145, 156
Amitraz (Mitaban), 159, 164, 171
Amoeba, of snakes, 214
Amyloid degeneration, 10
Amyloidosis, 10
Anaplasia, 57

Anastomosis, 32
Ancylostoma braziliense, 202
Ancylostoma caninum, 202
Ancylostoma sp., 113
Ancylostoma tubaeforme, 202
Anemia, 127
Aneuploid, 94
Animal and Plant Health Inspection Service (APHIS), 111
Animals, used in research, 63
Anoplocephala magna, 193
Anoplocephala perfoliata, 193
Anoplura lice, 152–153
Anthelmintic, 130
Anthracosis, 16
Anthrax, 114–115
Antibody, 68–75
Antieoplastic agents, 103
Antigen, 68, 69f, 71–72
Antimicrobial agents, 103
Aplasia, 13
Aponomma sp., 167
Apoptosis, 10
Arachnids, 125–126, 130, 132–134
Argas sp., 169
Argyria, 16
Arrested larvae, 179
Arterial calcification, 15
Arteriosclerosis, 15
Arthropods, 132
Arthropods, as infectors, 80
Asa (king), 5
Asbestosis, 16
Ascaridia sp., 215

Words in italics are genus and species names. F indicates figure.

Ascarids
    of dogs and cats, 199–201
    of horses, 190
    as parasites, 113
    of swine, 185–186
Ascaris, 84
Ascaris suum, 185–186
Aspiculuris tetraptera, 212
Astrocytoma, 60
Asymptomatic carrier, 79
Atmospheric pressure, and
        injuries, 103
Atrioventricular (AV) node, 38
Atrophy, 14
Attenuation, 82
Autoantibodies, 68–69
Autoimmune hemolytic anemia, 70
Autoimmunity, 68–71
Autopsy, and pathology, 6
Autoreactive T cells, 68–69
Autosomal recessive disorders, 92

B

Babesia bigemina, 128
Babesia canis, 156
Babesiosis, 156
Bacillus anthracis, 114
Bacteremia, 28, 83
Bacterial diseases, zoonotic, 113–117
Bacteriophages, 82
Baermann technique, 239–241
Basophils, 40, 41
Baylisacaris procyonis, 113
Benign tumor, 56, 57, 59, 60–61
Bilirubin, 17
Biopsy, 3
Biting lice, 152–153
Black bile (melancholy), 5
Blackhead, 217–218
Blacklegged ticks, 129–130,
        134, 137, 156
Blood
    chemistry profiles of, 42
    circulation of, 38–39
    fluid homeostasis in, 42–43
    as humor, 5
    makeup of, 40–41
Blood smear, 233
Blushing, 43
Body fluid
    blood as, 5, 38–43
    distribution of, 42–43
    hemodynamic disturbances in, 43–46
    imbalances in, 48–49
Boophilus annulatus, 128

Borrelia anserina, 169
Borrelia burgdorferi, 118, 127
Bots, 137–138, 143–144, 147, 152, 167
Bovicola equi, 137
Brown dog tick, 136–146
Brucella sp., 115
Brucellosis, 111, 115
Bubonic plague, 115
Buffy coat method, 232

C

Calcification, 14
Calcinosis circumscripta, 15
Calcinosis universalis, 15
Calliphora, 133
Calor (heat), 23
Cancer, 56
Cancer research, 63
Canine
    demodicosis, 171
    lymphosarcoma, 73
    panosteitis, 96
    piroplasmosis, 156.
    See also Dogs
Capillaria sp., 216
Carbon tetrachloride, 4–5
Carcinogens, 58, 102, 103, 106
Carcinoma, 61
Caseous necrosis, 12
Catarrhous (catarrhal)exudate, 27
Cats
    carcinomas in, 58f, 63–64, 102
    Ctenocephalides felis, 134, 137, 153
    diseases of, 248–250
    fleas in, 134
    Isopora felis, 209–210
    nutritional disorders of, 106
    Toxocara cati, 113, 199–201, 212
    Toxoplasma gondii, 118, 210–211
Cattle
    anthrax in, 114–115
    brucellosis in, 115
    diseases of, 255–259
    ectoparasites of, 122–131
    endoparasites of, 177–185
    grubs in, 137–138, 138
    ticks in, 128
Cecal worm, 215
Cellophane tape preparation,
        226–227
Cells
    degeneration of, 9, 10
    growth, healing of, 9, 30
    injuries to, 4, 9–11, 18, 19
    lesions of, 3–4

    membranes of, 9
    morphology of, 11
    nonself, 68
    pathological pigmentation in,
        16–17
    response to injury, 12–14
Cellulitis, 27
Cerebrospinal nematoiasis (larva
        migrans), 113
Ceroid, 16
Cestodes, 126
Chemical carcinogenesis, 58–59
Chemical hemostatis, 46
Chemotaxis, 25–26
Chewing lice (mallophaga), 130
Cheyletiella, 226
Cheylitiella parasitovorax, 163
Chickens
    ectoparasites of, 168–170
    endoparasites of, 215–218
    mites of, 170
Chigger, 170
Chirodiscoides caviae, 163
Chlamydia psittaci, 116
Chlamydiosis, 116
Chondroma, 61
Chondrosarcoma, 61
Chorioptes, 131, 137
Chromosome mutations, 92, 93–94
Chronic generalized passive
        congestion, 45
Chronic infection, 79
Chronic inflammation, 25, 28–30
Circulatory system, 38–39
Cistudinomyia cistudinis, 167
Clinical pathology, 6
Clostridium tetani, 80, 116, 130
Clot, 46
Coagulase, 84
Coagulation, 46
Coagulation necrosis, 12
Cobalt, 106
Coccidia, 184, 193, 209–210, 217
Coccidiosis, 184, 210
Cochliomyia hominivorax, 137
Collagen, 29, 31–32, 47
Colostrum, 68
Common ascarid, of poultry, 215
Companion animals, 16, 63, 104, 106,
        152, 171
Complement, 70
Complete blood count (CBC), 28,
        41–42
Congenital hereditary traits, 91
Congestion, 43–45

Contact carrier, 79
Contact inhibition, loss of, 57
Contagious vs. infectious disease, 81
Convalescent carrier, 79
Cooperia sp., 180
Cross-reactivity, 69
Cryptosporidium, 184, 194, 212
Ctenocephalides canis, 153
Ctenocephalides felis, 134, 137, 153
Cutaneous lymphosarcoma, 73f
Cutaneous myiasis, 167
Cuterebra, 130, 152, 160–161, 166
Cyst, 177
Cytogenetic disorders, 93–94
Cytogenetic tests, 94
Cytotoxic hypersensitivity, 70

D

Davainea proglottina, 216
Deer liver fluke, 183
Deer tick, 129–130, 156
Dehydration, 48–49
Delayed hypersensitivity, 70–71
Demodetic mange, 130, 157–158
Demodetic mange mite, 164
Demodex, 130–131, 137, 171, 227
Demodex aurati, 164
Demodex canis, 157
Demodex criceti, 164
Demodex meroni, 164
Demodicosis, 131, 157
Deoxyribonucleic acid (DNA), 4, 9
Dermacentor albipictus, 129
Dermacentor andersoni, 129
Dermacentor variabilis, 133, 156
Dermanyssus gallinae, 170
Dermatophyte testing media (DTM), 232–233
Descriptive Classification System, 61
Diagnosis, 3
Dichlorvos, 192, 195
Dictyocualus sp., 180–181
Differential diagnosis, 5
Differential WBC count, 41–42
Diffuse inflammation, 24
Dipetalonema reconditum, 205–206, 229–231
Diphyllobothrium, 84
Dipylidium caninum, 153, 207
Direct smear, for fecal exam, 227–228
Dirofilaria immitis, 204–205, 206, 219, 229–231
Disease, cellular basis of, 9
Disease production, 82–83
Diseases, image guide for, 276–284

Disseminated intravascular coagulation (DIC), 33
Dogs
    Ctenocephalides canis, 153
    diseases of, 250–253
    ectoparasites of, 152–160
    endoparasites of, 199–212
    Isopora canis, 209–210
    Toxocara canis, 113, 199–201
    See also Canine
Dolor (pain), 23
Dominant genes, 92
Double pore tapeworm, 207
Dysplasia, 14
Dystrophic calcification, 14

E

Ear cytology, 225–226
Ear mange mites, of rodents, 164–165
Ear mites, 159–160, 164–165
Ecchinococcus granulosus, 182
Ecchymoses, 45, 46f
Echidnophaga gallinacea, 168
Echinococcus granulosus, 208–209
Echinococcus multilocularis, 208–209
Ectoparasites, of companion animals, 152
Ectoparasites, of large animals
    of cattle, 134–141
    of horses, 135–137
    of sheep, 131–133
    of swine, 133–135
Ectoparasites, of small animals
    of chickens, turkeys, birds, 168–170
    of dogs and cats, 152–160
    of ferrets, foxes, mink, 166–167
    of rabbits, rodents, 160–165
Ectotherms, 167
Edema, 48
Eimeria leuckarti, 193
Eimeria sp., 184, 190, 213, 217
Eimeria stiedae, 213
Electrical injuries, 104
Embolus, embolism, 47
Emigration, cell, 25
Endogenous pigmentation, 16–17
Endoparasites, of large animals
    of horses, 190–193
    of laboratory, farm animals, 175–177
    of ruminants, 178–185
    of swine, pet pigs, 185–1890
Endoparasites, of small animals
    of dogs and cats, 199–212
    of ferrets, 212
    of poultry, birds, 215–218

    of reptiles, 214
    of rodents, rabbits, 212–213
Endoplasmic reticulum, 9, 10f
Endothelium, 47
Endotoxins, 83
Entamoeba invadens, 214
Entrapment, 82
Environmental diseases, 102–104
Enzyme-linked immunosorbent assay (ELISA), 87, 201
Eosinophils, 40, 41
Epistaxis, 45
Erosion ulcers, 28
Erythema, 171
Erythrocytes, 40
Etiology, 3
Euploid, 94
Exogenous pigmentation, 16
Exotoxins, 83
Experimental pathology, 6
Extrinsic pathway, 47
Exudate, 23, 25, 26, 27

F

Face fly, 136
Farm animals, endoparasites of, 175–177
Fasciola hepatica, 182–183
Fascioliodes magna, 183
Fat necrosis, 12
Fat-soluble vitamins, 104–105
Fatty degeneration, 10
Fecal egg count, 237–241
Fecal flotation, 178f, 228
Felicola subrostratus, 153
Feline. *See* Cats
Feline infectious peritonitis (FIP), 74
Feline leukemia, 73
Ferrets, endoparasites of, 212
Fever, 28
Fibrin, 46
Fibrinolysin, 84
Fibrinopurulent exudate, 27
Fibrinous exudate, 27
Fibroma, 60
Filter technique, for heartworm, 229–231
Fine needle aspiration (FNA), 226
Flea allergy dermatitis (FAD), 155
Fleas
    of chickens, turkeys, birds, 168–169
    of dogs and cats, 154–156
    of ferrets, foxes, mink, 166
    of horses, 137
    of swine, 134
Flesh flies, 161, 167

Flies
    of cattle, 134–138, 135–136
    of chickens, turkeys, birds, 168
    of dogs and cats, 152
    of ferrets, foxes, mink, 166
    of horses, 135–136
    of rabbits, rodents, 160–161
    of reptiles, 167
    of sheep, 131–132, 134
Fluid imbalances, 43, 48–49
Flukes, 126, 175, 177, 182–183
Fly strike, 152
Focal inflammations, 24
Fomite, 79
Fowl ticks, 169
Foxes, ectoparasites of, 166–167
Fragile X syndrome, 94
Francisella tularensis, 162
Frontline spray (Fipronil), 160
Frosch, Paul, 6
Functio laesa, 23
Fungal culture, 232–233
Fur mites, 163

G

Gangrene, 12
Gangrenous necrosis, 12
Gapeworm, 215–216
Gasterophilus sp., 135
Gene mutations, 92
Genetic etiology, 3
Genetics, disorders of, 91–95
Genome mutations, 92
Giardia
    and absorption, 84–85
    in birds, 218
    in dogs, 211, 213
    in horses, 194
    illustrated, 178f, 212f
    means of infection, 80
    in rodents, 213
    in ruminants, 185
    testing for, 118, 228
Giardia lamblia, 118, 211–212
Giardiasis, 118
Gliricola porcelli, 153
Global warming, 101
Goats, 177–185, 255–259
Gopher-tortoise tick, 167–168
Gram's stains, 225
Granulation tissue, 30
Granulocytes, 40
Granulomatous inflammation, 28–30
Guinea pig fur mite, 163
Guinea pig lice, 162
Guinea pigs, diseases of, 244–246

Gulf Coast tick, 128–129, 134, 137
Gyrodactylus salaris, 175
Gyropus ovalis, 162

H

Haematobia irritauns, 136
Haematopinus asini, 137
Haemodipsus ventricosus, 161
Haemonchus sp., 178–180
Hamsters, 164
Healing, of injured tissue, 30–32
Heartworm, 204–205, 219, 229–231
Heavy metal poisoning, 103
Heinz bodies, 50, 51f
Hemacytometer use, 234
Hematemesis, 45
Hematin, 17
Hematocrit, 40
Hematoidin, 17
Hematoma, 45
Hematuria, 45
Hemodynamic changes, 25
Hemodynamic disturbances, 43–46
Hemoglobin derivatives, 16–17
Hemopericardium, 45
Hemoperitoneum, 45
Hemoptysis, 45
Hemorrhage, 45
Hemorrhagic exudate, 27
Hemosiderin, 17
Hemostasis, 46
Hemothorax, 45
Heparin, 41, 69
Hepatoma, 61
Hereditary traits, 91
Hermaphrodites, 94
Hetakeris gallinae, 215
Heterakis, 85
Heterotopic bone formation, 15
Heteroxygous, 91
High egg passage vaccine (HEP), 82
Hip dysplasia, 96–97
Hippocrates, 5–6
Hippocratic Method, 6
Hippocratic Oath, 6
Histamine, 26, 41, 69, 70f
Histomonas, 85
Histomonas meleagridis, 217–218
Hog louse, 134
Homoxygous, 91
Hookworm, 113, 202
Hopopleura sp., 162
Horn fly, 136
Horses
    botflies of, 135–136
    diseases of, 255–259

ectoparasites of, 135–137
endoparasites of, 190–194
tapeworms of, 193, 194f
Host-parasite relationship, 86
Humors, cardinal, 5–6
Hyaline degeneration, 10
Hyalinosis, 10
Hyalomma aegyptium, 167–168
Hyaluronidase, 84
Hyatid tapeworm, 182, 208–209
Hydropic degeneration, 10
Hymenolepis tapeworms, 213
Hyostrongylus rubidis, 188
Hypercalcemia, 14–15
Hyperemia, 43–44
Hyperkalemic periodic paralysis
    (HYPP), 97
Hyperplasia, 13
Hypersensitivity reactions, 69–71
Hyperthermia, 103
Hyperthermic injuries, 103
Hypertrophy, 13
Hypoderma bovis, 137–138, 138
Hypoderma lineatum, 137–138, 138
Hypodermosis, of cattle, 138
Hypoplasia, 13
Hypothermia, 103
Hypovolemic shock, 47

I

Iatrogenic response, 81
Iguana tick, 167
Immediate hypersensitivity, 69
Immune carrier, 79
Immune complex hypersensitivity, 70
Immune function disturbances, 71–72
Immunity, 68
Immunodeficiencies, secondary, 71–72
Immunological diseases, 68–73
Immunosuppression, 86
Immunosuppressive agents, 103
Incision, surgical, 30–32
Infarction (infarct), 43, 47
Infecting organism, 79
Infection
    and contagion, 81
    defined, 79
    organisms of, 79
    pathogen properties, 82
    sources of, 79–81
Infectious disease, 79
Infectious vs. contagious disease, 81
Inflammation
    acute, 25–28
    chronic and granulomatous, 28–30
    defined, 23

exudate-based, 27–28
gross appearance of, 23
and healing of injured tissue, 30–32
location-based, 26, 28
onset and duration, 24–25
reaction of tissue to, 23f, 24
signs of, 23
systemic reactions to, 28
Injuries, physical, 103–104
Injury, adaptation to, 4
Insects, 125
Insterstitial compartment, 42
Interferons, 27
Intestinal strongyles, 180
Intracellular compartment, 42
Intracellular killing, 71
Intraintestinal compartment, 42
Intravascular compartment, 42
Intrinsic pathway, 47
Invermectin, 130, 133, 135, 136, 159,
    163, 192, 195, 219
Inversion, of chromosomes, 94
Ions, 9
Ischemia, 11, 12
Isochromosome formation, 94
Isopora, 228
Isopora canis, 209–210
Isopora felis, 209–210
Isopora rivolta, 209–210
Isopora suis, 190
Ixodes dammini, 127
Ixodes pacificus, 127, 129
Ixodes scapularis, 127, 129, 156

J

Jenner, Edward, 68
Johne's disease, 29f, 34

K

Karyolysis, 11
Karyorrhexis, 11
Karyotypic changes, 57
Keloid, 32
Kidney worm, of pigs, 188–189
Knemidocoptes mutans, 170
Knott's technique, modified, 229
Koch, Robert, 6

L

Labile cells, 13
Laboratory animals, endoparasites of,
    175–177
Larva migrans syndrome, 113
Leptospirosis, 116
Lesions, 3–4, 24
Lethal cellular injury, 9

Leukemia, 60
Leukocytes, 40–41
Leukocytosis, 28
Leukopenia, 28
Lice
    of chickens, turkeys, birds, 168
    of dogs and cats, 152–153
    of horses, 137
    as pests, 130–131
    of rabbits, rodents, 161–162
    of swine, 134
Lingonathus setosus, 152–153
Lipochromes, 16
Lipofuscin, 16
Liquefaction necrosis, 12
Lithiasis, 15–16
Liver cancer, 4–5
Liver fluke, of ruminants, 182–183
Lockjaw, 116
Loeffler, Friedrich, 6
Lone star tick, 128, 134, 137
Lucila, 133
Lugol's iodine, 177, 179f, 227
Lung flukes, 207
Lungworms
    of lizards, 214
    of pigs, 188
    of ruminants, 180–181
    of snakes, 214
Lyme disease, 118, 127, 130
Lymphatic obstruction, 48
Lymphatic system, 27
Lymphocytes, 41
Lymphokines, 27
Lysomes, 9

M

Macrophages, 30–31
Malignant tumor, 56, 57–58, 60–61
Mallophaga, 152–153
Malnutrition, 71–72
Mammary tumor neoplasia, 33
Mange, 130, 133
Mange mite, of guinea pigs, 164
Marek's disease, 73
Mast cells, 69, 70f
Material Safety Data Sheets
    (MSDS), 112
Mechanical force injuries, 103
Mediators, chemical, 26, 44
Melanin, 16
Melanoma, 60
Melena, 45
Melophagus ovinus, 131
Menacanthus stramineus, 168
Mendelian disorders, 92–93

Menopen gallinae, 168
Metaplasia, 14
Metastasis, 57–58, 60
Metastatic calcification, 14–15
Metastrongylus sp., 188
Methemoglobin, 17
Mice, diseases of, 242–244
Microfilariae, 203
Microscope use, 223–224
Microscopic examination, 3f
Microsporum canis, 232
Microsporum gypseum, 232
Milk allergy, 69
Mineralization, pathological, 14–15
Mink, ectoparasites of, 166–167
Mitaban, 159, 164, 171
Mites
    of cattle, 130–131
    of chickens, turkeys, birds, 169–170
    described, 133–134
    of dogs and cats, 156–160
    of ferrets, foxes, mink, 166–167
    of horses, 137
    of rabbits, rodents, 162–165
    of reptiles, 168
    of sheep, 133
    of swine, 134–135
Mitochondria, 9, 10f
Modified McMaster technique,
    237–239
Molecular diagnosis, of genetic
    disorders, 94–96
Moniezia sp., 181
Monoclonal tumor cells, 57
Monocytes, 41
Morphology, of tumors, 61–63
Mosquitoes, 118, 160, 203
Mouse, nude, 63
Mouse mange mite, 164
Mucoid degeneration, 10
Muellerius capillaris, 180–181
Multifactorial disorders, 93
Multifocal inflammation, 24
Musca autumnalis, 136
Mutations, 92
Mycobacterium bovis, 102, 114
Mycobacterium tuberculosis, 113–114
Myiasis, 137, 152
Myobia musculi, 162–163
Myocardial infarction, 47
Myocoptes musculinus, 164
Myocoptes romboutsi, 164

N

Necropsy, 3, 6
Necrosis, 9, 10–12

Nematodes
  described, 126
  of dogs and cats, 199–206
  of horses, 190–193
  of laboratory, farm animals,
      175, 176f
  of poultry, birds, 215–216
  of reptiles, 214
  of rodents, rabbits, 212
  of ruminants, 178–181
  of swine, pet pigs, 185–189
Nematodirus sp., 180
Neoplasia, 14, 56, 72–73. *See also*
      Tumors
Neorickettsia helminthoeca, 85
Neoschoengastin americana, 170
Neutrophils, 40, 41, 70
New methyline blue stain, 225
Nodular worm, of pigs, 187
Nonclassical inheritance, 94
Nonpathogenic parasites, 86
Nonself cells, 68
Normovolemic shock, 47–48
Northern fowl mite, 169–170
Noscomial infection, 81
Nothing per os (NPO), 87
Notoedres cati, 159
Notoedres muris, 164–165
Notoedres notoedres, 164–165
Notoedric mange mite, 159
Nucleus, 9, 10f
Nude mouse, 63
Nutritional disease, 84–85,
      104–106

O

Obesity, 106
Occupational Safety and Health Act
      (OSHA), 110–112
Oesophagostomum sp., 187
Oestrus ovis, 132
Onchocerca larvae, 85
Ophidascaris sp., 214
Ophionyssus natricis, 168
Opsonization, 71
Ornithonyssus bacoti, 165
Ornithonyssus sylviarum, 169–170
Ornithosis, 116
Osteoma, 60
Ostertagia sp., 178–180
Otobius megnini, 137
Otodectes, 225
Otodectes cynotis, 159–160, 166
Oxygen, cellular, 9
Oxyuris equi, 192–193

P

Packed cell volume (PCV), 40
Papilloma, 61
Paragonimus Kellicotti, 207
Paranoplocephala mimillana, 193
Parascaris equorum, 185, 190
Parasites
  defined, 84
  image guide for, 260–276
  inert, 86
  infections of, 84–85
  tolerance of, 86
  zoonotic, 113
Paratuberculosis, 34
Parenchyma, 60
Parvovirus, 87
Passalurus ambiguus, 212
Passive immunity, 68
Pasteur, Louis, 6
Pasteurella multocida, 165
Pathogenesis, 3
Pathogenicity, 82
Pathological mineralization, 14
Pathology, 3–4, 6
Pediculosis, 153
Peracute infection, inflammation, 24
Peracute infection,inflammation, 79
Permanent cells, 13
Petechiae, 45, 46f
Pet pigs, 185–190
Phaenicia, 133
Phagocytosis, 26
Phlegm, 5
Phlegmon, 28
Phormia, 133
Physical injury, 103–104
Phytotoxin, 83
Pig ascarid, 185–186
Pigmentation, pathological, 16–17
Pigs. *See* Pet Pigs, Swine
Pig whipworm, 187–188
Pinworms, 192–193, 212
Planaria, 126
Plasma, 40
Plasma proteases, 26–27
Plasmodium, 85
Platelet plug, 46
Platelets, 40, 41
Platynosomum concinum, 126
Plumbism, 16
Polymerase chain reaction (PCR)
      analysis, 94
Polyplax sp., 162
Polyps, 63
Porphyria, 17

Porphyrins, 17
Poultry. *See* Chickens, Turkeys
Practical Classification System,
      60–61
Prepatent period, 200, 204
Primates, diseases of, 253–255
Proglottids, 175
Proliferation, of cells, 9, 12
Prostaglandins, 27
Protein-losing enteropathy, 34
Proteolytic enzymes, 70
Protostrongylus rufescens, 180–181
Protozoal diseases, 118
Protozoans
  of dogs and cats, 209–212
  of horses, 193
  of laboratory, farm animals, 177
  as parasites, 126
  of poultry, other birds, 217–218
  of reptiles, 213–214
  of rodents, rabbits, 214
  of ruminants, 184–185
  of swine, pet pigs, 190
Pruritus, 155
Pseudohermaphrodites, 94
Pseudomembranous exudate, 27
Psittacosis, 116
Psoroptes cuniculi, 165
Psorptes, 130, 131, 137
Public health, responsibility for, 111,
      112
Purpura, 45
Purulent exudate, 27
Pus, 12
Pustules, 158
Pyknosis, 11
Pyrogens, 28

R

Rabbit louse, 161
Rabbits, 161–165, 212–213, 246–248
Rabies, 117
Racoon roundworm, 113
Radfordia affinis, 162–163
Radfordia ensifera, 163
Radiation injuries, 104
Raillietina sp., 216
Rat and mouse lice, 162
Rat fur mite, 163
Rats, diseases of, 242–244
Recessive genes, 92
Recombinant DNA, 91
Red blood cells, 40, 70
Regeneration, 30
Regional Classification System, 61

Repair, tissue, 30
Reptiles, parasites of, 167–168, 213–214
Resolution, 30
Retrovirus, 73
Reverse genetics, 91
Rhabdias sp., 214
Rhabdovirus, 104
Rheumatoid arthritis, 70
Rhipicephalus sanguineus, 155
Ribolucleic acid (RNA), 9
Rickettsia typhi, 162, 165
Ring chromosomes, 94
Ringworm dermatophytosis, 119–120
Rocky Mountain spotted fever, 156
Rocky Mountain wood tick, 140, 148
Rodent botfly, 160–161
Rodenticide toxicity, 18
Rodents, 161–165, 212–213. *See also*
    Mice, diseases of; Mouse, nude;
    Rats, diseases of
Romanowsky stains, 224–225
Roundworms, 126, 195
Rubor (redness), 23
Ruminants, diseases of, 255–259
Ruminant tapeworm, 181

S

Safety, procedures for, 110–118
Salmonellosis, 116
Sarcocystitis, 185, 211
Sarcophaga sp., 161
Sarcoptes, 141, 146, 148
Sarcoptes scabei, 134, 158–159, 166, 170
Sarcoptic mange mite, 158–159
Sarcoptic mange (scabies), 130, 158
Scabies, 130, 158
Scaly-leg mite, 170
Scarring, 28, 32
Schistocytes, 33
Screwworm fly, 137
Secondary infection, 81
Selamectin (Revolution), 159, 160
Self-antigens, 68–69
Septicemia, 28, 83
Serosanguineous exudate, 27
Serotonin, 26
Serous exudate, 27
Severe combined immunodeficiency (SCID), 95
Sex-linked inheritance, 92
Shaft louse, 168
Sheep, 131–133, 177–185, 255–259
Shock, 47
Side effects, 102–103

Siderosis, 16
Silicosis, 16
Single-gene disorders, 92–93
Sinoatrial node (SA), 38
Skin scraping, 227
Skin turgor, 48, 49f, 49t
Snake mites, 168
Sodium pump, 9, 10
Spinose ear tick, 127–128, 134
Squamous cell carcinoma, 58f, 61, 64, 102
Stable cells, 13
Stable fly, 135–136
Staining techniques, 224–225
Starling equilibrium, 48
Steatosis, 10
Stephanurus dentatus, 188–189
Steroids, 29
Stick-tight flea, 168–169
Stomach worm, of pigs, 188
Stomoxys calcitrans, 135–136
Stroma, 59, 60, 61
Strongyles, 191–192
Strongyloides ransomi, 189
Strongyloides stercoralis, 203
Strongyloides westeri, 193
Strongylus edentatus, 191
Strongylus equinus, 191
Strongylus vulgaris, 191–192
Subacute inflammation, 25
Sublethal cellular injury, 9
Sucking lice (Anoplura), 132, 152–153
Superinfection, 81
Suppurative exudate, 27
Swine, 133–135, 185–190, 255–259.
    See also Pet pigs
Synergism, 81
Syngamus trachea, 215–216
Syphacia sp., 212

T

Tabanids, 134–135, 152
Tabanus sp., 134
Taenia pisiformis, 207–208
Taenia saginata, 126
Taenia solium, 189–190
Taenia sp., 181–182
Taenia taeniaeformis, 207–208
Tapeworms
    diverting nutritive substances,
        84, 85f
    in dogs and cats, 207–209
    in horses, 193
    in laboratory, farm animals, 175, 176f
    in poultry, other birds, 216

    in reptiles, 214
    in rodents, rabbits, 213
    in ruminants, 181–182
    in swine, pet pigs, 189–190
    taeniid, 181–182, 207–208
Teratoma, 60
Tetanus (Lockjaw), 116
Thermolability, 84
Threadworms, 189, 193, 203
Thrombocytes, 41
Thrombocytopathy, 41
Thrombocytopenia, 45
Thrombus, 44, 47
Tick-borne diseases, 118
Ticks
    in cattle, 127–130
    in chickens, turkeys, birds, 169
    described, 132–133
    in dogs and cats, 127, 155–156
    in ferrets, foxes, mink, 166
    in horses, 137
    in rabbits, rodents, 162
    in reptiles, 167–168
    in swine, 134
Tissue
    classification system for, 60
    healing of, 30–32
    and hypersensitivity reaction, 69–71
    in pathology, 3–5
    vascular changes in, 25–26
TNM classification system, 61
Tobacco smoking, 102
Tolerance, of parasites, 86
Toxascaris leonina, 191, 199–201, 212
Toxemia, 28
Toxicity, 50
Toxin formation, production,
        83–84, 85
Toxin-induced immunosuppression, 72
Toxocara canis, 113, 199–201
Toxocara cati, 113, 199–201, 212
Toxoid, 84
Toxoplasma gondii, 118, 210–211
Toxoplasmosis, 112, 118
Transgenics, 91
Trauma, of parasites, 84
Trematodes
    described, 126, 175
    of dogs and cats, 207
    illustrated, 177f
    as parasites, 126
    in ruminants, 182–184
Treponema, 81
Trichinella spiralis, 186–187
Trichodectes canis, 152–153

Trichomonads, 217
Trichomonas gallinae, 217
Trichomonas gallinarum, 217
Trichophyton mentagrophytes, 232
Trichostrongylus sp., 178–180
Trichuris, 85
Trichuris suis, 187–188
Trichuris vulpis, 203
Tritrichomonas, 213
Tritrichomonas foetus, 184–185
Trixacarus caviae, 164
Trophozoite, 177
Tropical rat mite, 165
Trypanosoma brucei
     gambiense, 85
Trypanosoma cruxi, 85
Tuberculosis, 113–114
Tumors, 23, 56–63
Turkey chigger, 170
Turkeys, 168–170, 215–218

U

Ulcerative colitis, 71
Ulcers, 28
Uncinaria stencephala, 202
Undernutrition, 104

Undulant fever, 115
Urinalysis, 234–236

V

Vaccines, 82, 83f
Vascular permeability, 25
Vasoconstriction, 46
Vasodilatation (vasodilation), 25
Vectors, 80, 85
Veterinarians, 6
Veterinary vs. human medicine, 43
Viral diseases, 117
Virology, 6
Virulence, 87
Virus-induced immune disorders, 72
Visceral larval migrans (VLM), 201
Von Willebrand's disease, 41

W

Warble, 127
Wasting disease, 106
Water intoxication, 49
Water-soluble vitamins, 105
Whipworm, 187–188, 203
White blood cells, 40–41, 70
Winter tick, 129, 137

Wohlfahrtia vigil, 161, 166
Woolsorter's Disease, 114–115
World Health Organization, 101
Wuchereria, 85

X

X-linked disorders, 92, 93f

Y

Yellow bile (choler), 5
Yersina pestis, 115, 165

Z

Zoonoses
    and animal bites, 110
    bacterial diseases, 113–117
    defined, 110
    parasitic diseases, 113
    protozoal diseases, 118
    tick-borne diseases, 118
    Toxocara sp., 201
    Toxoplasma gondii, 210–211
    transmission of, 112
    viral diseases, 117
Zootoxins, 83

IMPORTANT! READ CAREFULLY: This End User License Agreement ("Agreement") sets forth the conditions by which Cengage Learning will make electronic access to the Cengage Learning-owned licensed content and associated media, software, documentation, printed materials, and electronic documentation contained in this package and/or made available to you via this product (the "Licensed Content"), available to you (the "End User"). BY CLICKING THE "I ACCEPT" BUTTON AND/OR OPENING THIS PACKAGE, YOU ACKNOWLEDGE THAT YOU HAVE READ ALL OF THE TERMS AND CONDITIONS, AND THAT YOU AGREE TO BE BOUND BY ITS TERMS, CONDITIONS, AND ALL APPLICABLE LAWS AND REGULATIONS GOVERNING THE USE OF THE LICENSED CONTENT.

1.0 SCOPE OF LICENSE

1.1 <u>Licensed Content.</u> The Licensed Content may contain portions of modifiable content ("Modifiable Content") and content which may not be modified or otherwise altered by the End User ("Non-Modifiable Content"). For purposes of this Agreement, Modifiable Content and Non-Modifiable Content may be collectively referred to herein as the "Licensed Content." All Licensed Content shall be considered Non-Modifiable Content, unless such Licensed Content is presented to the End User in a modifiable format and it is clearly indicated that modification of the Licensed Content is permitted.

1.2 Subject to the End User's compliance with the terms and conditions of this Agreement, Cengage Learning hereby grants the End User, a non-transferable, nonexclusive, limited right to access and view a single copy of the Licensed Content on a single personal computer system for non-commercial, internal, personal use only. The End User shall not (i) reproduce, copy, modify (except in the case of Modifiable Content), distribute, display, transfer, sublicense, prepare derivative work(s) based on, sell, exchange, barter or transfer, rent, lease, loan, resell, or in any other manner exploit the Licensed Content; (ii) remove, obscure, or alter any notice of Cengage Learning's intellectual property rights present on or in the Licensed Content, including, but not limited to, copyright, trademark, and/or patent notices; or (iii) disassemble, decompile, translate, reverse engineer, or otherwise reduce the Licensed Content.

2.0 TERMINATION

2.1 Cengage Learning may at any time (without prejudice to its other rights or remedies) immediately terminate this Agreement and/or suspend access to some or all of the Licensed Content, in the event that the End User does not comply with any of the terms and conditions of this Agreement. In the event of such termination by Cengage Learning, the End User shall immediately return any and all copies of the Licensed Content to Cengage Learning.

3.0 PROPRIETARY RIGHTS

3.1 The End User acknowledges that Cengage Learning owns all rights, title and interest, including, but not limited to all copyright rights therein, in and to the Licensed Content, and that the End User shall not take any action inconsistent with such ownership. The Licensed Content is protected by U.S., Canadian and other applicable copyright laws and by international treaties, including the Berne Convention and the Universal Copyright Convention. Nothing contained in this Agreement shall be construed as granting the End User any ownership rights in or to the Licensed Content.

3.2 Cengage Learning reserves the right at any time to withdraw from the Licensed Content any item or part of an item for which it no longer retains the right to publish, or which it has reasonable grounds to believe infringes copyright or is defamatory, unlawful, or otherwise objectionable.

4.0 PROTECTION AND SECURITY

4.1 The End User shall use its best efforts and take all reasonable steps to safeguard its copy of the Licensed Content to ensure that no unauthorized reproduction, publication, disclosure, modification, or distribution of the Licensed Content, in whole or in part, is made. To the extent that the End User becomes aware of any such unauthorized use of the Licensed Content, the End User shall immediately notify Cengage Learning. Notification of such violations may be made by sending an e-mail to infringement@cengage.com.

5.0 MISUSE OF THE LICENSED PRODUCT

5.1 In the event that the End User uses the Licensed Content in violation of this Agreement, Cengage Learning shall have the option of electing liquidated damages, which shall include all profits generated by the End User's use of the Licensed Content plus interest computed at the maximum rate permitted by law and all legal fees and other expenses incurred by Cengage Learning in enforcing its rights, plus penalties.

6.0 FEDERAL GOVERNMENT CLIENTS

6.1 Except as expressly authorized by Cengage Learning, Federal Government clients obtain only the rights specified in this Agreement and no other rights. The Government acknowledges that (i) all software and related documentation incorporated in the Licensed Content is existing commercial computer software within the meaning of FAR 27.405(b)(2); and (2) all other data delivered in whatever form, is limited rights data within the meaning of FAR 27.401. The restrictions in this section are acceptable as consistent with the Government's need for software and other data under this Agreement.

7.0 DISCLAIMER OF WARRANTIES AND LIABILITIES

7.1 Although Cengage Learning believes the Licensed Content to be reliable, Cengage Learning does not guarantee or warrant (i) any information or materials contained in or produced by the Licensed Content, (ii) the accuracy, completeness or reliability of the Licensed Content, or (iii) that the Licensed Content is free from errors or other material defects. THE LICENSED PRODUCT IS PROVIDED "AS IS," WITHOUT ANY WARRANTY OF ANY KIND AND CENGAGE LEARNING DISCLAIMS ANY AND ALL WARRANTIES, EXPRESSED

OR IMPLIED, INCLUDING, WITHOUT LIMITATION, WARRANTIES OF MERCHANTABILITY OR FITNESS FOR A PARTICULAR PURPOSE. IN NO EVENT SHALL CENGAGE LEARNING BE LIABLE FOR: INDIRECT, SPECIAL, PUNITIVE OR CONSEQUENTIAL DAMAGES INCLUDING FOR LOST PROFITS, LOST DATA, OR OTHERWISE. IN NO EVENT SHALL CENGAGE LEARNING'S AGGREGATE LIABILITY HEREUNDER, WHETHER ARISING IN CONTRACT, TORT, STRICT LIABILITY OR OTHERWISE, EXCEED THE AMOUNT OF FEES PAID BY THE END USER HEREUNDER FOR THE LICENSE OF THE LICENSED CONTENT.

8.0 GENERAL

8.1 <u>Entire Agreement.</u> This Agreement shall constitute the entire Agreement between the Parties and supercedes all prior Agreements and understandings oral or written relating to the subject matter hereof.

8.2 <u>Enhancements/Modifications of Licensed Content.</u> From time to time, and in Cengage Learning's sole discretion, Cengage Learning may advise the End User of updates, upgrades, enhancements and/or improvements to the Licensed Content, and may permit the End User to access and use, subject to the terms and conditions of this Agreement, such modifications, upon payment of prices as may be established by Cengage Learning.

8.3 <u>No Export.</u> The End User shall use the Licensed Content solely in the United States and shall not transfer or export, directly or indirectly, the Licensed Content outside the United States.

8.4 <u>Severability.</u> If any provision of this Agreement is invalid, illegal, or unenforceable under any applicable statute or rule of law, the provision shall be deemed omitted to the extent that it is invalid, illegal, or unenforceable. In such a case, the remainder of the Agreement shall be construed in a manner as to give greatest effect to the original intention of the parties hereto.

8.5 <u>Waiver.</u> The waiver of any right or failure of either party to exercise in any respect any right provided in this Agreement in any instance shall not be deemed to be a waiver of such right in the future or a waiver of any other right under this Agreement.

8.6 <u>Choice of Law/Venue.</u> This Agreement shall be interpreted, construed, and governed by and in accordance with the laws of the State of New York, applicable to contracts executed and to be wholly preformed therein, without regard to its principles governing conflicts of law. Each party agrees that any proceeding arising out of or relating to this Agreement or the breach or threatened breach of this Agreement may be commenced and prosecuted in a court in the State and County of New York. Each party consents and submits to the nonexclusive personal jurisdiction of any court in the State and County of New York in respect of any such proceeding.

8.7 <u>Acknowledgment.</u> By opening this package and/or by accessing the Licensed Content on this Web site, THE END USER ACKNOWLEDGES THAT IT HAS READ THIS AGREEMENT, UNDERSTANDS IT, AND AGREES TO BE BOUND BY ITS TERMS AND CONDITIONS. IF YOU DO NOT ACCEPT THESE TERMS AND CONDITIONS, YOU MUST NOT ACCESS THE LICENSED CONTENT AND RETURN THE LICENSED PRODUCT TO CENGAGE LEARNING (WITHIN 30 CALENDAR DAYS OF THE END USER'S PURCHASE) WITH PROOF OF PAYMENT ACCEPTABLE TO CENGAGE LEARNING, FOR A CREDIT OR A REFUND. Should the End User have any questions/comments regarding this Agreement, please contact Cengage Learning at delmar.help@cengage.com.

Minimum System Requirements:

PC:

* Operating System: Windows 2000 w/SP4, XP w/SP2, Vista
* Hard Drive space: [200MB]
* Screen resolution: 800 × 600 pixels
* CD-ROM drive

PC Setup Instructions:

1. Insert disc into CD-ROM drive. The program should start automatically. If it does not go to step 2.
2. From My Computer, double-click the icon for the CD drive.
3. Double-click the start.exe file to start the program.

Technical Support:

Telephone: 1-800-648-7450
8:30 AM – 6:30 PM Eastern Time
E-mail: delmar.help@cengage.com